# Basics and Trends in Sensitivity Analysis

## *Theory and Practice in R*

# Computational Science & Engineering

The SIAM series on Computational Science and Engineering publishes research monographs, advanced undergraduate- or graduate-level textbooks, and other volumes of interest to an interdisciplinary CS&E community of computational mathematicians, computer scientists, scientists, and engineers. The series includes introductory volumes aimed at a broad audience of mathematically motivated readers interested in understanding methods and applications within computational science and engineering, monographs reporting on the most recent developments in the field, and volumes addressed to specific groups of professionals whose work relies extensively on computational science and engineering.

SIAM created the CS&E series to support access to the rapid and far-ranging advances in computer modeling and simulation of complex problems in science and engineering, to promote the interdisciplinary culture required to meet these large-scale challenges, and to provide the means to the next generation of computational scientists and engineers.



**Series Volumes**

Da Veiga, Sébastien, Gamboa, Fabrice, Iooss, Bertrand, and Prieur, Clémentine, *Basics and Trends in Sensitivity Analysis: Theory and Practice in R*

Vidyasagar, M., *An Introduction to Compressed Sensing*

Antoulas, A. C., Beattie, C. A., and Güğercin, S., *Interpolatory Methods for Model Reduction*

Sipahi, Rifat, *Mastering Frequency Domain Techniques for the Stability Analysis of LTI Time Delay Systems*

Bardsley, Johnathan M., *Computational Uncertainty Quantification for Inverse Problems*

Hesthaven, Jan S., *Numerical Methods for Conservation Laws: From Analysis to Algorithms*

Sidi, Avram, *Vector Extrapolation Methods with Applications*

Borzì, A., Ciaramella, G., and Sprengel, M., *Formulation and Numerical Solution of Quantum Control Problems*

Benner, Peter, Cohen, Albert, Ohlberger, Mario, and Willcox, Karen, editors, *Model Reduction and Approximation: Theory and Algorithms*

Kuzmin, Dmitri and Hämäläinen, Jari, *Finite Element Methods for Computational Fluid Dynamics: A Practical Guide*

Rostamian, Rouben, *Programming Projects in C for Students of Engineering, Science, and Mathematics*

Smith, Ralph C., *Uncertainty Quantification: Theory, Implementation, and Applications*

Dankowicz, Harry and Schilder, Frank, *Recipes for Continuation*

Mueller, Jennifer L. and Siltanen, Samuli, *Linear and Nonlinear Inverse Problems with Practical Applications*

Shapira, Yair, *Solving PDEs in C++: Numerical Methods in a Unified Object-Oriented Approach, Second Edition*

Borzì, Alfio and Schulz, Volker, *Computational Optimization of Systems Governed by Partial Differential Equations*

Ascher, Uri M. and Greif, Chen, *A First Course in Numerical Methods*

Layton, William, *Introduction to the Numerical Analysis of Incompressible Viscous Flows*

Ascher, Uri M., *Numerical Methods for Evolutionary Differential Equations*

Zohdi, T. I., *An Introduction to Modeling and Simulation of Particulate Flows*

Biegler, Lorenz T., Ghattas, Omar, Heinkenschloss, Matthias, Keyes, David, and van Bloemen Waanders, Bart, editors, *Real-Time PDE-Constrained Optimization*

Chen, Zhangxin, Huan, Guanren, and Ma, Yuanle, *Computational Methods for Multiphase Flows in Porous Media*

Shapira, Yair, *Solving PDEs in C++: Numerical Methods in a Unified Object-Oriented Approach*

# Basics and Trends in Sensitivity Analysis

## *Theory and Practice in R*

SÉBASTIEN DA VEIGA
Safran Tech
Magny-les-Hameaux, France

FABRICE GAMBOA
Institut de Mathématiques de Toulouse
Université P. Sabatier
Toulouse, France

BERTRAND IOOSS
EDF Lab
Chatou, France

CLÉMENTINE PRIEUR
Université Grenoble Alpes
Grenoble, France

Society for Industrial and Applied Mathematics
Philadelphia

10 9 8 7 6 5 4 3 2 1



| | |
|---|---|
| *Publications Director* | Kivmars H. Bowling |
| *Executive Editor* | Elizabeth Greenspan |
| *Developmental Editor* | Mellisa Pascale |
| *Managing Editor* | Kelly Thomas |
| *Production Editor* | Louis R. Primus |
| *Copy Editor* | Bruce Owens |
| *Production Manager* | Donna Witzleben |
| *Production Coordinator* | Cally A. Shrader |
| *Compositor* | Cheryl Hufnagle |
| *Graphic Designer* | Doug Smock |

**Library of Congress Cataloging-in-Publication Data**

Names: Veiga, Sébastien da, author. | Gamboa, Fabrice, author. | Iooss, Bertrand, author. | Prieur, Clémentine, author. | Society for Industrial and Applied Mathematics, issuing body.
Title: Basics and trends in sensitivity analysis : theory and practice in R / Sébastien Da Veiga, Fabrice Gamboa, Bertrand Iooss, Clémentine Prieur.
Description: Philadelphia : Society for Industrial and Applied Mathematics, [2021] | Includes bibliographical references and index. | Summary: "This book provides an overview of global sensitivity analysis methods and algorithms, including their theoretical basis and mathematical properties"– Provided by publisher.
Identifiers: LCCN 2021014255 (print) | LCCN 2021014256 (ebook) | ISBN 9781611976687 (paperback) | ISBN 9781611976694 (ebook)
Subjects: LCSH: Sensitivity theory (Mathematics) | Global analysis (Mathematics) | R (Computer program language)
Classification: LCC QA402.3 .V35 2021 (print) | LCC QA402.3 (ebook) | DDC 003/.5–dc23
LC record available at *https://lccn.loc.gov/2021014255*
LC ebook record available at *https://lccn.loc.gov/2021014256*

# Contents

**Foreword** **vii**

**Preface** **ix**

**Notation** **xiii**

**Acronyms and Initialisms** **xv**

**1 Introduction** **1**
- 1.1 Computer experiments, uncertainty quantification, and sensitivity analysis . 1
- 1.2 Sensitivity analysis settings . . . . . 3
- 1.3 Examples of real-world problems . . . . . 4
- 1.4 General methodology . . . . . 11
- 1.5 Introduction to the R package sensitivity . . . . . 14
- 1.6 Overview of the methods and contents of this book . . . . . 14

**2 A first look at screening using R** **17**
- 2.1 Screening by discretized designs of experiments . . . . . 17
- 2.2 Sampling-based sensitivity analysis . . . . . 28
- 2.3 Derivative-based global sensitivity analysis . . . . . 36
- 2.4 Extensions to qualitative inputs . . . . . 43
- 2.5 Summary . . . . . 43

**3 Variance-based sensitivity measures** **45**
- 3.1 From functional ANOVA to Sobol' indices . . . . . 45
- 3.2 Sampling-based estimators . . . . . 58
- 3.3 Nonscalar outputs . . . . . 76
- 3.4 Further references . . . . . 90

**4 Spectral and metamodel-based estimation** **93**
- 4.1 Spectral approaches . . . . . 93
- 4.2 Metamodel-based approaches for Sobol' indices . . . . . 105
- 4.3 Further references . . . . . 115
- 4.4 Summary . . . . . 115

**5 Variance-based sensitivity measures with dependent inputs** **117**
- 5.1 State of the art: Difficulties and drawbacks . . . . . 117
- 5.2 Grouping . . . . . 119
- 5.3 Covariance-based sensitivity analysis . . . . . 121

5.4 Independent and full Sobol' indices . . . 126
5.5 Shapley effects . . . 132
5.6 Examples in R . . . 142
5.7 Summary . . . 146

**6 Beyond variance-based indices** **149**
6.1 Moment-independent importance measures . . . 149
6.2 Kernel-based sensitivity analysis . . . 156
6.3 Goal-oriented sensitivity analysis . . . 168
6.4 Robustness analysis in uncertainty quantification . . . 173
6.5 Summary . . . 185

**7 A case study in R: COVID-19 epidemic model** **187**
7.1 A compartmental model for COVID-19 . . . 188
7.2 Prior uncertainty quantification and sensitivity analysis . . . 195
7.3 Data calibration and posterior sensitivity analysis . . . 208

**A Sampling designs for model exploration in R** **229**
A.1 Introduction . . . 229
A.2 Basic sampling techniques . . . 230
A.3 Space-filling criteria . . . 235
A.4 Building space-filling designs . . . 239
A.5 Design classification using minimum spanning tree . . . 245

**B Reproducing kernel Hilbert space** **249**

**C Toy functions** **251**
C.1 Arctangent temporal function . . . 251
C.2 Bratley model . . . 251
C.3 HeterDisc function . . . 251
C.4 Ishigami function . . . 251
C.5 Linkletter decreasing function . . . 252
C.6 Morris function . . . 252
C.7 Sobol' $g$-function . . . 252

**D Software** **253**
D.1 Packages for sensitivity analysis in the R environment . . . 253
D.2 A generic platform for uncertainty treatment: OpenTURNS . . . 255

**Bibliography** **259**

**Index** **289**

# Foreword

I am really pleased to be invited to write the foreword to Bertrand, Clémentine, Fabrice, and Sébastien's book and delighted to see this new book on sensitivity analysis for several reasons.

First, to me at least, having another book devoted to sensitivity analysis and computer simulations based on mathematical models marks the maturing of the topic across the engineering disciplines and throughout the mathematical, computer experiments, and physical sciences. Sensitivity analysis is a crucial step in model building. Gaining insights into a model's behavior, its structure, and its response to changes in the model inputs is essential. With the constant increase in application demands and the continuous advances in theoretical research, more sophisticated models and techniques have been built to capture sophisticated facets of the problems under analysis. This book presents a comprehensive view of global sensitivity analysis as a modern discipline and how it can be applied in a large range of application areas. Of course, different application domains are characterized by models and data with different properties, and the authors' extensive practical experience is evident in their detailed and revealing views of a range of new theoretic and algorithmic results in sensitivity analysis, including many real-world applications. An attractive feature of the book is its discussion of some of the most important sensitivity analysis R software tools and its illustrations of these tools in practice in a way that I am sure will be attractive to readers and potential users of the methodology.

A second reason for my delight at the publication of this book is more personal and concerns the authors, with whom I have been personally involved during my active academic years in France. I was always impressed by their constant enthusiasm and their important and original contributions to the continued development of global sensitivity–related methods.

May 2020, Cape Town, South Africa

**Anestis Antoniadis**
*Emeritus professor, University Grenoble Alpes, France*
*Honorary research associate, University of Cape Town*

# Preface

*"Ce qui n'est pas légèrement difforme a l'air insensible; d'où il suit que l'irrégularité, c'est à dire l'inattendu, la surprise, l'étonnement sont une partie essentielle et la caractéristique de la beauté."*

Charles Baudelaire, Fusées

In many fields, such as environmental risk assessment, agronomic system behavior, aerospace engineering, and nuclear safety, mathematical models turned into computer code are used for making predictions and performing simulations when experiments would be too expensive or even infeasible. Modern computer models that simulate physical phenomena often input a high number of numerical parameters and physical variables and provide a number of outputs (scalars or functions). Inputs may include engineering or operating variables, variables that describe field conditions, and variables that include unknown or partially known model parameters.

For the development, analysis, and use of such computer models, *sensitivity analysis* is an invaluable tool. It involves studying how the uncertainty in the output of a model can be apportioned to different sources of uncertainty in its inputs. For instance, it may be used to determine the input variables that most contribute to some output behavior, detect noninfluential inputs, and uncover interaction effects within a model. Sensitivity analysis can also be used to rank the influence of inputs as well as to calibrate, verify, understand, and simplify models. Furthermore, it can be used to help validate computer code, guide research efforts, and justify system design safety.

The historical approach to sensitivity analysis is known as the *local* approach. Here, the impact of small input perturbations on a model's output is studied. The small perturbations occur around reference values (e.g., the mean). This approach—which is deterministic—consists in calculating or estimating the partial derivatives of the model at a specific point. To compute these derivatives, we use model adjoint–based methods, which allow the processing of models with a large number of input variables. Such methods are commonly used in solving large environmental systems, e.g., in climate modeling, oceanography, and hydrology. Many applied mathematicians and computer scientists have contributed to such workflows, in particular by developing adjoint models and automatic differentiation tools.

From the mid- to late 1980s onwards, to overcome the limitations of local methods (linearity and small variations), a new class of methods was developed under a statistical framework. In contrast to local sensitivity analysis, this was named *global* sensitivity analysis because it considered the whole possible range of variation in each input. Such approaches have for the most part been popularized by engineers and scientists in the UQ (uncertainty quantification) teams in U.S. national laboratories (Sandia and Los Alamos) and the SAMO (Sensitivity Analysis of Model Outputs) community,[1] coordinated by the European Commission Joint Research Centre in Ispra, Italy.

[1]For information on methods, publications, and software, see `https://ec.europa.eu/jrc/en/samo` and `https://en.wikipedia.org/wiki/Sensitivity_analysis`; see `http://www.gdr-mascotnum.fr/samo.html` for a list of past conferences.

Despite much work and the development of numerous methods for global sensitivity analysis, theoretical results had been lacking until the end of the 1990s. Subsequently, applied mathematicians and statisticians became more interested in the subject. In the United Kingdom, MUCM (Managing Uncertainty in Complex Models)[2] has brought together a large community developing Bayesian techniques for computer experiments. In the United States, a convergence between applied mathematicians and statisticians has led to the creation of the UQ group of the Society for Industrial and Applied Mathematics (SIAM),[3]

In France, the CNRS research group MASCOT-NUM (Stochastic Analysis Methods for Numerical Codes and Treatment[4]) was created to coordinate French research in the fields of uncertainty quantification, sensitivity analysis, and computer experiments. Numerous training programs and software developments have appeared, including several useful packages for the software R (in particular the sensitivity package, which is used in this book) and Open Source platforms (in particular OpenTURNS described in Appendix D.2) gathering the most useful UQ and SAMO algorithms. Collaborative work has also been encouraged between academic and industrial teams, leading, for example, to the COSTA BRAVA (Complex Spatio-temporal Analysis by Reduced Models and Sensitivity Analysis) project funded by the French national research agency (project ANR-09-COSI-015). This book aims to give an overview of results from the last fifteen years, some even coming from recent Ph.D. theses. Our main goal is to provide a unified view of most of the new theoretic and algorithmic results in and around sensitivity analysis, while keeping an eye on applications of the methods we describe.

This book aims to be a reference companion to all engineers, researchers, and students who use complex numerical models and have an interest in sensitivity analysis techniques. Mathematicians from the computer experiment community will be interested in recent results on sensitivity indices. We also expect the book to be of great use in industry and the computer modeling community. Bringing together all of the methods used to perform sensitivity analysis in one place should be a catalyst for better cultural dissemination of these methods into industry. To help the reader, this book offers many examples and programming examples in the R software and its sensitivity package. For reproducibility, all the codes are freely available at `https://bookstore.siam.org/cs23/bonus`.

### Acknowledgments

First of all, we are very grateful to our institutes and companies in providing support that allowed us to write this book. We warmly thank EDF R&D, Safran, Paul Sabatier Toulouse III University and Grenoble Alpes University. Support from Grenoble–Rhône Alpes Inria project/team AIRSEA and from the ANR-3IA Artificial and Natural Intelligence Toulouse Institute are also acknowledged.

Let us tell you about the genesis of this book. In early 2000, Sébastien was beginning his Ph.D. at IFPEN, Clémentine was a young assistant professor at INSA and at Paul Sabatier University, Bertrand was a junior research engineer at CEA, and Fabrice was a forty-year-old professor at Paul Sabatier University. At this time, computer code experiments and more precisely global sensitivity analysis was in France an emergent field in applied statistics mainly carried and pushed by researchers from the industrial world. Our work in sensitivity analysis and this book would never have existed without our meeting with these pioneers: Anestis Antoniadis, Nicolas Devictor, and François Wahl. Thank you so much for your scientific perspicacity! Anestis, thank you so much for your generosity and the pride we take in the preface to this book that you were kind enough to write. Thank you also for having promoted the idea of the creation of the GdR

[2]`http://www.mucm.ac.uk/`

[3]`https://www.siam.org/membership/activity-groups/detail/uncertainty-quantification`

[4]`http://www.gdr-mascotnum.fr`

CNRS MASCOT NUM. This research group has been a driving force in animating the scientific French community working on computer code experiments since 2008. We warmly thank all the people from this research group for the deep scientific synergies driven. Let us also mention the first project at the end of 2000 of writing a book on sensitivity analysis by Bertrand, Hervé Monod, Gilles Pujol, and Curtis Storlie. Due to job changes of all these coauthors, this project has been stopped, but the book outline has inspired the present one. Gilles was also the founder of the R sensitivity package (in 2006) that became a software success story and is one of the pillars of the present book.

We kindly thank Elisabeth Greenspan from SIAM for having proposed we write this book a long time ago and then for her gentleness, encouragement, and great patience. Part of this book has been written during some meetings in the four corners of France. We warmly thank the CIRM in Marseille and the LPSM in Paris (gracias a Nico y Gérard) for their hospitality; Véro for her house in *les Landes*; Soso for her house in *les Côteaux de Seine*; Nathalie for hosting Clémentine in Toulouse (and introducing her to Gizmos); the labs EDF Chatou, IMT, and LJK; and the containment imposed by the COVID pandemic, which allowed us to finish this book.

A special mention to our anonymous reviewers and our proofreaders Patrix Cattiaux, Vincent Chabridon, Guillaume Damblin, Josselin Garnier, Céline Helbert, Thierry Klein, and Agnès Lagnoux. Your time spent and your constructive remarks and suggestions on some parts of the first version of the book is sincerely appreciated. Thank you very much also for the sound scientific advice from Art Owen and Ralph Smith. Many thanks are also due to Kevin Bleakley for his strong efforts to correct our "Prisunic"' English in several chapters, as well as SIAM compositor Cheryl Hufnagle.

Many parts of the book come from contributions of past and current Ph.D. students: Benjamin Auder, Nazih Benoumechiara, Clément Bénard, Thomas Browne, Gaëlle Chastaing, Kevin Elie-Dit-Cosaque, Vincent Feuillard, Laurent Gilquin, María Belén Heredia, Adrien Hirvoas, Marouane Il Idrissi, Alexandre Janon, Henri Mermoz Kouye, Paul Lemaître, Amandine Marrel, Simon Nanty, Ekaterina Sergienko, Adrien Spagnol, Jérome Stenger, Jean-Yves Tissot, and Elena Volkova. We thank you very sincerely for having taught us so many things. We are very grateful to our close colleagues and collaborators, Elise Arnaud, Franck Barthe, Julien Bect, Eric Blayo, Agnès Boncannelé, Emanuele Borgonovo, Pierre Etoré, Jean-Claude Fort, Josselin Garnier, Titi Klein, Sergeï Kucherenko, Matieyendou Lamboni, Béatrice Laurent, Loïc Le Gratiet, Véronique Maume-Deschamps, Maëlle Nodet, Art Owen, Luc Pronzato, Olivier Roustant, Andrea Saltelli, Bruno Sudret, Roman Sueur, Jean-Claude Strong, Stefano Tarantola, Laurence Viry, and Olivier Zahm. Many ideas and results provided here come from some joint researches with you. Thank you so much for your good mood and your creativity. We also kindly thank our colleagues from EDF, INRAE, Safran Tech, and CEA: Aurélie Arnaud, Michaël Baudin, Géraud Blatman, Nicolas Bousquet, Claire Cannamela, Mathieu Couplet, Thibault Delage, Robert Faivre, Merlin Keller, Stéphanie Mahévas, Chu Mai, Michel Marquès, Jean-Marc Martinez, Gildas Mazo, Hervé Monod, Alberto Pasanisi, Guillaume Perrin, Anne-Laure Popelin, Christian Rey, Alejandro Ribés, Nicolas Son, and Elisabeta Vergu. Thank you to Annie, Cathy, Delphine, and Suzan for the administrative support and all hallway discussions.

To finish these thanks, inventory à la Prévert, thank you so much to Eléa, Lise, Véronique, Domitille, Gaétane, Guillem, Christophe, Annie, Olivier, Ika, Kattin, Marie, Beñat, Manuel, Emeric, Orlane, and Sophie and for your patience and undying love.

*"Ne méprisez la sensibilité de personne. La sensibilité de chacun, c'est son génie."*

Charles Baudelaire, Fusées

# Notation

| | |
|---|---|
| $n, N$ | Sample sizes |
| $d$ | Dimension of input vector |
| $k$ | Dimension of output vector |
| | |
| $G(\cdot)$ | Numerical model |
| $Y$ | Random output (scalar, vectorial or functional) |
| $\mathbf{X} = (X_1, \ldots, X_d)$ | Random input vector |
| $Y = G(\mathbf{X})$ | Input-output model relation |
| $\phi = \phi(Y) = \text{QoI}(Y)$ | Quantity of Interest (a statistics on $Y$) |
| | |
| $\boldsymbol{Y}^n = \left(Y^{(1)}, \ldots, Y^{(n)}\right)^T$ | Sample vector of $Y$ |
| $\boldsymbol{X}^n = \left(X_1^{(i)}, \ldots, X_d^{(i)}\right)_{i=1,\ldots,n}$ | Sample matrix of $\mathbf{X}$ (dimension $n \times d$) |
| | |
| $\mathbb{N}$ | Set of nonnegative integers |
| $\mathbb{N}_*$ | Set of positive integers |
| $\mathbb{Z}$ | Set of integers |
| $\mathbb{R}_+$ | Set of nonnegative real numbers $\mathbb{R}_{\geq 0}$ |
| $\mathcal{M}_1^+$ | Set of probability measures on $\mathbb{R}$ |
| $E_j$ | Metric space |
| $\mathcal{B}(E_j)$ | Borelian |
| | |
| $X_j \in (E_j, \mathcal{B}(E_j))$ | Random input variable |
| $E \subset \mathbb{R}^d$ | Variation domain of scalar inputs $\mathbf{X}$ |
| $\text{P}_\mathbf{X}, F_\mathbf{X}$ | Probability distribution of $\mathbf{X}$ |
| $p_\mathbf{X}, f_\mathbf{X}$ | Density function of $\mathbf{X}$ |
| $\mathbf{1}_A$ | Indicator function of the set $A$ |
| $\mathbb{L}^2(\text{P}_\mathbf{X})$ | $\{f : x \in \mathbb{R}^d \mapsto f(x) \in \mathbb{R},\, \int_\mathcal{X} f^2(x)\text{P}_\mathbf{X}(dx) < +\infty\}$ |
| $\|\cdot\|, \langle\cdot\rangle$ | (Euclidean or Hilbertian) norm and inner product |
| $\delta_x$ | Dirac measure centered on point $x$ |
| | |
| $\xrightarrow{\mathcal{D}}$ | Convergence in distribution |
| $\xrightarrow{\mathbb{P}}$ | Convergence in probability |
| $\xrightarrow{\text{a.s.}}$ | Convergence almost sure |
| | |
| $H^T$ | Transpose of $H$ ($H$ is any matrix) |
| $\text{Tr}(H)$ | Trace of $H$ ($H$ is any square matrix) |

| | |
|---|---|
| $\mu_X = \bar{X} = \mathbb{E}X$ | Mean of $X$ |
| $\operatorname{Var} X$ | Variance of $X$ |
| $\sigma_X = \sqrt{\operatorname{Var} X}$ | Standard deviation of $X$ |
| $\operatorname{Cov}(X, Y)$ | Covariance between $X$ and $Y$ |
| $x_\alpha = q^\alpha(X)$ | $\alpha$-quantile of $X$ |
| $p = \mathbb{P}(Y \in \mathcal{R})$ | Probability of failure ($Y$ lies in critical region $\mathcal{R}$) |
| $\hat{\beta}$ | Estimate of the statistical quantity $\beta$ |
| $\mathcal{U}(a, b)$ | Uniform distribution on the interval $[a, b]$ |
| $\mathcal{N}(\mu, \sigma^2)$ | Gaussian distribution of mean $\mu$ and variance $\sigma^2$ |
| $\Phi$ | Distribution function of the standard Gaussian distribution ($\mu = 0$, $\sigma^2 = 1$) |
| $\varphi$ | Density function of the standard Gaussian distribution |
| $[1 : d]$ | Set $\{1, \ldots, d\}$ |
| $\mathcal{P}_d = \mathcal{P}([1 : d])$ | Set of all subsets of $\{1, \ldots, d\}$ |
| $A \in \mathcal{P}_d$ | Subset of indices |
| $A \in \mathcal{P}_d$, $\|A\|$ | cardinality of subset $A$ |
| $A$, $B \in \mathcal{P}_d$, $A \subset B$ | $A$ subset of $B$ |
| $A$, $B \in \mathcal{P}_d$, $A \subsetneq B$ | $A$ subset of $B$, $A \neq B$ |
| $\prod$ | Cartesian product when applied to sets |
| $\bar{A} = \{i \in [1 : d] : i \notin A\}$ | Complementary set of A |
| $-i = \overline{\{i\}}$ | Complementary set of index $i$ |
| $-ij = \overline{\{i, j\}}$ | Complementary set of indices $i$ and $j$ |
| $\mathbf{X}_A = (X_j : j \in A)$, $A \in \mathcal{P}_d$ | Group of input variables |
| $\mathbf{X}_{\bar{A}} = (X_j : j \notin A)$ | Group of input variables |
| $\mathbf{X}_{-i} = (X_j : j \neq i)$ | Group of input variables |
| $\operatorname{Proj}_A$ | Orthogonal projector on the square integrable functions of $\mathbf{X}_A$ |
| $\Pi_A$ | The set of permutations of $A \in \mathcal{P}_d$ |
| $\Pi_n$ | The set of permutations of $\{1, \ldots, n\}$ |
| $\mathbf{X}$ and $\mathbf{X}'$ | Two independent copies |
| $\mathbf{X} = (\mathbf{X}_A, \mathbf{X}_{\bar{A}})$ | |
| $\mathbf{X}^{\sim A} = (\mathbf{X}_A, \mathbf{X}'_{\bar{A}})$ | |
| $V = \operatorname{Var} Y$ | Total variance |
| $\xi_j$ | First-order sensitivity of $X_j$ |
| $V_A$ | Partial variance associated to $A$ |
| $V_j$ | Partial variance for $A = \{j\}$ |
| $V_{\bar{j}}$ | Partial variance associated to $A = \overline{\{j\}}$ |
| $S_A$ | Sobol' sensitivity index associated to $A$ |
| $S_j$ | First-order Sobol' index of $X_j$ |
| $S_A^{\text{clos}}$ | Closed Sobol' sensitivity index associated to $A$ |
| $S_j^{\text{T}}$ | Total Sobol' index of $X_j$ |

# Acronyms and Initialisms

| | |
|---|---|
| ANOVA | Analysis of variance |
| a.s. | Almost surely |
| BEPU | Best-estimate plus uncertainty |
| cdf | Cumulative distribution function |
| CLT | Central limit theorem |
| CPU | Central processing unit |
| DELSA | Distributed evaluation of local sensitivity analysis |
| DGSM | Derivative-based global sensitivity measures |
| DOE | Design of experiment |
| Eq. | Equation |
| FANOVA | Functional analysis of variance |
| FAST | Fourier amplitude sensitivity test |
| Fig. | Figure |
| GOSA | Goal-oriented sensitivity analysis |
| GP | Gaussian process |
| HDMR | High-dimensional model representation |
| HSIC | Hilbert–Schmidt independence criterion |
| IBLOCA | Intermediate break loss of coolant accident |
| ICU | Intensive care unit |
| i.i.d. | Independent and identically distributed |
| IPM | Integral probability metrics |
| LHD | Latin hypercube design |
| LHS | Latin hypercube sample |
| MMD | Maximum mean discrepancy |
| MST | Minimum spanning tree |
| NDT | Nondestructive testing |
| NPV | Net present value |
| OAT | One at a time |
| ODE | Ordinary differential equation |
| PCA | Principal component analysis |
| pdf | Probability density function |
| PDE | Partial differential equation |
| PLI | Perturbed law–based indices |
| PoD | Probability of detection |
| QMC | Quasi-Monte Carlo |
| QoI | Quantity of interest |
| QRBD | Quasi-random balance design |
| RB | Reduced basis |

| | |
|---|---|
| RBD | Random balance design |
| RKHS | Reproducing kernel Hilbert space |
| SA | Sensitivity analysis |
| SAE | Sum of absolute error |
| SB | Sequential bifurcation |
| SFD | Space-filling design |
| SRC | Standard regression coefficient |
| SRRC | Standard rank regression coefficient |
| UQ | Uncertainty quantification |

## Chapter 1

# Introduction

When dealing with a computer model which simulates real phenomena, global sensitivity analysis techniques aim to apportion the model's output uncertainty to uncertainty in its inputs. The results of this can then be used for model calibration, model validation, and decision-making processes, i.e., any processes where it is useful to know which variables contribute most to output variability. In many works, much attention has been paid to screening techniques [111] and variance-based sensitivity measures, also known as Sobol' indices [442]. Indeed, numerical model builders and users have shown great interest in tools like these, which take full advantage of the advent of high-powered computing and numerical methods; see [187, 110, 126] for industrial and environmental applications. Several texts [414, 415, 111, 127, 110, 395] have covered these topics rather extensively. However, none of these describes the many recent technical advances in sensitivity analysis (SA). This is the main goal of the book: to provide a unified view of most of the new theoretical and algorithmic results in and around SA while keeping an eye on applications of the methods described. For a more extensive view of practical issues related to SA, the reader could refer to the recent position paper of Razavi et al. [395]

In this introductory chapter, SA of the model output is first considered in the general framework of uncertainty management in the numerical simulation and computer experiment domains. Then Section 1.2 defines the four SA settings covering the main goals of users so that all of the SA methods described in the book can be linked to real applications. In order to outline the practical interest of SA in engineering and scientific studies, Section 1.3 provides a brief overview of seven examples from the authors' own experience. Following this, Section 1.4 more precisely describes the general SA methodology (with respect to the four SA settings) and the underlying mathematics. Section 1.5 introduces the R package sensitivity, which will be used throughout the book. Finally, the book's layout is summarized in Section 1.6.

## 1.1 ▪ Computer experiments, uncertainty quantification, and sensitivity analysis

Uncertainty studies of computer models are often carried out in order to take into account and better understand sources of imprecision when using numerical modeling. Arising mainly from the needs of industry, generic uncertainty management methodologies have been developed and are now widely used in practical settings [255, 110, 24]. All of these methods have now been grouped together under the name *uncertainty quantification* (UQ) [438]. UQ is also closely related to the broader subject area of the design, analysis, and modeling of computer experiments [421, 127, 241], which has been of exponentially growing interest across many scientific and engineering disciplines [157, 235].

A schematic diagram showing how UQ works is shown in Fig. 1.1. It was proposed and discussed in detail in de Rocquigny et al. [110] and gives a good summary of the main steps in UQ:

- Step A: Specify the model $Y = G(\mathbf{X})$ under study (with, in its simplest version, the random input variables $\mathbf{X} = (X_1, \ldots, X_d) \in \mathbb{R}^d$ and the output variable of interest $Y \in \mathbb{R}$) and the statistical quantity of interest (QoI) of the study.
- Steps B and B': Input UQ by direct or inverse methods in order to assign a probability density function (pdf) to each input, and model the probabilistic dependence between inputs.
- Steps C and C': Propagate the uncertainty of all inputs through the model, and quantify the influence of each on the QoI. Thus, SA is seen as an important part of the uncertainty propagation step.

If the numerical model $G(\cdot)$ is time-expensive, a metamodel can be a useful tool for performing steps B', C, and C'; this is a mathematical function that approximates the numerical model based on only a few calculations of $G(\cdot)$, thus providing a time-inexpensive predictor.

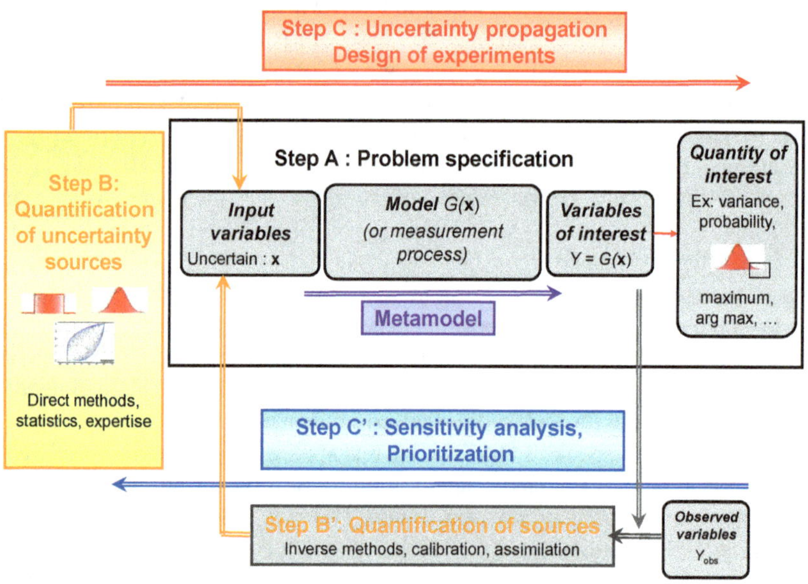


**Figure 1.1.** *The uncertainty management methodology.*

Metamodels are frequently used to simulate the behavior of an experimental system or time-expensive code based on a given number of output values. Under the name *response surface methodology*, it was originally proposed as a statistical tool for finding the operating conditions of a process at which certain responses were optimized [51]. Subsequent generalizations led to these methods being used to develop approximating functions of deterministic computer codes [118, 406, 243]. Essentially, constructing a metamodel means generating a fast-to-evaluate mathematical model that fits the initial data and has good predictive performance. They are thus efficient for UQ and SA methods requiring several thousand model calculations [127, 438].

Of course, close connections exist between metamodeling tools and statistical/machine learning techniques. Indeed, the latter consist of statistical methods that help one understand the functional relationships between variables. One of their main goals is the effective construction of an estimate of the function connecting the input variables to the output ones. Generally, no physical model is available, and the estimate is based only on the data. In recent years, many powerful statistical methods (regression trees, neural networks, etc.) have been studied and put into action on a huge class of problems, such as image processing and character recognition, among numerous others. A wide-ranging overview of statistical learning methods can be found in Hastie et al. [185].

In fact, it might be more correct to say that SA techniques can been thought of as cousins to variable selection problems in statistical learning. For example, there is a very strong connection between Sobol' sensitivity theory and importance score methods obtained when using resampling techniques like random forests [170, 499, 30, 8, 29].

There appear to be four major differences between theory developed for computer experiments and theory for dealing with experimental data—like in the machine learning paradigm:

- In an engineering or physical context, data are generally produced or representable by a physical or mathematical model linking the input variables to the output ones.
- In general, in machine learning (and especially neural networks), the statistician has access to a huge dataset. This is rarely the case in the physical modeling context, where data acquisition may be very expensive.
- In the computer experiment context, the data sample is fixed in advance, as the experimental design has to be defined by the user. Theoretical properties linking the design structure and a particular metamodel can then be considered in order to maximize the quality of the fitted metamodel.
- Computer experiments are usually deterministic (except in stochastic simulation cases), unlike real data samples that may be considered in machine learning, in which measurement errors are present but often unknown.

## 1.2 ▪ Sensitivity analysis settings

A clear description of the objectives of a study is essential before performing SA. Saltelli et al. [417] and Pappenberger et al. [357] have emphasized prioritizing the most influential inputs, identifying noninfluential inputs, mapping the output behavior as a function of the inputs, and calibrating certain model inputs using available information. Borgonovo [43] adds to this such things as understanding the model structure (especially in terms of interactions between inputs), analysis of the sign associated with changes (direction of variation of the model outputs), and stability (does some input perturbation change a model-supported decision?). With respect to such settings, many methods have been developed using various computational, mathematical, and statistical tools coming from data analysis, graphical analysis, information and decision theory, experimental design, Monte Carlo techniques, statistical learning methods, and so on. Several overviews of these methods have been provided in the literature [240, 142, 189, 110, 357, 209, 26, 499, 48, 213, 46, 395]. Note that in these works, an input is sometimes called a *factor* or *parameter*.

This book proposes an engineering point of view and considers the following four important SA settings:

1. The *model exploration* setting aims to understand the behavior of a model by trying to investigate the input-output relationship, e.g., starting with an initial visual analysis of the input-output sample.

2. The *factor fixing* setting aims to reduce the number of uncertain inputs by setting unimportant factors as constants. Unimportant factors are those that, if set to any particular value, do not lead to a significant reduction in output uncertainty.

3. The *factor prioritization* setting aims to identify the most important factors. For instance, in the case of independent inputs, the most important one is that which, if set to a specific constant, would lead to the greatest reduction in uncertainty in the QoI. Variance-based SA is the traditional approach here, but when looking for inputs important to a specific range of output values, methods need to be modified to take this into account.

4. The *input pdf robustness* setting aims to analyze variation in the QoI with respect to uncertainty in inputs' pdfs.

The *input pdf robustness* setting is a new proposal of the present book but can be connected to the stability setting of Borgonovo [43]. The *factor fixing* and *factor prioritization* settings were introduced in Saltelli and Tarantola [416] and Saltelli et al. [417], along with two other settings: a *variance cutting* setting—whose goal it is to find which inputs have to be set as constants to obtain a target value for the output variance—and a *factor mapping* setting—whose goal it is to determine which inputs are most responsible for producing values of the output in a specific region of interest. From the present book's point of view, these two settings are integrated into the factor prioritization one. Note also that the *model exploration* setting includes the *model structure* and *sign of change* settings described in Borgonovo [43].

## 1.3 ▪ Examples of real-world problems

SA concerns all applications where numerical simulation models are used to understand, analyze, and optimize real processes, e.g., in the environmental, business, social science, and engineering fields. This section contains several examples coming from the authors' direct experience.

### 1.3.1 ▪ Biogeochemical application

Marine biogeochemical models are now commonly integrated as modules within complex ocean circulation modeling systems. They are thus increasingly used across a range of applications. However, these models raise certain questions, especially regarding their tuning. Generally speaking, they are systems of nonlinear ordinary differential equations, each equation expressing the time evolution of a given state variable due to hydrodynamic effects (transport and diffusion) and fluxes between the various components of an ecosystem. One particularity is that the evaluation of these fluxes involves numerous parameters, typically several times more than the number of variables. A second is that the values of these parameters are often not very well known; in fact, they depend on the physical and biogeochemical context, while their available reference values were generally estimated in a particular situation (during a field trial) or in laboratory experiments (i.e., not in a real-world setting). The uncertainty in these values is thus generally quite large (see, e.g., [426] for a review of the identification of such parameters). In this context, SA may be valuable. It can help one better understand the model itself (*model exploration* setting) and identify which parameters are the most influential and thus must be carefully calibrated (*factor fixing* setting).

The model used in this particular example was MODECOGeL.[5] It was developed by Lacroix [256] for investigating biogeochemical activity in the Ligurian Sea, coupling a 1D hydrodynamic model of the mixed layer with a 12-component ecosystem model.

[5]MODECOGeL: MODèle d'ECOsystème du GHER (GeoHydrodynamics and Environment Research) et du LOV (Laboratoire d'Océanographie de Villefranche-sur-Mer).

This hydrodynamic model is a 1D version of GHER's primitive equations model [339]. The state variables are the horizontal velocity, the potential temperature, the salinity, and the turbulent kinetic energy. A full description of the model can be found in Lacroix and Nival [258] and Lacroix and Grégoire [257], where it was applied to simulate the behavior of the system during the FRONTAL[6] oceanographic expeditions from 1984 to 1988. The ecosystem model provides a 12-component ($C_i$, $i = 1, \dots, 12$) description of the ecosystem of the Ligurian Sea. The time evolution of each state variable is governed by the equation

$$\frac{\partial C_i}{\partial t} = \texttt{ADV}_i + \texttt{DIFF}_i + \texttt{SMS}_i \quad \text{with} \quad \texttt{SMS}_i = \sum_{j \neq i} \texttt{FLUX}(C_j \to C_i), \tag{1.1}$$

where $\texttt{ADV}_i$ and $\texttt{DIFF}_i$ are advection and diffusion terms (governed by the hydrodynamic model), and $\texttt{SMS}_i$ is the *source minus sink* term summing up the fluxes ($\texttt{FLUX}(C_j \to C_i)$) between the various components of the ecosystem (conservation implies that $\texttt{FLUX}(C_j \to C_i) = -\texttt{FLUX}(C_i \to C_j)$). Equation (1.1) is solved numerically (between the sea surface and a depth of 405 m) using a constant vertical discretization (1 m) and a constant time step (6 minutes). Outputs are saved daily at all depths.

Without getting into details (a more complete description of the ecosystem model can be found in Lacroix and Grégoire [257]), we simply emphasize the fact that the mathematical expression of these flux terms contains numerous parameters whose values are not known precisely.

The biogeochemical fluxes ($C_j \to C_i$) parametrized in MODECOGeL are summarized in Table B.6 in Prieur et al. [374]. Each flux depends on several parameters, which are listed in Table C.7 in Prieur et al. [374]. To give an idea of the role of each parameter in MODECOGeL, the biogeochemical fluxes are organized into several categories: primary production, secondary production, mortality, exudation, excretion, growth of bacteria, decomposition of particulate organic matter, and nitrification. A detailed description of these different categories can be found in Prieur et al. [374]. This variety of processes, many of them depending on several factors, explains why there are so many parameters in biogeochemical models. Moreover, the parametrization of each process as a joint function of the model state and parameters is often complex and nonlinear, with the consequence that it is usually impossible for a user to obtain an idea of the sensitivity of the whole system to the parameters. A rational and automatic approach is thus needed. To model the uncertainty, a probability distribution must be specified for each input parameter (see Table C.7 in Prieur et al. [374]).

Different quantities of interest (i.e., the output values $y$) can be defined according to the main scientific objectives of a sensitivity study. For example, it is possible to take an interest in, for any $i = 1, \dots, 12$, the values of $\max_t C_i(0, t)$ and $\text{argmax}_t C_i(0, t)$.

In the context of the above application, the inputs space's dimension is equal to the (large) number of parameters $d = 74$. For reasons such as this, most SA studies require a potentially prohibitive number of model evaluations. Prieur et al. [374] present an implementation using a grid computing environment, with computing resources distributed across three sites.

### 1.3.2 ▪ Flooding model

An assessment of the water level in sections of a watercourse is essential for several industries subject to flood risks as well for populations living near rivers or in flood-prone areas. Safety studies are undertaken in order to compare water levels which could be reached to the height of a dike protecting from flooding (Fig. 1.2). When the water level exceeds the height of the dike, flooding occurs.

[6]FRONTAL report, 1989. Résultats des missions "FRONTAL." Tome I: 1984–1985. Tome II: 1986–1988. Station zoologique, Villefranche-sur-mer, France

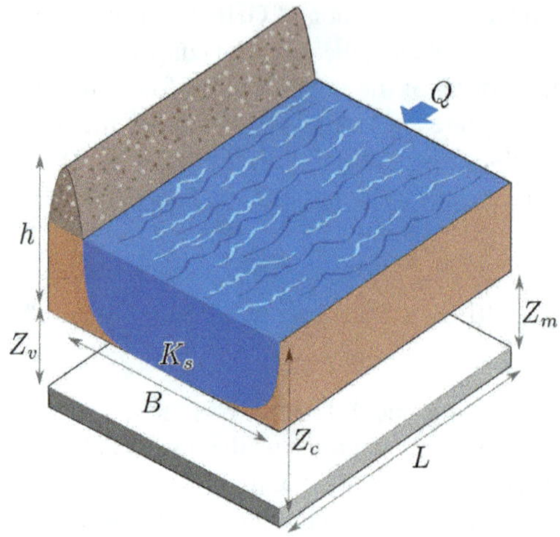


**Figure 1.2.** *The flood example: schematic view of a river and a dike. Courtesy of Merlin Keller (EDF R&D).*

With mathematical models, the water level can be evaluated as a function of the discharge and other physical parameters. This hydraulic phenomenon is governed by the Saint Venant shallow water equations (1871), relying the water level $H$ (measured with respect to the riverbed) at the abscissa $x$ and time $t$ to the discharge $Q$, the water section $s$, the lateral inflows $q_L$, the slope $I$, and the head losses $J$ due to friction between the water body and the riverbed:

$$\frac{\partial s}{\partial t} + \frac{\partial Q}{\partial x} = q_L,$$
$$\frac{\partial Q}{\partial t} + \frac{\partial (Q^2/s)}{\partial x} + gs\frac{\partial H}{\partial x} = gs(I - J).$$

From a simplified version of the Saint Venant equations, an analytical model can also be obtained; this has been used as the basis of several works—see [108, 109, 209, 126].

At EDF, the Mascaret and TELEMAC2D computer codes (resp., 1D and 2D solvers of the Saint Venant equations [168, 166]) allow engineers to simulate flood events. Petit et al. [367] and Roustant et al. [401] have applied SA techniques to the Mascaret model, simulating the flooding of a specific French river. They considered 37 inputs (geometric variables, river friction coefficients, slope perturbations, and discharge variables). The goal of these studies was to identify influential variables among the large number of inputs (*factor fixing* setting). Using TELEMAC2D, Goutal et al. [167] analyzed in-depth the sensitivity of spatially localized and simulated water levels with respect to six uncertain hydraulic model inputs (*factor prioritization* setting).

### 1.3.3 ▪ Optimization of maintenance investment planning

Energy production companies want to assess and optimize their strategic investment decisions for their electricity production assets using probabilistic and optimization methods involving *cost of maintenance* strategies. In order to quantify the technical and economic impact of a candidate maintenance strategy, EDF has developed the VME software [291], which allows for the computation of certain economic indicators using Monte Carlo simulation (see Fig. 1.3). The major output result of VME's Monte Carlo simulation process is the probability distribution of the

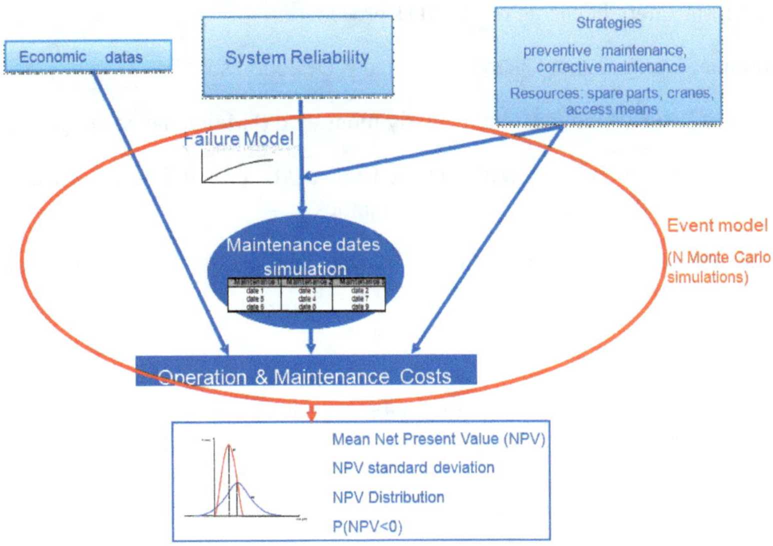


**Figure 1.3.** *Schematic diagram of EDF's maintenance strategy tool VME. Courtesy of Jérôme Lonchampt (EDF R&D).*

*net present value* (NPV) associated with the given maintenance strategy. From this distribution, certain indicators, such as the NPV's mean and standard deviation and the investment regret probability ($\mathbb{P}(\text{NPV} < 0)$), can easily be derived.

Uncertain input variables are numerous in such studies, including such things as the reliability data of components, economic indicators, duration of maintenance tasks, the operating times of facilities, and maintenance task dates [117]. In order to reduce uncertainty in the outputs, engineers need to understand the influence on NPV indicators of the input variables and interactions between them. The first objective is thus to quantify noninfluential inputs that can be set as constants without affecting output uncertainty (*factor fixing* setting).

Once the NPV indicators are obtained with sufficient accuracy, engineers are interested in optimizing the maintenance strategy, for instance, by determining the optimal investment dates, leading to the highest mean NPV and lowest investment regret probability. Due to the discrete nature of the events to be optimized, the method turns out to be based on genetic algorithms, which often require several thousand simulation runs. The main issue therefore becomes the computational cost of the stochastic simulator being optimized. Methods requiring a minimal number of simulation runs are of great interest, which is where metamodel-based methods associated with SA come in to play [334, 59].

### 1.3.4 ▪ Nondestructive examination of structures

The *probability of detection* (PoD) curve is a standard tool in a number of industries for evaluating the performance of *nondestructive testing* (NDT) procedures. The goal is to quantify the capacity of inspections to detect harmful flaws in a structure of interest. For example, in the aeronautic industry, regulations require damage tolerance assessments to be made to address the potential for failure due to critical material parts and manufacturing variability.

Figure 1.4 illustrates this tool when used for flaw detection. The PoD is plotted as a function of the main parameter of interest: the flaw size. Several indicators are extracted from this curve in order to perform safety licensing of the NDT procedure:

- the minimal flaw size that has to be detected ("Detection start");
- the sizes of flaws which must be regularly monitored ("Maintenance margin");
- the minimal size of flaws ($a90/95$) detected with a $90\%$ probability (associated with a $95\%$ confidence bound due to estimation uncertainties);
- the distance between the latter and the largest flaw size that retains structural integrity ("Safety margin").

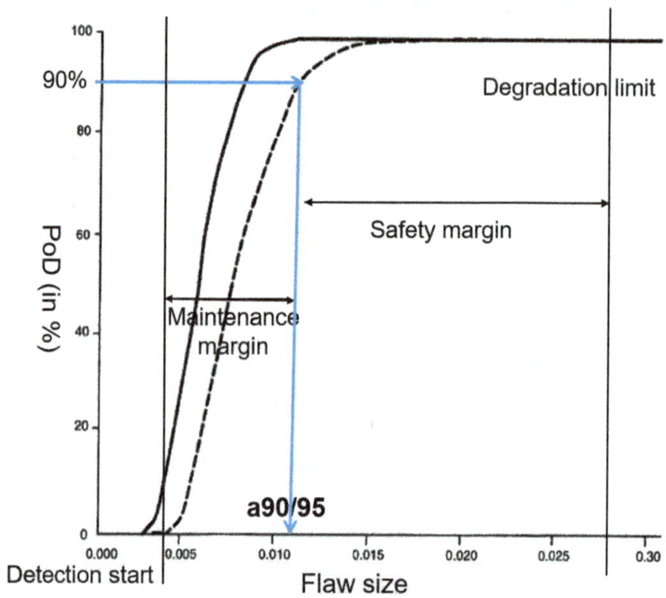


**Figure 1.4.** *A probability of detection curve (solid line) with its* $95\%$ *lower confidence bound (dashed line) and its decision indicators.*

Estimation of the PoD sometimes relies on deterministic computer models which simulate complex physical phenomena [63, 265]. This allows high costs to be avoided when calculating PoD curves for complex NDT procedures. In this framework, the uncertain input parameters of the computer code are modeled as random variables, with each assigned a pdf. In association with PoD, SA has proved to provide useful information to engineers trying to determine parameters that most influence PoD uncertainty (*factor prioritization* setting) [265].

However, most of the inputs' pdfs are not known precisely, so it is also necessary to gain in confidence with respect to this second level of uncertainty. This corresponds to having an *input pdf robustness* setting, which is considered in the PoD context in Iooss and Le Gratiet [208].

### 1.3.5 ▪ Management of oil reservoirs

A key challenge for oil producers is the efficient management of mature fields, with the specific goal of trying to improve future recovery. However, reservoir engineering studies typically involve a large number of parameters with uncertainty in them. Such parameters include geological ones (which can only be observed pointwise through drilling or indirectly through seismic campaigns), the flow properties of water and oil phases underground, and the position of faults in a reservoir.

For UQ of future production and for designing strategies for better recovery, it is thus crucial to identify which parameters have the highest impact on recovery. This explains why SA has become an important tool in reservoir engineering studies. It is worth mentioning that in general SA is coupled with a preliminary metamodeling step since reservoir engineering involves CPU-intensive fluid flow simulations.

UQ with a preliminary SA step for a mature field is detailed, for example, in Feraille and Marrel [132]. An illustration of SA for chemical-enhanced oil recovery with surfactant-polymer flooding can be found in Douarche et al. [116]. Both cases fall into the *factor prioritization* setting.

For the case of chemical-enhanced oil recovery, the objective is to inject polymers or surfactants to help increase recovery in mature reservoirs. But the process efficiency is severely impacted by many parameters, such as some polymer and surfactants properties, the reduction of oil saturation produced by these chemicals, or oil and water relative permeability. This is illustrated in Fig. 1.5.

SA can help to quantitatively analyze which parameters have the largest impact on the economic pertinence of deploying such chemical processes.

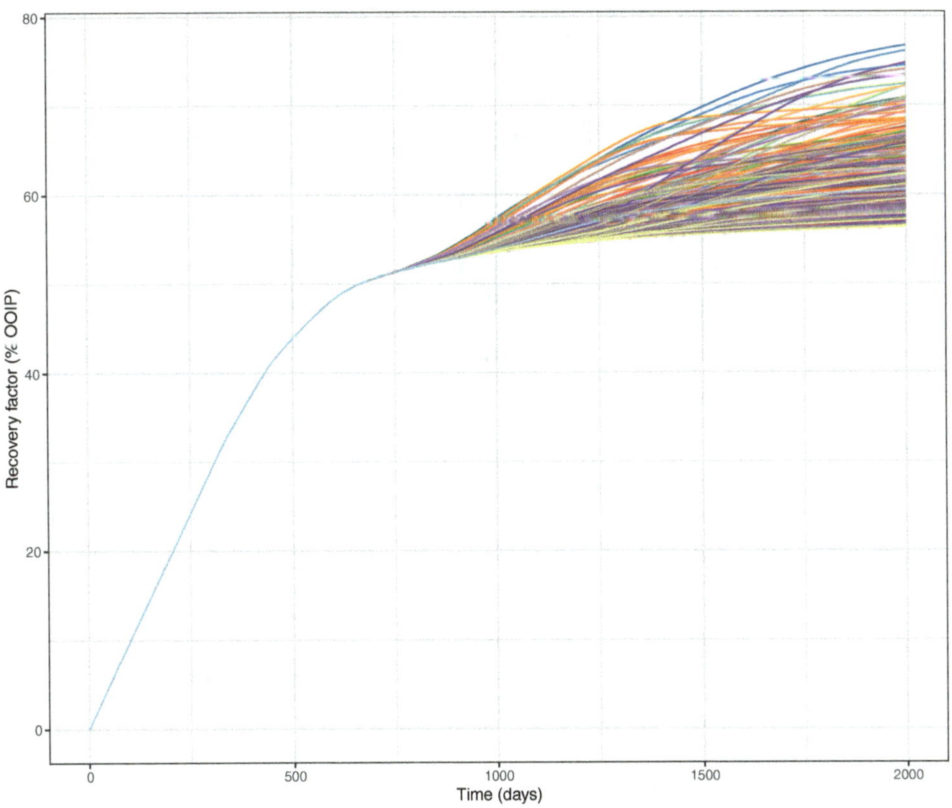


**Figure 1.5.** *Differences in recovery for different EOR (enhanced oil recovery) process parameters (based on Figure 11 from [116]).*

## 1.3.6 ▪ Thermal-hydraulic accident scenario simulation

The management of sources of uncertainty in the nuclear engineering field is of utmost importance, as it provides a complementary paradigm to assess the reliability and safety of nuclear facilities alongside standard deterministic safety analyses. The quantification of input uncertainties and their propagation through high-fidelity computer models—in order to estimate relevant statistical quantities of interest—constitutes the core phases of *best-estimate plus uncertainty* (BEPU) methods [342, 379]. This then typically leads to the computation of high-order quantiles for model outputs (temperature, pressure, etc.).

One example of this is simulation of an intermediate-break loss-of-coolant accident (IBLOCA) in a pressurized water reactor (see Fig. 1.6), which can be performed using a costly-to-evaluate thermal-hydraulic computer model. Indeed, as an experimental approach undertaken at a test facility only allows for the study of a limited number of situations, it needs to be bolstered by numerical simulations of the experimental loop [317]. A typical IBLOCA simulation may involve around a hundred uncertain input variables to be handled within the uncertainty modeling and propagation phases. Therefore, performing SA in a preliminary phase in the *factor fixing* setting is essential (see, e.g., the thermal-hydraulic application cases developed in Auder et al. [12] and Iooss and Marrel [210]).

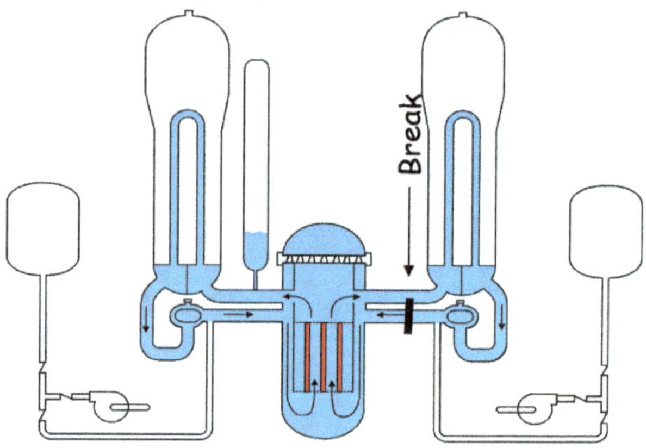


**Figure 1.6.** *Schematic view of a break in the hot leg of the primary circuit leading to an IBLOCA.*

Moreover, in a safety context, hypotheses have to be made in order to define input distributions; these may not always be easy to justify. This is especially the case in thermal-hydraulic accident scenario simulations, where model inputs are subject to epistemic uncertainty, meaning that they suffer from a lack of physical knowledge [341]. In all such cases, probabilistic models for these variables need to be established and associated with a certain level of accuracy or confidence. In this setting, SA may be useful to narrow down the list of input variables that most influence the QoI (e.g., 95% quantile of the model output). Even more importantly, SA allows us to quantify the impact of a lack of information about the input pdfs on the QoI (*robustness analysis* setting), as shown in Sueur et al. [467].

## 1.3.7 ▪ Loosening manufacturing tolerances

In industrial processes, parts are usually produced with small geometrical differences scattered around a nominal value. These unavoidable differences potentially influence the performance of objects being produced. To ensure adequate performance, manufacturing tolerance—i.e., the range of values allowed for in production—is kept tight enough that its impact can be supposed negligible. However, loosening certain key manufacturing tolerances might result in lower manufacturing costs while maintaining performance at a satisfactory level.

In particular, geometric tolerances for parameters with little influence on performance could be loosened; this strategy could then be encoded as an SA study. This point of view was taken in Vignat et al. [490] in the context of a helicopter engine's combustor, illustrated in Fig. 1.7. Around the compressor, several multiperforations are placed so as to refresh the combustor. The design and production of these multiperforations are driven by many parameters and associated manufacturing tolerances. In this example, a preliminary study showed that some of these parameters had very low impact on the combustor's performance and that loosening their tolerances would still lead to a satisfactory engine, while others were crucial and should not be loosened. More precisely, this was carried out by evaluating the impact of certain cost-driving manufacturing tolerances on the combustor's performance via SA (*factor prioritization* setting). The input variations were obtained by collecting real data on the production line.

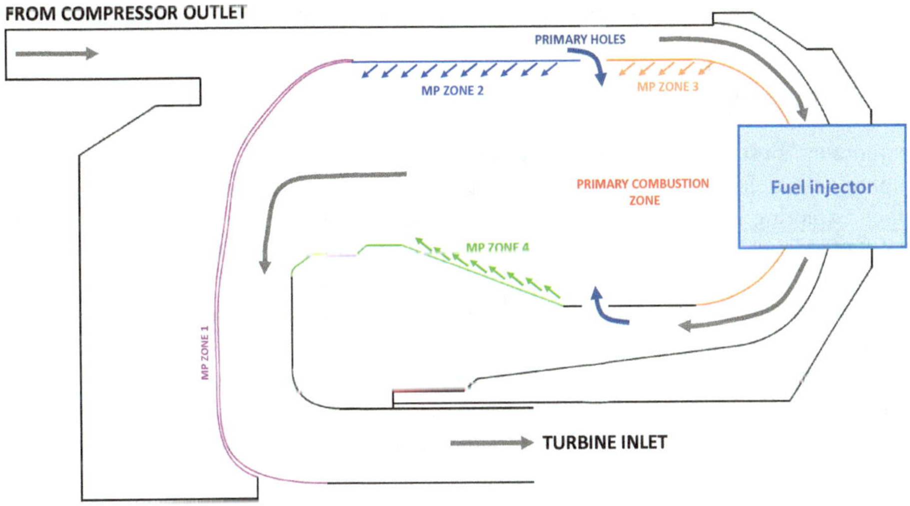


**Figure 1.7.** *Schematic view of a helicopter engine combustor. Courtesy of Julien Leparoux (SafranTech).*

## 1.4 ▪ General methodology

In this book, the model input vector is denoted $\mathbf{X} = (X_1, \dots, X_d) \in \mathbb{R}^d$. The inputs $X_i$ ($i = 1, \dots, d$) are assumed to be independent, except in Chapter 5 or when otherwise indicated. For the sake of simplicity, except in Section 3.3 of Chapter 3, our studies are restricted to a scalar output $Y \in \mathbb{R}$ of the computer code (or *model*) $G(\cdot)$:

$$Y = G(\mathbf{X}). \tag{1.2}$$

In the probabilistic setting, $\mathbf{X}$ is a random vector defined by a probability distribution, and $Y$, the model output of interest, is a random variable. The case where $\mathbf{X}$ has a nonfixed pdf will also be considered in Section 6.4 of Chapter 6.

Depending on the information required by the study and the SA settings (see Section 1.2), different classes of mathematical methods can be defined; see Iooss and Lemaître [209] for a previous classification attempt. This type of classification scheme can help users select the most appropriate method.

### 1.4.1 ▪ Model exploration

In an initial SA approach, one may simply want to better understand the direction of variation, regularity and complexity of model outputs, identify potential discontinuities, uncover possible interaction effects between inputs, etc.

This initial run could be based on a rough exploration of the model, with several well-located runs (in the domain of variation of the inputs). The results of this initial analysis of the input-output sample can then be visualized using tornado diagrams, scatterplots, cobweb plots, CUSONORO curves, etc. [255, 136]. Statistical models such as smoothing techniques can also be useful for visualizing the main effects of the inputs [461].

Of course, the whole range of SA techniques are designed to explore models; we describe three—more sophisticated—ones in the following sections. The use of a metamodel (Section 1.1), which allows new points to be predicted with minimal CPU time cost, then gives access to a full exploration of the input space, possibly with the help of visualization techniques [136].

### 1.4.2 ▪ Screening

An exploratory design is one which tries to transmit information to the model's outputs from the input domain of variation, often starting with total ignorance of the model's output behavior and then exploring it in the most efficient way possible. One of the first goals is to identify noninfluential inputs using only a small number of model runs while making realistic hypotheses on the model's complexity. This is called *screening* [111, 506].

This more sophisticated SA approach consists of detecting active and inactive variables and is particularly important in cases where a model has many inputs (e.g., several dozen). Screening techniques aim to establish a qualitative ranking of input factors with a minimal model evaluation budget. They are mainly associated with the *factor fixing* setting and are often a preliminary step before calibration—using available information (real observed outputs, constraints, etc.) or before building a metamodel.

### 1.4.3 ▪ Quantitative partitioning

In many cases, one wishes to more precisely quantify the effect of each uncertain input on the output of interest. In the probabilistic approach shown in Fig. 1.1, one has to investigate the joint pdf of the inputs. From an initial sample of input and output values (e.g., a Monte Carlo sample or another numerical experimental design which uses the input joint pdf), quantitative sensitivity indices can be obtained by analyzing the statistical relationship between the inputs and the output. This sampling-based approach, introduced in Helton et al. [189], then allows the uncertainty propagation and SA steps to be treated in the same way.

It is also possible to further exploit the stochastic framework—which furnishes specific tools such as conditional probability distributions and the decomposition of variance, leading to global sensitivity measures like Sobol' indices and generalizations thereof (see Chapters 3, 4, and 6). Then statistical tools such as sampling and asymptotic theory allow us to propose and study statistical estimates of global sensitivity measures.

In this context and for the *factor prioritization* setting, which aims to provide a quantitative sensitivity index for each input, many methods have been proposed in recent decades, including the following:

- Linear correlation coefficients and standard regression coefficients based on a linear model relating the inputs to outputs [417]. Simple extensions of these elementary measures consist in using the rank transformation of the input and output samples, leading to rank correlation coefficients and standard rank regression coefficients.

- General variance-based importance measures (Sobol' indices) which give, for any nonlinear model $G(\cdot)$ with finite variation (i.e., $\text{Var}(Y) < \infty$), the parts of the output variance due to each input and each interaction between inputs [442, 373].

- Derivatives-based global sensitivity measures which extend the deterministic local approach to a probabilistic global one (averaging the sensitivities given by the derivatives over the whole input space) [251, 249].

- Density-based importance measures [43], which are moment-independent metrics that complement the vision provided by variance-based ones.

- Dependence measures and kernel-based sensitivity indices [99, 107], which extend Sobol' indices to take into account higher-order information (than the expectation) about output behavior in order to provide more in-depth details.

In most real-world applications, SA is applied by using the output variance (or some other uncertainty measure) for the QoI, even if the real QoI of the study concerns a targeted region of the output variable's domain. In Chapter 6, SA techniques that consider more specific QoI will be described. In particular, it is important to use the correct SA methodology when the QoI is associated with rare events (e.g., an extreme quantile or failure probability).

### 1.4.4 ▪ Robustness analysis

One of the most critical hypotheses in uncertainty propagation studies involves the choice of distribution for uncertain input variables, a choice which propagates through the numerical model. In general, each initial pdf is selected based on various theoretical or heuristic rules (statistical inference, design or operation rules, expert judgment, etc.) and is then refined to a certain level of accuracy or confidence. Hence, bringing stringent justification to an overall approach requires quantifying the impact of pdf modeling assumptions on the QoI.

*Perturbed law-based indices* (PLIs)—different from standard SA methods—perform this type of robustness analysis [272, 208, 363], fully addressing the input pdf robustness question in SA. The PLI principle is to assess the influence of a perturbation on a parameter or moment of the input pdf on some QoI. Any QoI can be considered, e.g., the mean model output, its variance, a probability that the output exceeds a threshold, or a quantile of the output. The quantile case is studied in Sueur et al. [467] and Gauchy et al. [153], with their quantile-PLI applied in a thermal-hydraulic setting. One big advantage of their technique is that the number of model runs required to estimate the PLI is independent of the number of inputs.

Other solutions can be based on a probabilistic modeling of the parameters of the inputs' pdfs, followed by computing sensitivity indices of these epistemic parameters [327]. The robust Bayesian analysis [33, 204] is also a great ingredient in the study of the sensitivity of Bayesian answers to assumptions and uncertain calculation inputs. Such approaches can be adapted to the context of SA of model outputs. Finally, extraprobabilistic modeling of the inputs is also proposed by numerous authors, and certain sensitivity tools can be associated with this more complex representation [188].

## 1.5 ▪ Introduction to the R package sensitivity

In this book, some examples are run using commands available in the sensitivity package [206] in R. The R software [385], commonly used for statistical analyses, is a powerful tool for knowledge diffusion in the statistics community. Since it is open source and easily available, R is the software of choice for many statisticians in education and industry. For global SA, the sensitivity includes a collection of functions for factor screening, sensitivity measure estimation, and reliability SA of model outputs. Most of the functions are for models with scalar outputs only, though a few support multidimensional ones.

In this package, programming is object oriented and works with SA "objects." Each SA method has its own SA object. In this workflow, we run generic print() and plot() (or ggplot() from the ggplot2 package) methods on SA objects. Indeed, calls to these methods are found in most of the sensitivity package's functions.

The sensitivity package has been designed to work with either models written in R or external models such as large-scale computational codes. This choice is entered via the "model" input argument, present in all of the package's functions. The "model" argument is expected to be either a function or a predictor (i.e., an object with a predict function such as, for instance, lm()). This model is invoked once for the whole experimental design. It can also be left as NULL; the latter is referred to as the decoupled approach and is used with external computational codes that rarely run on the statistician's own computer. Examples of how to use each function in the sensitivity package can be found using R's built-in help support.

After launching R and installing the sensitivity package, loading the package requires the following command:

```
library(sensitivity)
```

To know more about the package's features, type

```
help(package="sensitivity")
```

A detailed list with explanations of all features in the sensitivity package is also provided in Section D.1.1.

Finally, important features of the sensitivity package are the useful ask() and tell() functions [86], which allow calls to the model to be decoupled from running the SA function. The ask() function interacts with the SA method in order to provide an answer to the question, Which simulations are needed?, while the tell() function interacts with the SA method in order to update the SA result's object by saying, "Here are the simulations." More details on these can be found in Collette et al. [86].

## 1.6 ▪ Overview of the methods and contents of this book

The purpose of this book is to give an overview of past and present SA methods and algorithms, including some very recent theoretical results. In some places, an engineering point of view is also taken in order to explain the usefulness of methods in practice. In addition, applications of several methods are run using R commands (see Section 1.5). Indeed, it is clear to us that in order for SA methods to be widely used by researchers, engineers, end-users, and students, simple-to-use software is fundamental. For reproducibility, all the codes used throughout the chapters are freely available at `https://bookstore.siam.org/cs23/bonus`.

Chapter 2 presents several screening techniques devoted to identifying influential and noninfluential inputs (coming under the *model exploration* and *factor fixing* settings). As most of these methods are not based on deep theoretical bases, the choice has been made to introduce them using examples and R code. Chapter 3 develops the theoretical aspects of Sobol' indices (*model exploration*, *factor fixing*, and *quantitative partitioning* settings) as well sampling-based estimators. Chapter 4 is devoted to the spectral and metamodel-based approaches for estimating Sobol' indices. However, when model inputs are statistically dependent, the methods of Chapter 3 do not apply. Chapter 5 presents ways to deal with variance-based measures in that context (*model exploration* and *factor fixing* settings). As the variance is known to only provide partial information on uncertainty, Chapter 6 explains how to go beyond variance-based sensitivity measures (all settings as well as *robustness analysis*). Finally, a case study in R related to a COVID-19 epidemic model (calibrated on the 2020 French infection data) is given in Chapter 7.

As shown in this book, SA remains a very active field of research. Real-world problems are often more complex than the simple models supposed in theoretical and methodological works. Moreover, due to their very new or incomplete nature, several other SA issues have not been discussed in this book. Bibliographical references and possible paths for future research are given at the end of each methodological section in sections called "Conclusions and Further References."

# Chapter 2

# A first look at screening using R

Screening methods are preliminary numerical experiments designed to identify inputs (also known as factors) with the largest influence on a particular model output. The aim of screening is to end up with a short list of important factors via a rapid (in terms of CPU time) exploration of a code's behavior. One assumption often used (and often verified) as a working hypothesis in screening design is that the number of inputs that are truly important to the model output is small by comparison to the total number of variables in the model. Since screening designs deal with models containing a large number of inputs, a trade-off has to be made between computational cost and information extracted.

Section 2.1 describes a few methods inspired by the design of experiment techniques, focusing on those which have been developed in the context of numerical experiments. Then Section 2.2 presents graphical tools along with regression analysis and feature selection techniques that can be used with ease on any sample of model inputs and outputs. Section 2.3 describes visualization tools and measures based on the model output's derivative (with respect to each input). All these sections consider the cases where the model input variables are quantitative. Then Section 2.4 provides a few references and discusses the adaptation of screening methods to deal with qualitative inputs.

## 2.1 ▪ Screening by discretized designs of experiments

### 2.1.1 ▪ Introduction

The most natural SA approaches come from the design of experiments (DOE) field [325], developed a century ago for physical experiments (not computer ones). This field contains several techniques that answer screening requirements (e.g., well-known fractional factorial designs) and can therefore be considered a global (statistical) SA approach. The inputs are assumed to belong to a compact subset $D = \prod_{j=1}^{d} [x_j^{\min}, x_j^{\max}]$ of $\mathbb{R}^d$, and methods are most often based on a discretization of the inputs into levels, following standard practice in classical design of experiments theory developed in the context of physical experiments. All of the designs presented in this section are built on discretized input spaces.

### 2.1.2 ▪ Discretized designs of experiments

#### OAT designs

The screening method most used in engineering is based on the notion of *one-at-a-time* (OAT) design, where each input is varied while the others are kept fixed [241]. One of the most natural ways to define an OAT is to choose as reference point $\mathbf{x}^* = \left((x_1^{\min} + x_1^{\max})/2, \ldots, (x_d^{\min} + x_d^{\max})/2\right)$, the center of $D$. The reference experiment is then defined as $G(\mathbf{x}^*)$. The OAT effects $\Delta_j^+ y$ (resp., $\Delta_j^- y$) for each input $X_j$, $j = 1, \ldots, d$, are computed as [48]

$$\Delta_j^+ y = G(x_1^*, \ldots, x_{j-1}^*, x_j^{\max}, x_{j+1}^*, \ldots, x_d^*) - G(\mathbf{x}^*),$$
$$\Delta_j^- y = G(\mathbf{x}^*) - G(x_1^*, \ldots, x_{j-1}^*, x_j^{\min}, x_{j+1}^*, \ldots, x_d^*).$$

This leads to a design which requires $2d + 1$ model runs. These elementary effects represent the differences between the perturbed and reference responses. They are then compared to assess which factors are most significant in affecting the response. If for a particular input, say $x_j$, $\Delta_j^+ y$ and $\Delta_j^- y$ are different, then the relationship between input $x_j$ and the output is not strictly linear.

The main problem with OAT designs is that they cannot provide information about interactions between inputs or precision on estimated effects. Moreover, the results are meaningful only if the model's input-output relationship can be adequately represented by a first-order polynomial in the inputs. If the model is affected by nonlinearities (as is often the case in practice), then changes in inputs would lead to drastically different "sensitivities" depending on the choice of reference point $\mathbf{x}^*$. In any case, in practice OAT design is often used as a verification procedure to manually test the effect of each input (see, for instance, the studies of Auder et al. [12]) and to detect software anomalies.

OAT design is one particular tool from a wide class of methods originating in classical DOE [325] theory, which aims to explore as well as possible, based on few well-designed experiments, a phenomenon described by an input-output model. More specifically, OAT designs are discretized designs in the sense that their inputs are assumed to take only a finite (in general small) number of values (called *levels*). They can be used to perform a rough version of SA. In the following section, we give a global overview of the tools and techniques used in DOE, referring the reader to some of the numerous reference books on the topic for more details.

### Factorial designs

Modern DOE has evolved from full factorial designs—which had the serious limitation of being constrained by the number of factors that could reasonably be investigated. Indeed, a full factorial design (see, e.g., [325]) consists in evaluating a model at all $L^d$ combinations of the discretized input values, with $d$ the input space dimension and $L$ the number of levels (if supposed to be the same for all factors). Full factorial designs are typically conducted at two levels only, low and high ($L = 2$), and are suitable only when there is a limited number of factors. Contrarily to OAT designs, full factorial designs allow one to estimate these interaction effects in a model of the form $G(\mathbf{x}) = \sum_{\mathbf{u} \subset [1:d]} \beta_{\mathbf{u}} \prod_{j \in \mathbf{u}} x_j$ by, for all $\mathbf{u} \subset [1:d]$,

$$\hat{\beta}_{\mathbf{u}} = \frac{1}{2}\left(\bar{G}_{\mathbf{x_u}=+1} + \bar{G}_{\mathbf{x_u}=-1}\right),$$

with $\bar{G}_{\mathbf{x_u}=+1}$ the mean of the responses of the model under the constraint $\prod_{j \in u} x_j = +1$ and $\bar{G}_{\mathbf{x_u}=-1}$ the mean of the responses of the model under the constraint $\prod_{j \in u} x_j = -1$.

Fractional factorial designs were then introduced to increase the number of factors that could reasonably be tested. Probably the most widely used ones are regular two-level fractional factorial designs, known as $2^{d-k}$ designs, with $1 \leq k < d$ an integer. As not all combinations of levels are tested, information on certain factor interactions is indistinguishable from that on others. This is called *aliasing* and is a known limitation of fractional designs. Table 2.1 gives the $2^{3-1}$ regular fractional factorial design constrained by the so-called defining relation $x_1x_2x_3 = +1$.

**Table 2.1.** $2^{3-1}$ *fractional factorial design constrained by* $x_1x_2x_3 = +1$.

| run | $x_1$ | $x_2$ | $x_3$ | $x_1x_2x_3$ |
|---|---|---|---|---|
| 1 | $-1$ | $-1$ | $+1$ | $+1$ |
| 2 | $-1$ | $+1$ | $-1$ | $+1$ |
| 3 | $+1$ | $-1$ | $-1$ | $+1$ |
| 4 | $+1$ | $+1$ | $+1$ | $+1$ |

This design involves only half of the full factorial design. From the defining relation $x_1x_2x_3 = +1$, it follows that $x_1 = x_2x_3$, $x_2 = x_1x_3$, and $x_3 = x_1x_2$; i.e., each main effect is aliased with a two-factor interaction. Thus, if one assumes a linear regression model without any interactions of order greater than or equal to three, $\beta_0+\beta_1x_1+\beta_2x_2+\beta_3x_3+\beta_{12}x_1x_2+\beta_{13}x_1x_3+\beta_{23}x_2x_3$, it is not possible to estimate separately the main effects and the interaction effects. It is possible to infer the submodel $\beta_0 + \gamma_1x_1 + \gamma_2x_2 + \gamma_3x_3$, where $\gamma_j$ contains both the main effect of $x_j$ and the interaction effect of $\mathbf{x}_{-j}$.

The main effect $x_1$ is said to be aliased with the second-order interaction $x_2x_3$, the main effect $x_2$ with the second-order interaction $x_1x_3$, and the main effect $x_3$ with the second-order interaction $x_1x_2$. More generally, the list of all aliased effects is encoded by the defining relation ($x_1x_2x_3 = +1$ in the above example), and the "sparsity-of-effects" principle legitimizes the assumption that the model is dominated by main effects and low-order interactions (second-order ones in the above example).

Here, the defining relation is made up of only one word, $x_1x_2x_3$, whose length is called the resolution. In general, however, it could involve several. The *resolution* is then defined as the minimal word length in the defining relation. This corresponds to the ability to estimate—separately from the design—the different main effects and low-order interactions. The most useful regular fractional designs are those with resolution III, IV, or V. Those below III are not useful because even some of the main effects will be aliased. Resolutions above V lead to a large number of runs, which is not necessarily justified since the additional cost mainly goes into the estimation of high-order interactions—which rarely occur in practice.

Note that regular fractional factorial designs lead to a number of runs equal to a power of two. It may be that the cost of such a design still remains unaffordable. In this case, it is possible to propose nonregular fractional factorial designs, which cannot be constructed by defining words. Nonregular designs are more flexible in their run size, but the price to pay is the difficulty of interpretation due to more complex aliasing relationships.

As a last resort, if each model run is overly time consuming, it is possible to propose screening for main effects with designs of size $n < d + 1$. For such designs, at least one main effect is aliased with one or more other main effects. The use of such supersaturated designs is thus justified only if there exists a sparsity of active factors and if the active main effects are important ones.

For more details on the construction and use of discretized DOE, we refer the reader to some reference books. The one by Montgomery [325] describes the main concepts of discretized DOEs, while Dean and Lewis [111] describes how to use them for screening purposes. The book by Kleijnen [241] focuses on developing them for use in analyzing numerical experiments (mainly for building polynomial metamodels).

At this stage, it is important to note that computer code experiments are different in several ways from real experiments. For the former, the input and output spaces are often high-dimensional, and, furthermore, deterministic computer codes do not produce any noise. As mentioned in Woods and Lewis [506], the most common source of error when screening factors in numerical models is induced by assuming a form of the surrogate model that is far too simple.

Sections 2.1.3 and 2.1.4 propose two screening approaches—still based on a discretization of the input space—that are more specifically designed for computer experiments.

### 2.1.3 ▪ The sequential bifurcation method

Group screening techniques [111] are an ensemble of methods that consist of constructing several groups of inputs, then identifying which of the groups are the most influential. This then reduces the dimension of the problem to the number of groups (instead of the number of inputs). Among these techniques, the *sequential bifurcation* (SB) method [35] is based on a dichotomic approach, sequentially separating the inputs into two groups. After each new model evaluation, SB tries to eliminate one of the two groups.

#### Main principles

For each of the inputs, two extreme levels are defined—the $+$ and $-$ levels—relative to the model outputs (the $+$ level means a larger output value than that corresponding to the $-$ level). The first step aggregates all inputs into a single group and tests whether or not that group has an effect by comparing the two extreme combinations. If the group has an important effect, then it is split into two subgroups. This is the "bifurcation" referred to in the name. The inputs of one subgroup take their $+$ level values, while the inputs of the other take their $-$ level values. From this input combination, a new model evaluation is performed and compared to the previous one to infer which group has the most influence. The noninfluential group is then eliminated by setting its inputs to arbitrary values in the following steps of the algorithm. These continue in a similar way: SB splits active subgroups into smaller ones and discards nonactive ones. At the end, only individual inputs remain, and all discarded inputs are considered noninfluential. The cost of running SB depends on the number of influential variables but also on the ranking strategy (i.e., the choice of inputs to be put in each subgroup). Figure 2.1 explains the principle of the method, where $Y[j]$ is defined such that $X_1, \ldots, X_j$ take their $+$ values and $X_{j+1}, \ldots, X_d$ their $-$ values.

**Remark 1.** *At each iteration of the SB algorithm, a division of the remaining group into two subgroups is performed. The idea is to put a similar number of inputs in each subgroup. No mathematical rule is proposed for the choice of inputs in each; this must be done using expert judgment. Moreover, inputs that are perceived to be the most influential should be gathered together in the same subgroup.*

The function sb() of the sensitivity package implements the SB method. A description of it can be obtained with

```
> library(sensitivity)
> help(sb)
```

The default usage for the first function call is simply

```
sb(d, sign = rep("+", d), interaction = FALSE)
```

where d is the number of factors, `sign` a vector of length d filled with "+" and "-" giving the signs of the factor effects assumed to be monotone ("+" if assumed increasing, "-" if assumed decreasing), and `interaction` a Boolean equal to `TRUE` if we suppose the model has interactions, `FALSE` otherwise. sb() returns a list of class "sb".

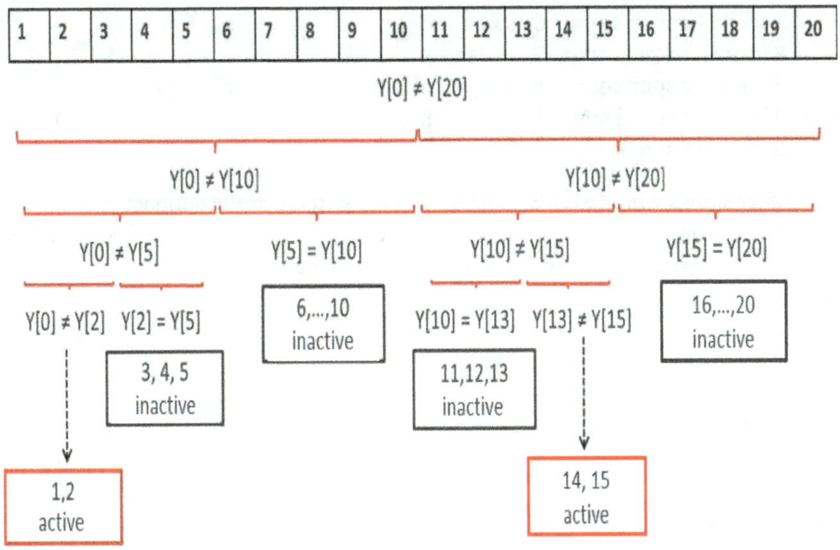


**Figure 2.1.** *Example of the sequential bifurcation method with* $d = 20$ *inputs. Courtesy of Claire Cannamela (CEA).*

Let us apply this function (with 5 iterations of the SB algorithm) to the Linkletter decreasing function (see Section C.5), found in the sensitivity package under the name linkletter.fun():

```
> library(sensitivity)
> d <- 10 # number of inputs
> sa <- sb(d) # An SB object has been created
> for (i in 1 : 5){
+   x <- ask(sa) # Give the new point to be computed by the model
+   y <- list()
+   for (j in names(x)){
+     xx <- matrix(x[[j]], nrow = 1)
+     cat(c("Step",i,"; Bifurcation input",j,"; Design=",xx,"\n"))
+     for (k in 1:d) if (xx[1,k] == -1) xx[1,k] <- 0
+     # because Linkletter fct is defined on [0,1] and not on [-1,1]
+     y[[j]] <- linkletter.fun(xx)
+     cat(c("\t Model response = ", y[[j]],"\n"))
+   }
+   tell(sa, y) # Update the SB object with the model output value
+ }

Step 1 ; Bifurcation input 0 ; Design= -1 -1 -1 -1 -1 -1 -1 -1 -1 -1
         Model response =  0
Step 1 ; Bifurcation input 10 ; Design= 1 1 1 1 1 1 1 1 1 1
         Model response =  0.3984375
Step 2 ; Bifurcation input 5 ; Design= 1 1 1 1 1 -1 -1 -1 -1 -1
         Model response =  0.3875
```

```
Step 3 ; Bifurcation input 3 ; Design= 1 1 1 -1 -1 -1 -1 -1 -1 -1
         Model response =  0.35
Step 4 ; Bifurcation input 2 ; Design= 1 1 -1 -1 -1 -1 -1 -1 -1 -1
         Model response =  0.3
Step 5 ; Bifurcation input 1 ; Design= 1 -1 -1 -1 -1 -1 -1 -1 -1 -1
         Model response =  0.2
```

The step number shows which SB iteration is under way (there are two model evaluations in Step 1), and the bifurcation input is the number $j$ at which the bifurcation is applied. The final result gives the effect of each group (along with the corresponding number of inputs in it) sequentially created by the SB algorithm:

```
> print(sa)
```

```
Groups:
  group    effect
1     1 0.2000000
2     2 0.1000000
3     3 0.0500000
4   4-5 0.0375000
5  6-10 0.0109375
```

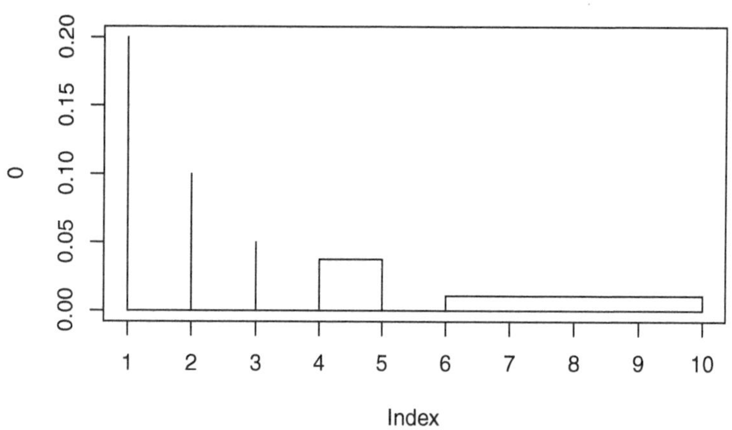


**Figure 2.2.** *The effects for inputs and groups of inputs obtained after* 5 *iterations of the SB method for the Linkletter decreasing function.*

The following command produces Fig. 2.2, which allows us to visualize this final result:

```
> plot(sa)
```

By typing

```
> names(sa)
```

```
[1] "p"              "sign"          "interaction" "i"              "y"
```

we observe that sa has several arguments: all of the input arguments detailed before, as well as the vector of bifurcations i and the vector of outputs y. Their values can be obtained with the commands

```
> sa$i
[1]  0 10  5  3  2  1
> print(sa$y, digits = 4)
[1] 0.0000 0.3984 0.3875 0.3500 0.3000 0.2000
```

### Extensions

The most important assumptions of the SB method are the monotonicity of the effects between the two levels of each input and knowledge of the sign of the effect of each input on the output. Second-order interactions can be easily taken into account using a slightly more sophisticated version of the SB algorithm [35]. A review of the variants and extensions of the SB method can be found in Kleijnen [241], for instance, the cases of random simulations (see also [494, 495]) and multiple outputs (see also [434]), as a method for the validation of the assumptions behind SB (see also [433]). Finally, Ankenman et al. [5] have introduced a stopping rule for the SB method as an adaptation of SB to identifying inputs influencing the variability rather than the mean.

## 2.1.4 ▪ Morris screening designs

A method introduced by Morris [329] allows inputs to be grouped into three categories:

1. Inputs with negligible effects.
2. Inputs with linear effects and without interactions.
3. Inputs with nonlinear effects and/or interactions. In this category, nonlinear effects cannot be distinguished from interactions.

The method consists in discretizing the input space for each factor, then drawing randomly a given number $r$ of OAT designs, each composed of $d+1$ points. Based on these $r$ OAT designs, the elementary effect of each factor is estimated, as described in more detail below.

Note that the Morris method, unlike those seen previously in the chapter, only applies to numerical inputs, not to categorical or mixed inputs.

### Elementary effects in the Morris method

Each input is first rescaled to $[0,1]$, then discretized as $L$ levels $\{0, \frac{1}{L-1}, \dots, 1\}$. This turns the original input space into the space $[0,1]^d$ discretized as a $d$-dimensional grid $\{0, \frac{1}{L-1}, \dots, 1\}^d$. Let $(\Delta_1, \dots, \Delta_d)$ be a vector of positive multiples of $1/(L-1)$ defined by the user. The $j$th elementary effect $d_j$ is the variation of the model when perturbed in the $j$th direction, i.e.,

$$d_j = \epsilon_j \frac{G(\mathbf{x} + \epsilon_j \Delta_j \mathbf{e}_j) - G(\mathbf{x})}{\Delta_j}, \tag{2.1}$$

where $\mathbf{e}_j$ is the $j$th vector in the canonical basis and $\epsilon_j$ is either 1 or $-1$ (randomly selected). Of course, the point $\mathbf{x} + \epsilon_j \Delta_j \mathbf{e}_j$ has to remain within $[0,1]^d$. If this is not the case, the direction of the perturbation is reversed by taking $-\epsilon_j$.

A specific OAT design is constructed using the following steps:

- Random choice of the initial point $\mathbf{x}$ in the discretized grid.

- Loop from 1 to $d$:
  - random choice without replacement of $j$ in $[1:d] = \{1, \ldots, d\}$;
  - random choice of $\epsilon_j$ in $\{-1, +1\}$ (verify that $\mathbf{x} + \epsilon_j \Delta_j \mathbf{e}_j \in [0,1]^d$; swap the value of $\epsilon_j$ if not);
  - the initial point $\mathbf{x}$ is replaced by $\mathbf{x} + \epsilon_j \Delta_j \mathbf{e}_j$.

A particularity of an OAT design with the Morris method is that each new start point $\mathbf{x}$ is taken as the previous perturbed point. Figure 2.3 gives an example of a Morris design with $d = 2$, $L = 6$, $r = 3$, and $\Delta = 3/5$, obtained with the code

```
> set.seed(825367) # To reproduce the same results
> library(sensitivity)
> sa <- morris(model = NULL, factors = 2, r = 3,
+          design = list(type = "oat", levels = 5, grid.jump = 3))
> plot(x = sa$X[,1], y = sa$X[,2], xlab = "X1", ylab = "X2",
+        cex = 1, col = c(rep(1,3),rep(2,3),rep(3,3),rep(4,3)))
```

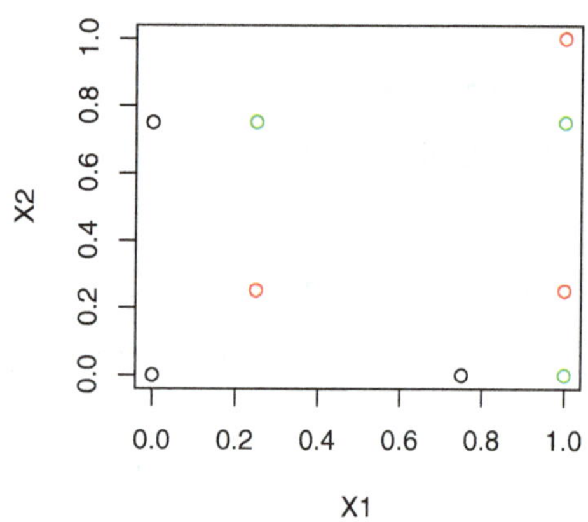


**Figure 2.3.** *Morris experimental design with $d = 2$, $L = 6$, $r = 3$, and $\Delta = 3/5$. Each color corresponds to one of the $r = 3$ trajectories of the complete Morris design.*

By evaluating the model on a random draw of the design $\mathbf{x}$, we get a realization of each elementary effect $d_j$, $j = 1, \ldots, d$. Each realization can be interpreted as a local sensitivity measure around the start point of the design. Morris [329] suggested as global sensitivity measures the quantities

$$\mu_j = \mathbb{E} d_j,$$
$$\sigma_j = \sqrt{\operatorname{Var} d_j}.$$

These global sensitivity measures can be estimated from $r$ random draws of the OAT Morris design $\mathbf{x}$,

$$\widehat{\mu_j} = \frac{1}{r} \sum_{i=1}^{r} d_j^{(i)},$$

$$\widehat{\sigma_j} = \sqrt{\frac{1}{r} \sum_{i=1}^{r} \left( d_j^{(i)} - \widehat{\mu_j} \right)^2},$$

where $\left(d_j^{(i)}\right)_{i=1\ldots r}$ is the corresponding $r$-sample of the elementary effect defined in Eq. (2.1). The interpretation is then summarized in Table 2.2. We say that $|\widehat{\mu}_j|$ is small if it is small relative to $\widehat{\sigma}_j$. What values of $\widehat{\sigma}_j$ are considered small and large are determined by the choice of an arbitrary threshold. Most of the time, the threshold is chosen in order to have a clear separation between influential and noninfluential factors.

**Table 2.2.** *Interpretation of Morris statistics.*

| | $\widehat{\sigma}_j$ small | $\widehat{\sigma}_j$ large |
|---|---|---|
| $\lvert\widehat{\mu}_j\rvert$ small | negligible effects | nonlinearities and/or interactions |
| $\lvert\widehat{\mu}_j\rvert$ large | only linear effects | nonlinearities and/or interactions |

**Remark 2.** *For each model input $X_j$, the sensitivity measure $\mu_j^* = \mathbb{E}|d_j|$, based on the absolute value of $d_j$, was introduced in [65]. Its aim is to avoid the cancellation effect, which can appear on $\mu_j$ with nonmonotonic functions.*

### Morris design algorithm

When the number of inputs is large, the elementary effects can only be calculated for a reduced number $N = r \times (d+1)$ of points in order to keep the number of model calls reasonable. In each trajectory of a Morris DOE, two successive simulations only differ by one factor (OAT design).

Morris design is random, and each trajectory forms a random walk in $[0,1]^d$. It is made up of a $(d+1) \times d$ matrix in which consecutive lines are identical except in one coordinate (e.g., the $j$th, where they differ by $\pm\Delta_j$). The matrix of the corresponding experimental design $B^*$ can be represented by the expression

$$B^* = \left( J_{d+1,1}x^* + \frac{\Delta}{2}((2B - J_{d+1,d})D^* + J_{d+1,d}) \right) P^*,$$

with

- $J_{d+1,1}$ (resp., $J_{d+1,d}$) a $(d+1) \times 1$ (resp., $(d+1) \times d$) matrix filled with ones;
- $B$ a $(d+1) \times d$ matrix whose element indexed by $1 \leq i \leq d+1$ and $1 \leq j \leq d$ is null if $i > j$ and equal to one otherwise;
- $D^*$ a $d \times d$ diagonal matrix composed of equiprobable $+1$'s and $-1$'s;
- $P^*$ a $d \times d$ random permutation matrix;
- $x^*$ a randomly chosen point on the grid.

The points of the experimental design (rows of the matrix) have to remain on the grid; the algorithm has to guarantee this happens with, e.g., a rejection sampling.

### Choosing the parameters

It is usual practice to choose a constant $\Delta_j = \Delta$ for all $j = 1, \ldots, d$. The choice of $r$ is important because it modulates the quality of the input space exploration and the convergence of sensitivity measures. This method has—in terms of model evaluations—a cost of $N = r \times (d+1)$. Saltelli et al. [417] propose setting the value of $r$ between 4 and 10.

The choice of parameters $L$ and $\Delta$ is also important. The underlying idea of the Morris method is to take a rough grid when discretizing the domain and make large jumps on this grid in each OAT design. It appears that for this kind of heuristic approach, with small $r$ in order to avoid overly many model calls, a value of $L$ between 6 and 11 with $\Delta = 1/(L-1)$ is a reasonable choice.

**Remark 3.** *If a large value of r ($>20$) is possible, users of the Morris method tend to take a low value for $\Delta$, meaning that local variations are studied. In that case, the Morris method is close to derivative-based global SA (see Section 2.3.3), which should then be preferred, as it is more solid from a theoretical point of view provided that regularity assumptions are satisfied.*

### Example

The morris() method is available in the sensitivity package. The following code shows how to apply it to the Morris function with $d = 20$ (see Section C.6):

```
> set.seed(12345) # To reproduce the same results
> library(sensitivity)
> sa <- morris(model = morris.fun, factors = 20, r = 10,
+            design = list(type = "oat", levels = 5, grid.jump = 2))
```

The obtained object sa contains—among other things—Morris design (sa$X) and model output values (sa$y) on this design. The number of model runs is given by

```
> length(sa$y)
```

```
[1] 210
```

The command print(sa) allows computing statistics of interest. It performs the same calculations as if we had entered

```
> mu <- apply(sa$ee, 2, mean)
> mu.star <- apply(sa$ee, 2, function(x) mean(abs(x)))
> sigma <- apply(sa$ee, 2, sd)
> print(cbind(mu,mu.star,sigma))
```

```
             mu  mu.star     sigma
X1    4.1706332 80.641742 87.668714
X2    2.7693230 62.157400 73.762803
X3   49.1180003 50.354187 51.817402
X4   23.3114112 65.616183 71.856369
X5   29.4589979 34.641199 48.671922
X6   23.1129337 45.507701 52.261663
X7   22.1142862 22.114286 26.249250
X8   45.3991509 45.399151  3.934418
X9   39.2051963 39.205196  4.248169
X10  37.6276567 37.627657  6.764721
X11  -3.9293663  5.592258  5.437169
X12   0.3773796  2.760687  4.393889
```

```
X13  1.7768889  3.425364  4.030624
X14 -2.9067298  5.193887  6.121095
X15  2.6841162  3.299804  3.646100
X16  3.3931808  5.738365  5.363439
X17 -4.5225101  6.494019  6.359742
X18 -1.5241302  3.453061  4.815955
X19  4.2896356  4.704044  3.080657
X20  3.4434433  3.971589  4.379384
```

One possible way to present the results consists in plotting each input factor $X_j$ ($j = 1, \ldots, d$) in terms of its coordinates $(\mu_j, \sigma_j)$ or $(\mu_j^*, \sigma_j)$, or even $(\mu_j, \mu_j^*, \sigma_j)$ in 3D [380]. For the Morris test function, Fig. 2.4 shows (via the $(\mu_j^*, \sigma_j)$ graph) the detection of potentially influential ($X_1$ to $X_{10}$ have large $\mu^*$) and noninfluential ($X_{11}$ to $X_{20}$ have small $\mu^*$) inputs. Because of their small values of $\sigma$, we conclude that inputs $X_8$, $X_9$, and $X_{10}$ have a linear influence on the model output, while $X_1$ to $X_7$ have a nonlinear influence or interaction effects. Figure 2.4 is obtained with the following command:

```
> plot(sa, xlim = c(0,85), cex = 1.5, cex.axis = 1.5, cex.lab = 1.5)
```

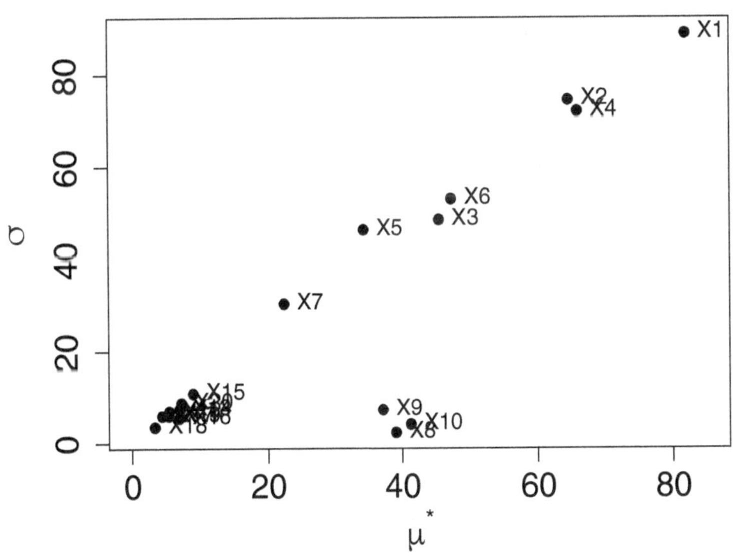


**Figure 2.4.** *Plotted results for the Morris method for the Morris test function ($d = 20$ inputs, $r = 10$ OAT experiments, $k = 5$ levels, with a grid step of $\Delta = 2/5$).*

### Extensions

Some modifications to the Morris method lead to a more effective sampling strategy, meaning a better exploration of the space for uncertain input factors; see [65]. Finally, other extensions of the initial Morris method have been introduced for second-order effect analysis [64, 92, 130], for estimation of Morris measures with any type of design [380, 420], and for dependent inputs [154]. Fédou and Rendas [130] have also formalized the Morris method in the framework of graph theory.

### 2.1.5 ▪ Further references

When the number of inputs is large, it is sometimes necessary to use methods which require fewer model runs than model inputs. Group screening (which reduces the input space's dimension by grouping inputs into a small number of groups) and supersaturated designs (for which the number of inputs is larger than the sample size) are then useful [111].

## 2.2 ▪ Sampling-based sensitivity analysis

The methods described in the previous sections are all based on very specific sample types, in particular those based on rough discretizations of each input. In this section, we consider so-called sampling-based approaches for SA [189, 414] where samples can be of any type, though Monte Carlo or space-filling designs (see Section A) are preferable. First, we take a look at data visualization techniques. We then develop indices based on linear model coefficients and feature selection techniques.

### 2.2.1 ▪ Visualization techniques

As an initial exploratory approach, graphical techniques can be useful in order to visualize and then understand relationships between model output and input variables. Let $\mathbf{X}^n = \left(X_1^{(i)}, \ldots, X_d^{(i)}\right)_{i=1,\ldots,n}$ be an $n$-sample matrix of the inputs $\mathbf{X}$, and let $\mathbf{Y}^n = \left(Y^{(1)}, \ldots, Y^{(n)}\right)^T$ an $n$-sample vector of the output $Y$. Scatterplots (visualizations of point clouds of the sample $(\mathbf{X}^n, \mathbf{Y}^n)$ for $Y$ versus $X_j$, $i = j, \ldots, d$) are one way to do this but in a visually subjective manner. The following code provides a visualization of a sample of size 200 from the HeterDisc function ($d = 4$; see Section C.3 with independent $X_j \sim \mathcal{U}(0, 20)$):

```
> set.seed(12345) # To reproduce the same results
> d <- 4
> n <- 200
> X <- matrix(runif(n*d, 0, 20), nrow = n, ncol = d)
> colnames(X) <- c("X1", "X2", "X3", "X4")
> Y <- heterdisc.fun(X)

> par(mfrow=c(2,2))
> for (j in 1:d){
+   plot(X[,j], Y, xlab = colnames(X)[j])
+   lines(lowess(X[,j], Y), col = 2, lwd = 2)
+ }
```

The results are shown in Fig. 2.5. The red lines correspond to LOESS smoothers, which use locally weighted polynomial regression [84]. These allow the mean trend of the output to be visualized with respect to each input by estimating the conditional mean of each scatterplot. In the SA context, numerous smoothing techniques (see [184] for a general overview of parametric and nonparametric methods) have been studied; see, for instance, Storlie and Helton [461] (polynomial regression, moving averages, locally weighted regression, smoothing splines, etc.) and Da Veiga et al. [102] (local polynomial regression). Using the generalized additive models [184] also allows one to apply smoothing in dimension greater than one.

Cobweb plots [255], also known as parallel coordinate plots, can be used to visualize the simulations as a set of trajectories. Here we use the parcoords package for the HeterDisc function,

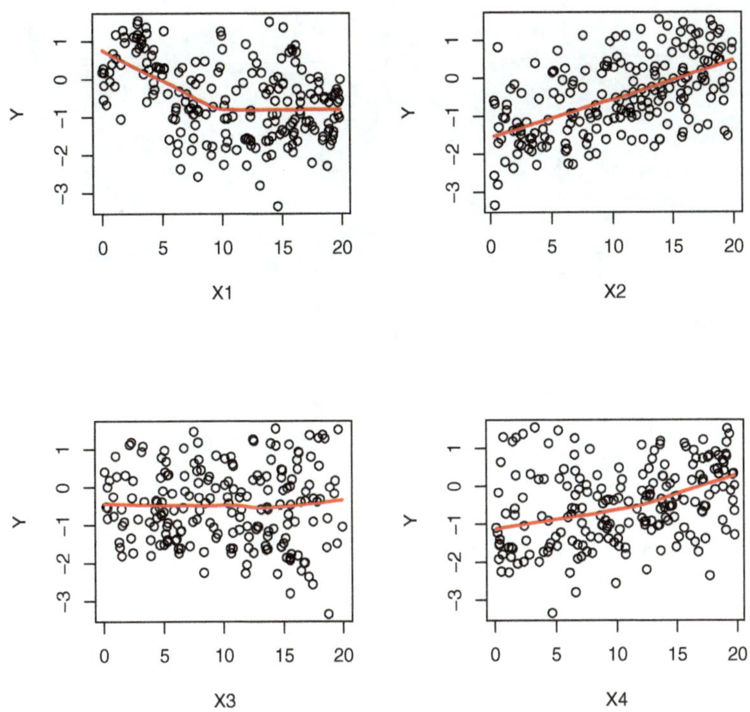


**Figure 2.5.** *Scatterplots of the model output as a function of each input (from a sample from the HeterDisc function).*

which leads to Fig. 2.6, where all simulations are displayed:

```
> library(parcoords)
> parcoords(data = as.data.frame(cbind(X,Y)), rownames = FALSE,
+           brushMode = "1D-axes-multi", alphaOnBrushed = 0.2)
```

A powerful feature of these plots comes from their interactivity (one can click on them to bring up specific details). For example, in Fig. 2.7, we have highlighted samples with low output values. This suggests that there is a potential interaction between $X_2$ and $X_3$ since only combinations with low values of $X_2$ and high values of $X_3$ lead to low output values.

Finally, *cumulative sum of normalized reordered output* (CUSONORO) plots, proposed by Plischke [371], are particularly useful for detecting trends and changes in trends in the data. The CUSONORO function for the input $X_j$ corresponds to the partial mean of the output (see [46]) for the values of $X_j$ less than each possible value in its domain. More precisely, for $u \in [0, 1]$, this function is given by

$$c_j(u) = \frac{u}{\sigma_Y}\mathbb{E}[Y - \mathbb{E}(Y)|X_j \leq F_{X_j}^{-1}(u)].$$

That is, it corresponds to the average mean of the standardized output when the input being considered is less than its quantile of order $u$. We have $c_j(0) = c_j(1) = 0$ and a positive (resp., negative) CUSONORO function if the output is decreasing (resp., increasing) as a function of $X_j$. If $X_j$ is symmetrically distributed and the dependence of $Y$ on $X_j$ is linear, then the CUSONORO function is centered and parabolic. Nonlinear output behavior is suspected if the (positive or negative) peak of the CUSONORO function is not in the center of the plot.

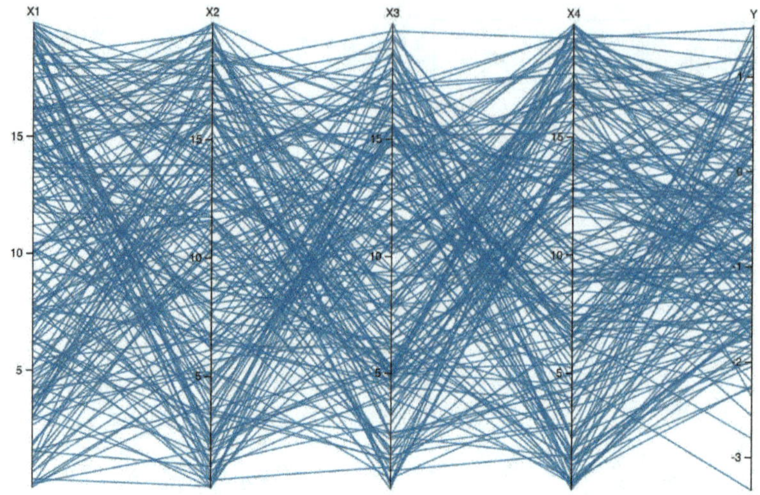


**Figure 2.6.** *Cobweb plot of a sample from the HeterDisc function.*

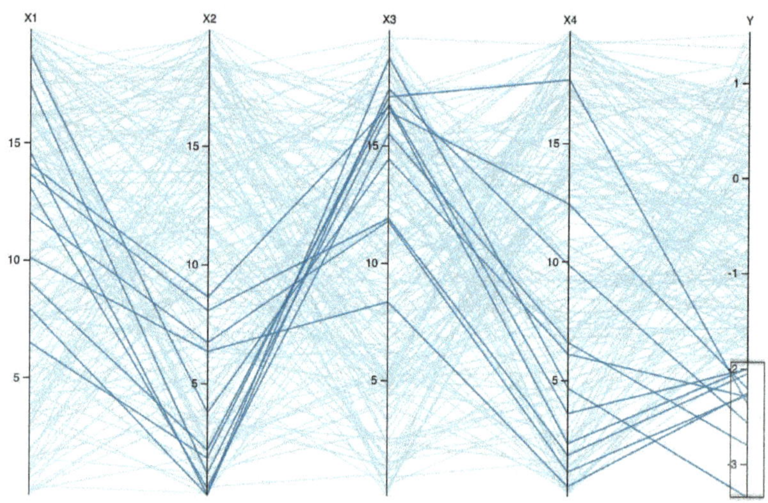


**Figure 2.7.** *Cobweb plot of a sample from the HeterDisc function filtered for low output values.*

The following code constructs the CUSONORO function for a sample of size 200 of the HeterDisc function ($d = 4$; see Section C.3), which is shown in Fig. 2.8:

```
> c <- (1:n)/n
> Yc <- ( Y - mean(Y) ) / ( n * sd(Y) )
> z <- cumsum( Yc[order(X[ ,1])])
> plot(c, z, type = "l", ylim = c(-0.3,0.3),
+       xlab = "Input quantiles", ylab = "Deviation from mean")
> abline(h = 0, lty = 2)
```

```
> for(i in 2:d) {
+   z <- cumsum( Yc[order(X[ ,i])])
+   lines(c, z, col = i)
+ }
> legend(0.8,-0.1,legend = c("X1","X2","X3","X4"),bty = "n",
+          lty = rep(1,4),col = c(1,2,3,4),cex = 1.2)
```

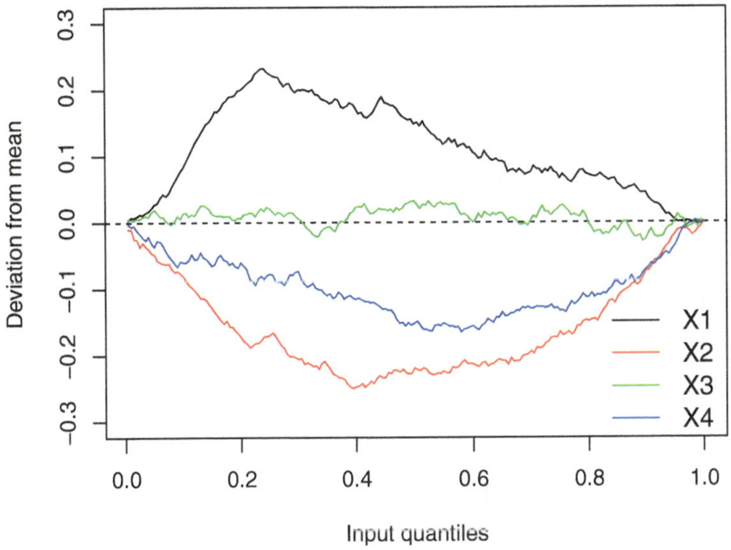


**Figure 2.8.** *CUSONORO plot for a sample of the HeterDisc function.*

Figure 2.8 shows that $X_1$ and $X_2$ are the most influential inputs, with a clear nonlinear trend for $X_1$ and a linear one for $X_2$. No trend is detected for $X_3$, while an increasing trend with slight nonlinearity is detected for $X_4$. These results accompany well what was seen in the scatterplots of Fig. 2.5.

## 2.2.2 ▪ Standard linear and rank regression techniques

If a sample of inputs/outputs $(\mathbf{X}^n, \mathbf{Y}^n) = \left(X_1^{(i)}, \dots, X_d^{(i)}, Y^{(i)}\right)_{i=1,\dots,n}$ is available, it is possible to fit the additive linear model

$$Y = \sum_{j=0}^{d} \beta_j X_j + \epsilon,$$

where $X_0 = 1$, $\beta = (\beta_0, \dots, \beta_d)^t \in \mathbb{R}^{d+1}$ is the effects vector, and $\epsilon \in \mathbb{R}$ is the model's error. Ordinary least squares can then easily estimate the effects. This allows the behavior of $Y$ to be explained given the values of $\mathbf{X}$, provided that the sample size $n$ is sufficiently large ($n > d$ at least). The associated sensitivity indices are called *standard regression coefficients* (SRC),

$$\text{SRC}_j = \beta_j \sqrt{\frac{\text{Var}\, X_j}{\text{Var}\, Y}},$$

where $\beta_j$ is the linear regression coefficient associated with $X_j$. If the linear hypothesis is valid and the input variables are independent, each $\text{SRC}_j^2$ expresses the part of the output variance explained by the input $X_j$.

If the linear hypothesis is rejected, a rank transformation can be performed whereby the sample $(\mathbf{X}^n, \mathbf{Y}^n)$ is transformed into a sample $(\mathbf{R}_X^n, \mathbf{R}_Y^n)$ by replacing the values by their ranks in each column of the matrix. Then a linear regression on the ranks gives the *standard rank regression coefficient* (SRRC), which is valid under monotonic hypotheses [414].

Computing the SRC and SRRC can be done using the function src() in the sensitivity package. On the sample of size 200 from the previous section (HeterDisc function), this gives

```
> library(sensitivity)
> src(X, Y)

Call:
src(X = X, y = Y)

Standardized Regression Coefficients (SRC):
      original
X1 -0.50709196
X2  0.60345220
X3  0.02356868
X4  0.31741793

> src(X, Y, rank=TRUE)

Call:
src(X = X, y = Y, rank = TRUE)

Standardized Rank Regression Coefficients (SRRC):
      original
X1 -0.50092142
X2  0.59115418
X3  0.02527736
X4  0.33003999
```

The estimation of these sensitivity indices is subject to uncertainty due to the limited sample size. Analytical formulas could be applied in order to estimate this uncertainty [82], but a simple bootstrap-based procedure is used instead in the sensitivity package. This leads to Fig. 2.9, where the error bars correspond to 95% confidence intervals, via the following commands:

```
> sa <- src(X, Y, nboot = 100)
> library(ggplot2) ; ggplot(sa)
```

SRC (resp., SRRC) indices are based on a linear (resp., monotonic) relationship between the output and the inputs, coded in R by

```
> model <- lm(Y~X1+X2+X3+X4, data = data.frame(X,Y))
```

Statistical techniques can then be used to support the linear (resp., monotonic) hypothesis, e.g., the classical coefficient of determination $R^2$ (resp., $R^{2*}$) or the predictivity coefficient (also called the Nash–Sutcliffe model efficiency) $Q^2$ (resp., $Q^{2*}$),

$$Q^2 = 1 - \frac{\sum_{i=1}^{m}\left[Y'^{(i)} - \widehat{Y}(\mathbf{X}'^{(i)})\right]^2}{\sum_{i=1}^{m}\left[Y'^{(i)} - \sum_{i=1}^{m} Y'^{(i)}\right]^2},$$

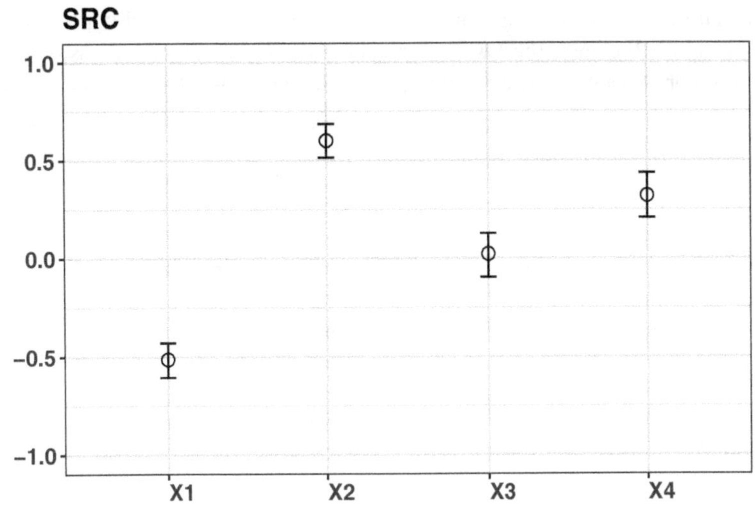


**Figure 2.9.** *Standard regression coefficients of the HeterDisc function.*

where $(\mathbf{X}'^{(i)}, Y'^{(i)})_{i=1,\dots,m}$ is an input-output test sample of size $m$ (not used to fit the model) and $\widehat{Y}(\cdot)$ is the predictor of the linear regression model. The value of $Q^2$ corresponds to the percentage of output variability explained by the linear regression model (a value equal to 1 means a perfect fit on the test sample).

From our previous linear regression model, we can get the $R^2$ with the command

```
> summary(model)$r.squared
```

```
[1] 0.6390936
```

This shows that the linear model is far from valid in this case.

**Remark 4.** *In the case of dependent inputs, SRC is no longer a valid importance measure for an input. In the global SA literature, the partial correlation coefficient (PCC) has been proposed as a substitute [414, 189, 209]. However, PCC is more a measure of the linearity between the output and one input.*

*In the linear regression analysis literature, other measures have been developed [225, 175]. For example, the semipartial correlation coefficient [225, 202] quantifies the additional explanatory power of each input but presents drawbacks for highly correlated inputs. Most interesting measures, based on the so-called general dominance analysis, find ways to partition the $R^2$ among the $d$ inputs. This is done by the LMG [284] (acronym coming from the initials of the authors' names, i.e., "Lindeman–Merenda–Gold"; see also Section* 5.5.7*), which is given by the function* lmg() *in the* sensitivity *package. Other game-theoretic concepts [131] allow to define interesting measures such as the* proportional marginal variance decomposition *(included in the function* e-mvd() *in the* sensitivity *package).*

*However, the main challenge is that these measures cannot be applied on models with large number of inputs. In such cases, Johnson's relative weights [224] are interesting alternatives (see a sensitivity analysis example in Clouvel et al. [85]).*

The linear regression model introduced above makes it possible to identify noninfluential variables by looking for small regression coefficients. To go further and even automate this selection, a natural extension consists in using a linear regression model which enforces sparsity in the regression coefficients: the *lasso*. Instead of solving the classical least-squares problem, the lasso estimator is obtained by adding a sparsity-inducing penalty,

$$\underset{(\beta_0,\ldots,\beta_d)}{\arg\min} \sum_{i=1}^{n} \left( Y^{(i)} - \beta_0 - \sum_{j=1}^{d} \beta_j X_j^{(i)} \right)^2 + \lambda \sum_{j=1}^{d} |\beta_j|,$$

where $\lambda \geq 0$ controls the level of sparsity. When $\lambda = 0$, the SRCs are recovered, while $\lambda \to \infty$ leads to a constant model. In practice, its value is set using cross-validation—the approach used in the glmnet package and also used here to illustrate the lasso for SA:

```
> library(glmnet)
> # Compute lasso estimate for several values of lambda
> lassomodel <- cv.glmnet(X,Y)
> # Select the best one through cross-validation
> # and exclude intercept
> coefsbestlassomodel <- coef(lassomodel,s = "lambda.1se")[2:5]
> # Renormalize them just like SRC ones
> print(coefsbestlassomodel * apply(X,2,sd)/sd(Y))

        X1         X2         X3         X4
-0.3917919  0.4987874  0.0000000  0.2237342
```

Here, note that the third variable has been automatically filtered out by the lasso model. Unfortunately, this approach does not help improve the low quality of the linear model in this test case. To try to overcome this, several generalizations of the lasso have been proposed. First, second-order interaction effects can be included via specific penalty functions involving hierarchical considerations [281, 37]. More general nonlinear generalizations have also been introduced, including sparse additive models [394, 321], the COSSO [283], and sparse polynomial chaos [41]. Some of these approaches are presented in Section 4.2.

### 2.2.3 ▪ Feature selection without a model

In machine learning, feature selection aims to identify relevant features (in a large set) with respect to a prediction task. The goal is to detect irrelevant or redundant features which may increase the prediction variance without reducing its bias. In fact, this closely resembles the factor screening goal in global SA. The main difference is that in global SA, input variables are usually assumed to be independent, whereas in feature selection, redundant features, i.e., highly dependent factors, are precisely those that need to be filtered out. This apparently naive distinction actually makes feature selection an interesting alternative to screening when some of the input variables are correlated. However, it is important to note that it is also a powerful option even in the independent case. We do not intend to give an exhaustive review of feature selection techniques here; instead, we simply give details of a few approaches which make use of the dependence measures introduced later in Chapter 6 (Sections 6.1 and 6.2).

The literature is abundant on feature selection and involves numerous approaches. Model-based techniques include, for example, the lasso and the sparse additive models mentioned above. Another line of work is model-free methods, where only dependence measures are used to identify relevant features. For illustration purposes, let us first introduce the concept of max-dependency [362]. Denote by $X_1, \ldots, X_p$ the set of available features, by $Y$ the target output to be predicted, and by $D(\cdot,\cdot)$ any measure quantifying the dependence between two random vectors. The max-dependency scheme for feature selection involves finding $m$ features $X_{i_1}, \ldots, X_{i_m}$ which have jointly the largest dependency with $Y$; i.e., the following optimization problem is solved:

$$\max_{\{i_1,\ldots,i_m\}\subset[1:p]} D(\{X_{i_1}, \ldots, X_{i_m}\}, Y).$$

However, the dependence measure $D(\cdot,\cdot)$ needs to be robust with respect to the number of dimensions, which is hard to achieve in practice when the number of samples is less than $m$. Consequently, marginal computations which only involve $D(X_k, Y)$ terms are usually used instead. The max-relevance criterion [362] serves in this context as a proxy to max-dependency, where the optimization problem is now

$$\max_{\{i_1,\ldots,i_m\}\subset[1:p]} \frac{1}{m}\sum_{k=1}^{m} D(X_{i_k}, Y). \tag{2.2}$$

But when the features are dependent, it is likely that this criterion will select redundant features. To limit the possibility of this, one can add a min-redundancy condition, expressed as

$$\min_{\{i_1,\ldots,i_m\}\subset[1:p]} \frac{1}{m^2}\sum_{k,l=1}^{m} D(X_{i_k}, X_{i_l}). \tag{2.3}$$

The final scheme, combining (2.2) and (2.3), is called minimal-redundancy-maximal-relevance (mRMR) and given by

$$\max_{\{i_1,\ldots,i_m\}\subset[1:p]} \left[\frac{1}{m}\sum_{k=1}^{m} D(X_{i_k}, Y) - \frac{1}{m^2}\sum_{k,l=1}^{m} D(X_{i_k}, X_{i_l})\right].$$

On the other hand, solving this is computationally infeasible when $m$ and $p$ are large due to the combinatorial nature of the problem. Usually instead, approximate solutions are found by iterative procedures, where features are added one at a time to the subset $X_{i_1}, \ldots, X_{i_m}$ (forward selection). Forward and backward procedures for mRMR were investigated by [362], where $D(\cdot,\cdot)$ was set as the mutual information (see Section 6.2.1). Similarly, forward and backward approaches, where mutual information is replaced with a kernel-based dependence measure, were introduced by [448]. Links between such feature selection methods and SA are investigated in Da Veiga [99] (see Section 6.2.1).

Let us illustrate this approach on the Sobol' $g$-function (see Section C.7) with $d = 30$ inputs and independent $X_j \sim \mathcal{U}(0,1)$). Only the first two inputs have a strong influence, and $X_9$ to $X_{30}$ are dummy variables. As the inputs are independent, a forward approach to solving mRMR simply boils down to computing the mutual information between each input and the output. This can be done with the sensitivity package using the sensiFdiv() function (see Section 6.1.2 for details):

```
> library(ks)
> set.seed(12345) # To reproduce the same results
> d <- 30
> n <- 200
> X <- matrix(runif(n*d,0,1), nrow = n, ncol = d)
> colnames(X) <- paste0("X",1:d)
> smi <- sensiFdiv(model = sobol.fun, X = X, fdiv = "KL")
> library(ggplot2) ; ggplot(smi)
```

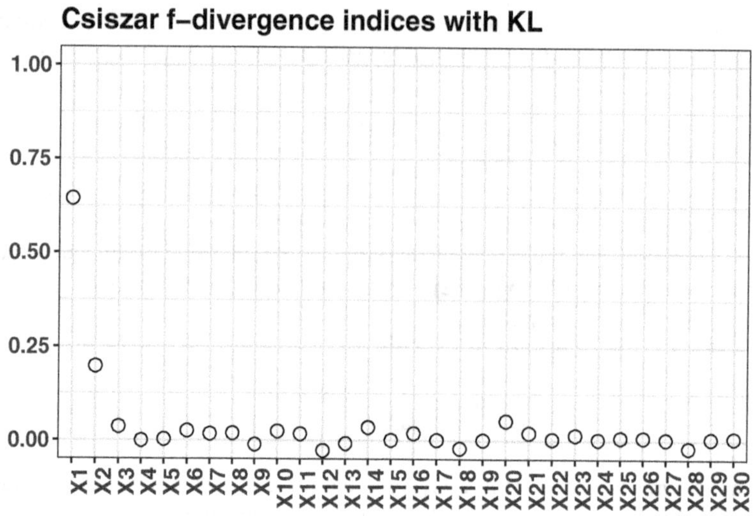


**Figure 2.10.** *Mutual information for the g-function with* 30 *inputs.*

Figure 2.10 clearly shows that only the first two inputs are detected as influential.

### 2.2.4 ▪ Further references

Other screening techniques involving statistical tests have been studied in detail in Kleijnen and Helton [242] and Helton et al. [189]. The general idea is to place a grid over the scatterplot for the output versus each input. A number of statistical tests can then be performed in order to determine whether the distribution of points across the grid cells is nonrandom. If so, this indicates that the input has an effect on the output.

This kind of subdivision scheme of the input domain or output range has also been proposed by Pannier and Graf [355]. Their sectional global sensitivity measures consist of computing the same sensitivity measures in each subdomain/subrange in order to compare the inputs' influence as a function of the chosen input subdomain or output subrange. A similar and more general idea has been developed by Fruth et al. [145], with the so-called support indices, who explore the local sensitivity behavior of an input over its support.

In the class of methods used in feature selection without models, statistical tests of independence have also been studied, in particular for kernel-based dependence measures [173, 513]. Also of note, Yamada et al. [510] merge the lasso and kernels in an interesting way. Finally, the distance correlation–based measure [473] introduced by Da Veiga [99] in SA has also been extended to the dependent-input case by Zhou et al. [514].

## 2.3 ▪ Derivative-based global sensitivity analysis

The historical approach to SA, known as the local approach, is deterministic. In it, the impact of small input perturbations around input nominal values on the model output is studied. By assuming that the function $G$ is differentiable with respect to each of its inputs, all of the partial derivatives of the model at this nominal point can be computed to provide sensitivity values. To be comparable, these local sensitivities may also need to be normalized. This section looks at various ways to use these derivatives for SA.

### 2.3.1 ▪ Local sensitivity analysis

For a long time, SA was performed with a local approach by computing perturbations of the model output due to small input perturbations around a nominal point. This deterministic approach consists in calculating sensitivity indices based on the model's partial derivatives at specific points of the input variable space [487, 340]. From these basic principles, sophisticated adjoint methods were also developed to treat large-scale differential systems [60, 61] with large numbers of input variables (from thousands to billions). Such approaches are particularly well suited to uncertainty analysis, SA, and inverse problems in environmental systems in domains like climatology, oceanography, and hydrogeology [359].

Let $\mathbf{x}^*$ be an initial set of model input data. In many works, the partial derivative $\frac{\partial G(\mathbf{x})}{\partial x_j}\big|_{\mathbf{x}=\mathbf{x}^*}$ is referred to as the "sensitivity" of the response $y^* = G(\mathbf{x}^*)$ to input parameter $x_j$. This type of approach is called local SA. As the model inputs can have different units, the local sensitivity $\eta$ is often normalized with respect to the local values $x_j^*$ and $y^*$ (if nonzero):

$$\eta_j = \frac{x_j^*}{y^*} \frac{\partial G(\mathbf{x})}{\partial x_j}\bigg|_{\mathbf{x}=\mathbf{x}^*}. \tag{2.4}$$

These indices are often called the elasticities of the model, for instance, in structural reliability and economics [270, 43]. Some additional details and points of view on local sensitivity indices (also called differential importance measures) can be found in Borgonovo and Plischke [48] and Borgonovo [43].

The $d$ partial derivatives $\frac{\partial G}{\partial x_j}\big|_{\mathbf{x}=\mathbf{x}^*}$ may be computed numerically by varying each input variable—one at a time (OAT; see also Section 2.1.2)—within a small interval around a nominal value and determining the corresponding effect on the output variable:

$$\frac{\partial G(\mathbf{x})}{\partial x_j} \approx \frac{G(x_1, \dots, x_j + \Delta x_j, \dots, x_d) - G(\mathbf{x})}{\Delta x_j}, \quad 1 \leq j.$$

In practice, they can be computed by a finite difference scheme using a finite perturbation $\Delta x_j^*$. The choice of this perturbation is known to be tricky: incorrectly estimated derivatives if too large, numerical instability if too small. This procedure, occasionally called the *brute-force method*, requires $d+1$ model computations. If central differences are used, the number of model computations required could be as great as $2d$. Although this method is conceptually simple to use and requires no additional model development, it is relatively computationally expensive and requires a trial-and-error process when selecting the parameter perturbations $\Delta x_j$ (a bad choice can lead to poor estimation of sensitivity over the whole domain if the response is nonlinear).

Nodet and Vidard [340] reviewed the theoretical and methodological tools of variational SA, which is based on these local derivatives. In particular, to compute the derivatives, finite differences can be avoided by the use of an adjoint-based method. As its computational cost is independent of the number of input variables, this allows models with a large number of inputs to be processed. Software can be used to automatically derive an adjoint model from an initial numerical model [174]. This subject is beyond the scope of the book; here we simply illustrate the use of derivatives for SA with R .

In R, the following functions can be used to compute the model derivatives:

- From an R expression (analytical formula for a model) and a list of variables for which the model has to be derived, the deriv() function performs symbolic differentiation and provides the formulas of the derivative model that are able to be evaluated.

- The numericDeriv() function numerically evaluates the gradient of an expression.

- The numDeriv package contains several functions for calculating the gradient, Jacobian, and Hessian of an R function representing the model (when there is only one input). It also includes various tuning parameters for the numerical schemes used (see also the numderiv() function of the pracma package).

Except for specific cases, models that are studied in SA cannot be expressed analytically and depend on several inputs. In the sensitivity package, the delsa() function allows derivatives to be computed (via a finite difference scheme) with respect to several inputs for a general R model. For the Morris function ($d = 20$; see Section C.6), we can compute the derivatives at the central point of the domain $[0, 1]^d$ with the following commands:

```
> library(sensitivity)
> d <- 20
> X0 <- matrix(rep(0.5,d), nrow = 1) # The domain central point
> der1 <- delsa(model = morris.fun,perturb = 1.01,
+                   X0 = X0, varprior = rep(1,d))
> print(der1$deriv[,1:d], digits = 2)

 [1] -25.06 -22.03   5.71 -24.24   5.44 -12.30  14.18  46.75  38.15  35.57
[11]  -2.00  -3.16   2.21  -0.72   0.95  -0.89  -6.71   3.32   4.03   1.82
```

The derivatives are output for each input. The argument `perturb` = 1.01 means that a 1% perturbation has been used around the central point when computing the finite differences. The argument `varprior` is necessary but not used at this stage.

Remember that derivative values have to be normalized in order to compare them with each other, e.g., with respect to the input-output values (see Eq. (2.4)). In place of this, another possibility is to multiply the derivative value with respect to each input by that input's standard deviation. This procedure gives importance measures for each input variable that can be interpreted in terms of global sensitivities (because it integrates the range of uncertainty in each input). It is obtained directly from a first-order Taylor expansion (linearization) of $G$ around the nominal point $\mathbf{x}^*$:

$$G(\mathbf{x}) \approx G(\mathbf{x}^*) + \sum_{j=1}^{d} \frac{\partial G(\mathbf{x})}{\partial x_j}\bigg|_{\mathbf{x}=\mathbf{x}^*} (x_j - x_j^*).$$

For $d$ independent inputs of variance $\operatorname{Var} X_j$ ($j = 1, \ldots, d$), the variance of $Y = G(\mathbf{x})$ is then approximated by

$$\operatorname{Var} Y \approx \sum_{j=1}^{d} \left( \frac{\partial G(\mathbf{x})}{\partial x_j}\bigg|_{\mathbf{x}=\mathbf{x}^*} \right)^2 \operatorname{Var} X_j. \tag{2.5}$$

Then, for this linearization of the model output in a neighborhood of $\mathbf{x}^*$, Eq. (2.5) divides up the model output variance into the contribution from each input. This leads to

$$\xi_j = \left( \frac{\partial G(\mathbf{x})}{\partial x_j}\bigg|_{\mathbf{x}=\mathbf{x}^*} \right)^2 \frac{\operatorname{Var} X_j}{\operatorname{Var} Y}. \tag{2.6}$$

From here on, the indices $\xi_j$ ($j = 1, \ldots, d$) will be known as first-order sensitivities.

By only using first-order derivatives computed at one point, nonlinear and interaction effects between inputs cannot be measured. In the following sections, a derivative-based exploration of the input domain is proposed in order to obtain further information in a global SA setting.

### 2.3.2 ▪ Distributed evaluation of local sensitivity analysis

Local SA has been shown to be efficient for various problems [60, 487] but only provides information in a neighborhood of the point $\mathbf{x}^*$. If the model derivatives can be computed, it is natural to propose derivative-based exploration methods in order to move toward the global SA context. Rakovec et al. [391] have proposed such tools, which they call distributed evaluation of local sensitivity analysis (DELSA) and which are available in the delsa() function of the sensitivity package. The idea is to choose a sample size, a design type, and a perturbation percentage (for evaluating derivatives at each sample point via finite differences). The design type can be a regular grid, a Sobol' sequence (see Section A.4), or a sample provided by the user. The perturbation percentage is chosen to be the same at all sample points. The variance of each input is also required because the delsa() function directly evaluates the sensitivities in Eq. (2.6). Note that the regular grids and Sobol' sequences are uniform distributions of the inputs between their minimal and maximal values (defined by the user). If Monte Carlo sampling or nonuniform distributions have to be applied to some inputs, it is then necessary to provide a valid sample as input to the delsa() function.

The following code sets up visualization of a sample of size 100 (a Sobol' sequence) of sensitivities for the HeterDisc function ($d = 4$; see Section C.3):

```
> d <- 4
> delsa <- delsa(model = heterdisc.fun, perturb = 1.01,
+                 par.ranges = replicate(d,c(0,20),simplify = FALSE),
+                 samples = 100, method = "sobol")
```

The command

```
> print(delsa)
```

returns statistical summaries for the sensitivities with respect to each input. However, by using the generic plot() function, the delsa() function also provides graphics containing more complete information.

Figure 2.11 helps us to visualize the dispersion of the first-order sensitivities of each input ($X_1$ to $X_4$ are renamed V1 to V4). While the first-order sensitivities of V2 are relatively uniformly distributed, those of V3 are concentrated around small values, and those of V1 and V4 are split into two groups (of equal size): one with null values and the other with high values (especially V1).

Figure 2.12 does not show any specific links between the model output values and first-order sensitivities for any of the inputs (except for V4, which shows larger first-order sensitivities for small output values). However, these plots clearly illustrate that the group of large first-order sensitivities are predominantly for V1 and V2, showing that the model is highly sensitive to these inputs. Finally, a separation of the first-order sensitivities into two different groups still applies for V1 and V4. At this level of discrimination, one can conclude that the model has two different regimes.

Figure 2.13 shows the scatterplots of the first-order sensitivities, known as D-scatterplots (for derivative scatterplots) in Borgonovo et al. [46], using unnormalized derivatives instead of first-order sensitivities. An analysis of these plots shows that the discontinuity arises at V1 $= 10$; V1 has no effect below this threshold, while it has a strong effect above. Thus, V1 governs the model's two regimes. Therefore, it is possible to deduce that V4 has an interaction effect with V1 because for one of the two regimes, V4 has no effect.

All of these conclusions from the DELSA analysis provide complementary details to the conclusions from Figs. 2.5 and 2.6, without derivatives.

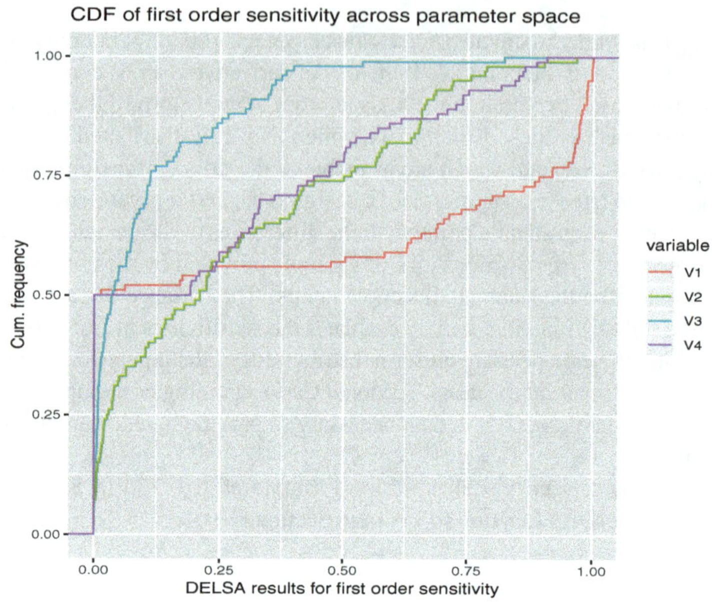


**Figure 2.11.** *Cumulative distribution function of a sample of first-order sensitivities for each input of the HeterDisc function.*

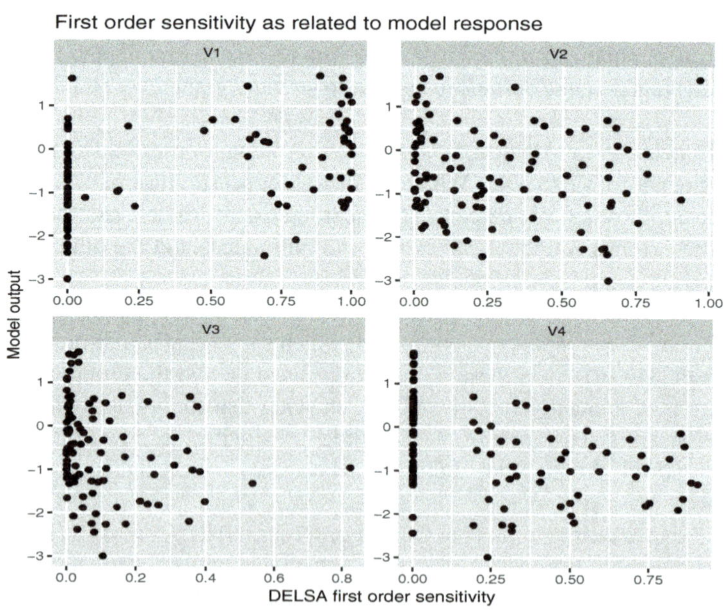


**Figure 2.12.** *Scatterplots of the model output as a function of the first-order sensitivities of each input (from a sample of the HeterDisc function).*

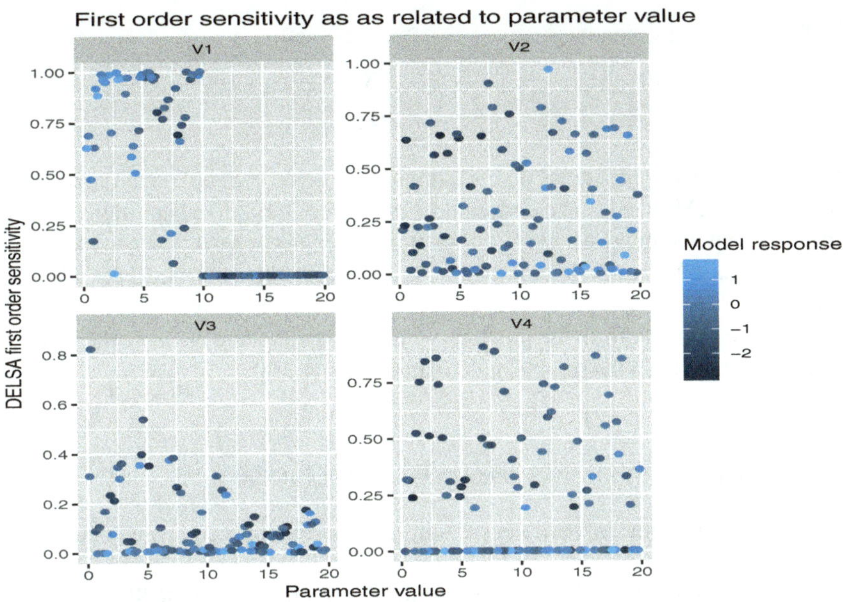


**Figure 2.13.** *Scatterplots of the first-order sensitivities of each input as a function of the same input (from a sample of the HeterDisc function).*

## 2.3.3 ▪ Derivative-based global sensitivity measures

As for importance measures for information contained in a sample of output and inputs values (see Sections 2.2.2 and 2.2.3), one can try to summarize the information contained in a sample of outputs and derivatives of the output with respect to the inputs. Sobol' and Gershman [443] have proposed what are known as derivative-based global sensitivity measures (DGSMs), which use the second moment of the model derivatives as an importance measure. Then they provide indices based on the global energy of the partial derivatives of the function $G$. We remark that the DGSM method can be seen as a natural extension of the Morris method (see Section 2.1.4), which is based on a rather rough estimation of the main effects using only a few model evaluations.

The DGSM method is motivated by the fact that a high value of the derivative of the model output with respect to some input variable means that a big variation in model output is expected when varying the value of the variable. This has been widely studied by Kucherenko et al. [251], Sobol' and Kucherenko [444, 445], Patelli et al. [361], Lamboni et al. [259], Kucherenko and Iooss [249], and Roustant et al. [402, 401]. In all of these studies, the DGSM method has been shown to be very efficient in the screening setting.

Let us first assume that $\frac{\partial G(\mathbf{X})}{\partial x_j}$ is square integrable for $j = 1, \ldots, d$. Then, for $j = 1, \ldots, d$, the DGSMs are defined by

$$\begin{aligned} \nu_j &= \mathbb{E}\left(\frac{\partial G(\mathbf{X})}{\partial x_j}\right)^2 \\ &= \int \left(\frac{\partial G(\mathbf{x})}{\partial x_j}\right)^2 \mathrm{P}_{\mathbf{X}}(d\mathbf{x}), \end{aligned} \tag{2.7}$$

where $\mathrm{P}_{\mathbf{X}}$ is the probability distribution of $\mathbf{X}$. A classical estimator for the DGSM is the empirical estimator

$$\widehat{\nu}_j = \frac{1}{n} \sum_{i=1}^{n} \left( \frac{\partial G(\mathbf{X}^{(i)})}{\partial x_j} \right)^2.$$

Experimental convergence properties of this estimator are given in Sobol' and Kucherenko [444]. Computation of DGSMs can also be performed with Monte Carlo–like algorithms such as Latin hypercube sampling and quasi-Monte Carlo sampling (see Section A). Kucherenko et al. [251] have shown that quasi-Monte Carlo outperforms Monte Carlo when the model has a low effective dimension.

Continuing the example from the previous section, it is simple to estimate the DGSM $\nu_j$ for each of the four inputs from the sample of derivatives of size 100:

```
> colMeans(delsa$deriv^2)

[1] 0.161774760 0.013155200 0.003249341 0.005000000
```

The results show the minimal influence of the third and fourth inputs compared to that of the first and second.

Of course, if the inputs have different probability distributions, their DGSMs cannot be compared directly. For a uniformly distributed input $X_j$, Sobol' and Kucherenko [444] have shown that $\nu_j$ has to be multiplied by a normalization factor $(b-a)^2/\pi$ (with $a$ and $b$ the minimal and maximal bounds of the uniform distribution of $X_j$). In Section 3.1.4, formulas for the normalization constant are given for any input probability distribution. These results are derived from inequalities developed between DGSMs and variance-based sensitivity indices.

For investigating interactions, Friedman and Popescu [143] and Roustant et al. [402] use the integral of squared mixed partial derivatives. For second-order interactions, they introduce (assuming similar conditions on the higher-order derivatives of $G$)

$$\nu_{i,j} = \int \left( \frac{\partial^2 G(\mathbf{x})}{\partial x_i \partial x_j} \right)^2 \mathrm{P}_{\mathbf{X}}(d\mathbf{x})$$

and, more generally, for $I \subseteq [1:d]$,

$$\nu_I = \int \left( \frac{\partial^{|I|} G(\mathbf{x})}{\partial \mathbf{x}_I} \right)^2 \mathrm{P}_{\mathbf{X}}(d\mathbf{x}),$$

with $\partial \mathbf{x}_I = \prod_{i \in I} \partial x_i$, where $|I|$ is the cardinality of $I$. In analogy to DGSMs, when $|I| \geq 2$, the quantities $\nu_I$ are called crossed DGSMs.

It has been shown that the DGSM method can be used for screening as well as for factor prioritization and is of great interest when an adjoint model of $G(\cdot)$ is available [402, 401]. The information this provides is strongly linked to Sobol' indices but comes from a different point of view.

### 2.3.4 ▪ Further references

Fruth et al. [145] have proposed to use more deeply the derivatives by exploring the local sensitivity behavior of an input over its support by defining support indices. Their first-order (resp., total) support indices is the square of the conditional expectation (resp., the conditional expectation of the square) of the model derivative. Then the expectation of a total support index is a DGSM. This approach allows to determine regions within the input domain for which the model variation is large.

It would seem also important to mention the connection here to active subspaces, which are an emerging set of tools for identifying unimportant directions in a model's input parameter space (see, e.g., [87] and references therein). There is also the recent work by Lewis et al. [275] which introduces a gradient-free active subspace construction method that avoids the need to sample from the gradient (which may not always be available). It works by constructing a rough approximation to the gradient matrix by employing the concept of *elementary effects* from Morris screening procedures (see Section 2.1.4).

## 2.4 ▪ Extensions to qualitative inputs

In this chapter, different techniques have been illustrated by restricting the examples to cases where the model input variables are quantitative. However, in practice one may encounter applications where some of the input variables are qualitative (also called categorical), i.e., have a finite number of unordered levels. Some of the aforementioned approaches generalize easily to this qualitative framework, while others need to be modified. Notes on this for each set of techniques are given below:

- Discretized design of experiments. Design of experiments techniques, introduced in Section 2.1.2 but not considered in depth in this book, are well set up to deal with qualitative inputs. Let us consider whether the techniques described in Section 2.1 still apply. The basic sequential bifurcation method relies on the discretization of inputs into two values. Therefore, it can be easily extended to bimodal qualitative inputs if the direction of variation of the model output is known when the level of a qualitative input changes. However, it cannot be extended in the case of inputs with more than two levels. The same problem appears for the Morris method, which is not applicable for qualitative (unordered) inputs. However, the sequential bifurcation and Morris methods easily apply in the case of ordinal input variables since these inputs are already discretized as ordered levels.

- Sampling-based techniques. Visualization techniques are easily extended to the qualitative setting since cobweb plots are unchanged and scatterplots can be replaced with, for example, boxplots or violin plots. Standard linear techniques still apply with the help of one-hot encoding. Moreover, in this context, sparsity involves a generalization of the lasso penalty known as the group lasso [511]. Nonlinear sparse models have also been developed for qualitative factors, e.g., the adaptive COSSO [460]. For feature selection techniques without models, the mutual information definition still holds, but estimation may need to be adapted [399]. As for kernel-based measures, they necessitate a kernel designed specifically for qualitative variables, ranging from the simple Dirac kernel to more sophisticated ones [2]. Finally, the case of ordinal input variables can easily be taken into account for sampling-based techniques, e.g., by starting with a rank transformation of the sample.

- Derivative-based techniques. A derivative of the model output with respect to a qualitative (or ordinal) input does not make any sense. Therefore, these techniques do not apply.

## 2.5 ▪ Summary

As a synthetic view of all the screening techniques described in this chapter, Fig. 2.14 provides a classification grid aiming to position them in a practical use viewpoint. Each technique is coarsely located as a function of its computational cost (in terms of the required number of model evaluations) and its underlying hypothesis on the model regularity or complexity. For example, a resolution III (denoted RIII) design of experiments (see Section 2.1.2) allows one to correctly estimate the first-order effects of a linear model which has no interaction, while a

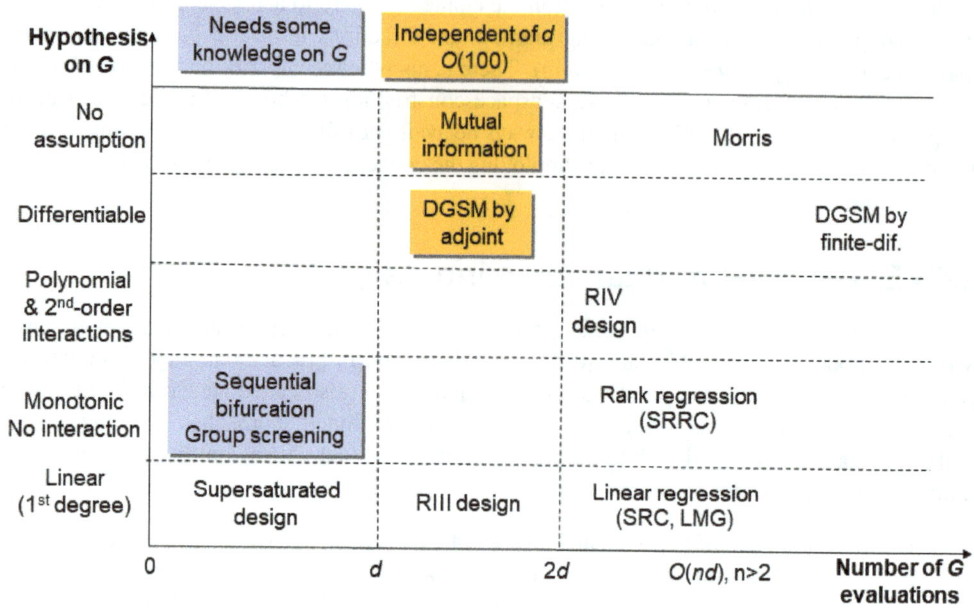


**Figure 2.14.** *Classification grid of screening methods in terms of the required number of model evaluations and their underlying hypothesis on the model.*

resolution IV (denoted RIV) design allows one to correctly estimate the first-order effects of a linear model which can have some second-order interactions. Another example is that the model has to be differentiable if the screening technique uses a sample of model derivatives, as in the estimation of DGSM or in the use of the DELSA method (see Section 2.3).

The costs of most of the methods depend linearly (or approximately linearly) on the number of inputs $d$. The methods whose costs are $d$-independent appear with an orange box. Moreover, the techniques requiring some knowledge on the model behavior are identified by a purple box. For example, this is the case for the sequential bifurcation method (see Section 2.1.3), which requires the sense of variation of the output with respect to each input.

# Chapter 3
# Variance-based sensitivity measures

This chapter deals with quantitative global SA, where the uncertainty of the model output is measured by second-order statistics, namely, the variance. From the functional analysis of variance decomposition, the following section (Section 3.1) introduces the variance-based importance measures, which are the so-called Sobol' indices. Except for very simple models, these indices cannot be analytically derived, and various algorithmic schemes have been proposed for their estimation. In this chapter, the standard sampling-based methods is described in Section 3.2, while other approaches are detailed in Chapter 4. Asymptotic and nonasymptotic properties are given for the simplest estimators based on the use of Monte Carlo samples. These two first sections are restricted to the case of a single scalar output. In Section 3.3, we develop ideas for dealing with multidimensional and functional outputs. All throughout this chapter, several estimation algorithms are illustrated by using the R software on simple analytical functions taken from Section C.

## 3.1 ▪ From functional ANOVA to Sobol' indices

This section describes the functional analysis of variance decomposition (FANOVA), which gives a theoretical basis for the definition of Sobol' indices as well the derivation of their properties.

### 3.1.1 ▪ Functional ANOVA

One of the main tools in global SA is the FANOVA decomposition. To understand in a nutshell this decomposition, let us focus on a simple functional toy model with two finitely supported inputs. More precisely, let $X_1$ and $X_2$ be independent balanced Bernoulli random variables. Now let $G$ be a real-valued application on $\{0,1\} \times \{0,1\}$. In other words, $G$ is a vector consisting of the four values of the function $G(0,0)$, $G(0,1)$, $G(1,0)$, $G(1,1)$. A very rough summary of this vector is given by its mean $G_\emptyset$. Further, if one wish to ignore the second (resp., first) variable, a better summary is then $G_{\{1\}}$ (resp., $G_{\{2\}}$), which is the deviation from the global mean of the average obtained by summing, after normalization, over the second (resp., first) variable. Roughly speaking, $G_{\{1\}}$ (resp., $G_{\{2\}}$) provides the correction to the global mean when one wishes to study the function $G$ ignoring the second (resp., first) variable. At a first glance, it is then tempting to provide for $G$ the interpretable summary $G_\emptyset + G_{\{1\}} + G_{\{2\}}$, where $G_\emptyset$ resumes the global behavior and $G_{\{1\}}$ (resp., $G_{\{2\}}$) the deviation to this global behavior when only $x_1$ (resp., $x_2$) is moving. Of course, if one wishes to reconstruct exactly the function $G$ from these summaries, the residual part $G_{\{1,2\}} = G - (G_\emptyset + G_{\{1\}} + G_{\{2\}})$ is missed. Intuitively,

$G_{\{1,2\}}$ represents the part of $G$ that cannot be explained by either global or marginal effects. As discussed below, the previous additive decomposition of $G$ has the very nice property of providing uncorrelated interpretable contributions. Let us now formalize a little bit more on this toy model the general decomposition. One way writes

$$G(X_1, X_2) = G_\emptyset + G_{\{1\}}(X_1) + G_{\{2\}}(X_2) + G_{\{1,2\}}(X_1, X_2) \tag{3.1}$$

with

$$\begin{aligned} G_\emptyset &= \frac{1}{4}(G(0,0) + G(0,1) + G(1,0) + G(1,1)), \\ G_{\{1\}}(x) &= \frac{1}{2}(G(x,0) + G(x,1)) - G_\emptyset, \\ G_{\{2\}}(x) &= \frac{1}{2}(G(0,x) + G(1,x)) - G_\emptyset, \quad (x \in \{0,1\}) \end{aligned}$$

and

$$G_{\{1,2\}}(x_1, x_2) = G(x_1, x_2) - G_\emptyset - G_{\{1\}}(x_1) - G_{\{2\}}(x_2), \quad (x_1, x_2 \in \{0,1\}).$$

It can be easily checked that the four functions $(G_A)_{A \subset \{1,2\}}$ satisfy the orthogonality property $\mathbb{E}[G_A(X_1, X_2) G_B(X_1, X_2)] = 0$ when $A$ and $B$ are distinct subsets of $\{1,2\}$. This of course implies that the variance of $G(X_1, X_2)$ may be decomposed as

$$\operatorname{Var} G(X_1, X_2) = \operatorname{Var} G_{\{1\}}(X_1) + \operatorname{Var} G_{\{2\}}(X_2) + \operatorname{Var} G_{\{1,2\}}(X_1, X_2).$$

In summary, a general finite function of two independent balanced Bernoulli random variables may be decomposed as an orthogonal sum of constant, marginal, and joint functions. As a matter of fact, we recognize here the classical orthogonal ANOVA decomposition for two binary factors in the very special case of a balanced design without noise on the measurement; see, e.g., [427]. Of course, from the general ANOVA decomposition of the linear model theory, this noiseless decomposition easily extends to the broader case where we have at hand a general vector $\mathbf{X}$ having independent components with finite support.

The previous decomposition is stated in the slightly much broader case of a square integrable function of a random vector $\mathbf{X}$ with independent components valued in abstract state spaces. In other words, the previous ANOVA decomposition will be performed in a very general setting where the inputs are assumed to be independent but may take their values in very general spaces. This decomposition will allow one to then define general sensitivity indices called Sobol' indices. These indices are energy ratios built on the ANOVA decomposition and are related to the so-called $F$-test (see [427]) in classical ANOVA. We will make the following important independence assumption on the input of the system under study. We will see in Theorem 3.3 that this assumption allows to write the ANOVA decomposition in the general frame described here.

**Assumption $\mathcal{A}_1$.** *Each component $X_i$, $i = 1, \dots, d$, is assumed to be valued in an abstract Polish measurable space $(E_i, \mathcal{B}(E_i))$. Here, $\mathcal{B}(E_i)$ denotes the Borel $\sigma$-field associated to $E_i$. Recall that a Polish space is a complete separable metric space. For any nonempty subset of indices $A \in \mathcal{P}_d$ ($\mathcal{P}_d = \mathcal{P}([1:d])$, the set of all subsets of $[1:d] = \{1, \dots, d\}$), we define*

$$(E_A, \mathcal{E}_A) = (\Pi_{i \in A} E_i, \otimes_{i \in A} \mathcal{B}(E_i)).$$

*Moreover, we introduce $(E, \mathcal{E}) = (E_{[1:d]}, \mathcal{E}_{[1:d]})$. Let $\mathrm{P}_{\mathbf{X}}$ denote the probability distribution of the random vector $\mathbf{X}$. The components $X_i$ of the random vector $\mathbf{X}$ are assumed to be independent; thus we have $\mathrm{P}_{\mathbf{X}} = \prod_{i=1}^{d} \mathrm{P}_{X_i}$ with $\mathrm{P}_{X_i}$ the probability distribution of $X_i$.*

We then denote by $\mathbb{L}^2(\mathrm{P}_{\mathbf{X}})$ the set of all measurable functions $f$ on $(E, \mathcal{E})$ such that $\mathbb{E}[f^2(\mathbf{X})] < +\infty$. The space $\mathbb{L}^2(\mathrm{P}_{\mathbf{X}})$ is a Hilbert space with inner product defined by

$$\langle f, g\rangle = \mathbb{E}[f(\mathbf{X})g(\mathbf{X})]\ \forall f, g \in \mathbb{L}^2(\mathrm{P}_{\mathbf{X}}).$$

In the following, $\mathrm{P}_{\mathbf{X}}$-*almost surely* is abbreviated by simply writing *a.s.*

**Remark 5.** *Note that this framework is not limited to real-valued random inputs $X_i$ ($E_i = \mathbb{R}$). It also includes more complex structures (images, signals, functions, etc.). Of course, in these more complicated frameworks, strong measure theory is required to define properly the random quantities occurring in the model. We refer the reader to Tjur [485] and Billingsley [38] for such constructions.*

To illustrate the previous definitions and assumption, let us discuss briefly an example involving a functional input.

**Example 3.1 (heat flow in a uniform rod).** Consider the heat equation describing the evolution of the temperature $T(t, s)$ as a function of the position $s \in [0, L]$ ($L > 0$) and the time $t \geq 0$ in a uniform rod with length $L$:

$$\frac{\partial T}{\partial t} = X_1 \frac{\partial^2 T}{\partial s^2},\ T(t, 0) = T(t, L) = X_2,\ T(0, s) = X_2 + X_3(s). \tag{3.2}$$

Let $t_0 > 0$ be given, and set $G(X_1, X_2, X_3) = T(t_0, L/2)$. This means that we are interested in the temperature at time $t_0$ in $L/2$ as a function of $X_1, X_2, X_3$. We have $d = 3$, the diffusivity coefficient $X_1$, and the border condition $X_2$; both lie in $E_1 = E_2 = \mathbb{R}_+$, while the random nonnegative process $X_3(s)$ lies in $E_3$, the set of all nonnegative continuous functions on $[0, L]$ vanishing at 0 and $L$. Under Assumption $\mathcal{A}_1$, the quantities $X_1, X_2, X_3$ on which the solution of Eq. (3.2) depends are random but stochastically independent. Notice further that $\mathcal{P}_3$ may be extensively written as

$$\mathcal{P}_3 = \{\emptyset, \{1\}, \{2\}, \{3\}, \{1, 2\}, \{1, 3\}, \{2, 3\}, \{1, 2, 3\}\}.$$

Notice that $\mathcal{P}_3$ contains exactly $8 = 2^3$ sets and that more generally $\mathcal{P}_d$ contains exactly $2^d$ sets. ■

We now aim at decomposing the functions in $\mathbb{L}^2(\mathrm{P}_{\mathbf{X}})$ on suitable *factor* subspaces. For any $A \in \mathcal{P}_d$, we denote by $\mathbb{L}^2_A$ the subspace of functions in $\mathbb{L}^2(\mathrm{P}_{\mathbf{X}})$ that are measurable on $(E_A, \mathcal{E}_A)$. Formally, for any $A \in \mathcal{P}_d$, set

$$\begin{aligned}\mathbb{L}^2(\mathrm{P}_{\mathbf{X}}) &= \{\text{measurable functions } f \text{ on}(E, \mathcal{E}) : \mathbb{E}[f^2(\mathbf{X})] < +\infty\},\\ \mathbb{L}^2_A &= \{f \in \mathbb{L}^2(\mathrm{P}_{\mathbf{X}}) : f \text{ is } (E_A, \mathcal{E}_A)\text{-measurable}\}.\end{aligned}$$

In particular, $\mathbb{L}^2_{[1:d]} = \mathbb{L}^2(\mathrm{P}_{\mathbf{X}})$, and $\mathbb{L}^2_\emptyset$ is the set of constant functions on $E$. Further, we also set $\mathbf{X}_A = (X_i : i \in A)$, and for any $\mathbf{x} \in E$, $\mathbf{x}_A = (x_i : i \in A)$. To alleviate the notation, we will also write $\mathrm{P}_A$ for $\Pi_{i\in A}\mathrm{P}_{X_i}$ and $\mathrm{P}_i$ for $\mathrm{P}_{\{i\}}$. Notice that an element of $\mathbb{L}^2_A$ may be written as $f(X_A)$, where $f$ is a measurable real function on $\mathbb{R}^{|A|}$ satisfying $\mathbb{E}[f(X_A)^2] < +\infty$ (here $|A|$ denotes the number of elements in $A$). A classical and very interesting operator acting on $\mathbb{L}^2(\mathrm{P}_{\mathbf{X}})$ with values in $\mathbb{L}^2_A$ is the orthogonal projector $\mathrm{Proj}_A$. For $f \in \mathbb{L}^2(\mathrm{P}_{\mathbf{X}})$, $\mathrm{Proj}_A f$ is the closest element in $\mathbb{L}^2_A$ of $f$ (closest means having the least $\mathbb{L}^2$ deviation). That is,

$$\inf_{h\in L^2_A} \mathbb{E}[(f(\mathbf{X}) - h(\mathbf{X}_A))^2] = \min_{h\in L^2_A} \mathbb{E}[(f(\mathbf{X}) - h(\mathbf{X}_A))^2] = \mathbb{E}[(f(\mathbf{X}) - (\mathrm{Proj}_A f)(\mathbf{X}_A))^2].$$

As $\mathrm{Proj}_A$ is a projection, we have

$$\mathrm{Proj}_A^2 f = \mathrm{Proj}_A(\mathrm{Proj}_A f) = \mathrm{Proj}_A f, \text{ so that } \mathrm{Proj}_A(f - \mathrm{Proj}_A f) = \mathrm{Proj}_A f - \mathrm{Proj}_A f = 0.$$

The last equality implies that $f - \mathrm{Proj}_A f$ is perpendicular to $\mathbb{L}_A^2$ so that, for any $h \in \mathbb{L}_A^2$,

$$\mathbb{E}[(f(\mathbf{X}) - (\mathrm{Proj}_A f)(\mathbf{X}_A))h(\mathbf{X}_A)] = 0.$$

Notice that this last property also characterizes $\mathrm{Proj}_A f$. That is, $\mathrm{Proj}_A f$ is the unique function $g$ of $\mathbb{L}_A^2$ such that $f - g$ is perpendicular to any member of $\mathbb{L}_A^2$. In other words,

$$\forall h \in \mathbb{L}_A^2, \langle h, g\rangle = \langle h, f\rangle \iff g = \mathrm{Proj}_A f. \tag{3.3}$$

Notice further that Eq. (3.3) also implies that $\mathrm{Proj}_A$ is self-adjoint. Indeed, for $f$ and $g$ square integrable functions, we may write

$$\langle \mathrm{Proj}_A f, g\rangle = \langle \mathrm{Proj}_A f, \mathrm{Proj}_A g\rangle = \langle \mathrm{Proj}_A g, \mathrm{Proj}_A f\rangle = \langle \mathrm{Proj}_A g, f\rangle = \langle f, \mathrm{Proj}_A g\rangle.$$

Last but not least, let $I$ denotes the identity operator on $\mathbb{L}^2(\mathrm{P}_\mathbf{X})$ ($If = f$ for all $f$). Then $I - \mathrm{Proj}_A$ is the orthogonal projector on $(L_A^2)^\perp$, the space of all square integrable functions that are orthogonal to all members of $L_A^2$. As a matter of fact, for $f \in \mathbb{L}^2(\mathrm{P}_\mathbf{X})$ and $h \in (L_A^2)^\perp$, we may write

$$\langle h, (I - \mathrm{Proj}_A)f\rangle = \langle h, f\rangle - \langle h, \mathrm{Proj}_A f\rangle = \langle h, f\rangle$$

and conclude using one more time Eq. (3.3). A classical result of probability theory (see, e.g., [485]) is that $\mathrm{Proj}_A f$ is the conditional expectation with respect to the $\sigma$-field $\mathcal{E}_A$. In other words, we have

$$(\mathrm{Proj}_A f)(\mathbf{X}_A) = \mathbb{E}[f(\mathbf{X})|X_A].$$

Now, under Assumption $\mathcal{A}_1$, the components of $\mathbf{X}$ are independent, and the computation of $\mathbb{E}[f(\mathbf{X})|X_A]$ becomes easier, as described in the following lemma.

**Lemma 3.2.** *Let $f \in \mathbb{L}^2(\mathrm{P}_\mathbf{X})$ and $A, B \in \mathcal{P}_d$. Recall that $\overline{A}$ denotes the complementary set of $A$. Then, under Assumption $\mathcal{A}_1$, we have, for almost $\mathrm{P}_\mathbf{X}$ all $\mathbf{x}$ in $E$,*

$$(\mathrm{Proj}_A f)(\mathbf{x}) = \mathbb{E}[f(\mathbf{X})|\mathbf{X}_A = \mathbf{x}_A] = \int_{E_{\overline{A}}} f(\mathbf{x})\mathrm{P}(d\mathbf{x}_{\overline{A}}), \tag{3.4}$$

$$(\mathrm{Proj}_B \mathrm{Proj}_A f)(\mathbf{x}) = (\mathrm{Proj}_{A\cap B} f)(\mathbf{x}) = (\mathrm{Proj}_A \mathrm{Proj}_B f)(\mathbf{x}). \tag{3.5}$$

***Proof.*** For $h \in \mathbb{L}_A^2$, as $E = E_A \times E_{\overline{A}}$ and by the independence assumption, we may write

$$\begin{aligned}\langle h, f\rangle = \mathbb{E}(hf) &= \int_{E_A \times E_{\overline{A}}} h(\mathbf{x}_A) f(\mathbf{x}) \mathrm{P}_{\overline{A}}(d\mathbf{x}_{\overline{A}}) \mathrm{P}_A(d\mathbf{x}_A)\\ &= \int_{E_A} h(\mathbf{x}_A)\left(\int_{E_{\overline{A}}} f(\mathbf{x})\mathrm{P}_{\overline{A}}(d\mathbf{x}_{\overline{A}})\right)\mathrm{P}_A(d\mathbf{x}_A).\end{aligned}$$

Here, the last equality comes from the Fubini theorem as $h(\mathbf{X}_A)f(\mathbf{X})$ is integrable. This implies Eq. (3.4) by using Eq. (3.3). Let now $h \in \mathbb{L}_{A\cap B}^2$, as $\mathbb{L}_{A\cap B}^2 \subset \mathbb{L}_A^2$ and $\mathbb{L}_{A\cap B}^2 \subset \mathbb{L}_B^2$. We have by Eq. (3.3)

$$\langle h, \mathrm{Proj}_B \mathrm{Proj}_A f\rangle = \langle h, \mathrm{Proj}_A f\rangle = \langle h, f\rangle.$$

Thus, we may conclude that Eq. (3.5) holds as $A \cap B = B \cap A$ and using Eq. (3.3). □

The following theorem appeared early in a work by Hoeffding [196] and is known in the global SA community as the Sobol' decomposition. It is called the Hoeffding decomposition in what follows. To our knowledge, it was first used as a statistical tool for functional variance analysis by Antoniadis [7]. It gives a nice orthogonal decomposition in the spirit of Eq. (3.1) but in a very general $\mathbb{L}^2(\mathrm{P}_\mathbf{X})$ framework.

**Theorem 3.3 (Hoeffding decomposition).** *Let $G \in \mathbb{L}^2(\mathrm{P}_\mathbf{X})$. Under Assumption $\mathcal{A}_1$, there exists a unique decomposition of $G$ in $\mathbb{L}^2(\mathrm{P}_\mathbf{X})$ of the form*

$$G(\mathbf{x}) = \sum_{A \in \mathcal{P}_d} G_A(\mathbf{x}_A), \quad \mathrm{P}_\mathbf{X} \ a.s., \tag{3.6}$$

*such that the two following properties hold:*

1. *$G_\emptyset$ constant a.s.*
2. *$\forall\, A \in \mathcal{P}_d,\ A \neq \emptyset,\ \forall\, i \in A,\ \int_{E_i} G_A(\mathbf{x}_A) \mathrm{P}_{X_i}(dx_i) = 0.$*

*The unique solution writes, $\forall\, A \in \mathcal{P}_d$,*

$$G_A(\mathbf{x}_A) = \sum_{B \subset A} (-1)^{|A|-|B|} \mathbb{E}[G(\mathbf{X}) | \mathbf{X}_B = \mathbf{x}_B] \quad a.s. \tag{3.7}$$

***Proof.*** Note first that the proof could be done using induction and the same arguments as in the bidimensional example developed at the beginning of this chapter. Nevertheless, a smarter elegant method developed in a general context is presented. Indeed, we follow the proof proposed by Kuo et al. [254] in the much more general context of a commuting projector collection (see Theorem 2.1 in [254]). Note that the results stated in this theorem are also very useful in proving the so-called anchored Hoeffding decomposition (see, e.g., [475]). It it also useful in Section 6.2 for some kernel generalizations in sensitivity analysis.

To begin with, let us first prove that there exists a decomposition of $G$ in the form of Eq. (3.6) such that properties 1 and 2 hold. Let $j \in [1 : d]$. To alleviate the notation, we will write $\mathrm{Proj}_{-j}$ for $\mathrm{Proj}_{\overline{\{j\}}}$. Notice that from Eq. (3.4), $(\mathrm{Proj}_{-j} G)(\mathbf{x})$ does not depend anymore on $x_j$ (a.s.). Moreover, from Eq. (3.5), for any $i, j \in [1 : d]$, $\mathrm{Proj}_{-i}$ and $\mathrm{Proj}_{-j}$ are commuting. We may write

$$I = \prod_{j=1}^{d} \left(\mathrm{Proj}_{-j} + (I - \mathrm{Proj}_{-j})\right).$$

Taking into account the commutation in expanding the last product leads to

$$I = \sum_{A \in \mathcal{P}_d} \left( \prod_{j \in A} (I - \mathrm{Proj}_{-j}) \right) \left( \prod_{j \in \overline{A}} \mathrm{Proj}_{-j} \right). \tag{3.8}$$

Considering the complementary set, one may easily check that $\bigcap_{j \in \overline{A}} \overline{\{j\}} = A$ so that, using once again Eq. (3.4), we have that $\left(\prod_{j \in \overline{A}} \mathrm{Proj}_{-j}\right) = \mathrm{Proj}_A$. Let us now define, for any $A \in \mathcal{P}_d$, $G_A = \left(\prod_{j \in A}(I - \mathrm{Proj}_{-j})\right) \mathrm{Proj}_A G$. It follows directly from Eq. (3.8) that, a.s.,

$$G(\mathbf{x}) = \sum_{A \in \mathcal{P}_d} G_A(\mathbf{x}).$$

By definition, we get $G_\emptyset = \text{Proj}_\emptyset G = \int_E G(\mathbf{x})\text{P}(d\mathbf{x})$ by Eq. (3.4); thus, $G_\emptyset$ is a.s. constant. Moreover, by Eq. (3.5), we have a.s.

$$G_A(\mathbf{x}) = \left(\prod_{j\in A}(I - \text{Proj}_{-j})\right)\text{Proj}_A G(\mathbf{x}) = \text{Proj}_A\left(\prod_{j\in A}(I - \text{Proj}_{-j})\right)G(\mathbf{x}),$$

so that $G_A(\mathbf{x})$ does not depend on $\mathbf{x}_{\overline{A}}$ and we may write $G_A(\mathbf{x}) = G_A(\mathbf{x}_A)$. Now, for $i \in A$ and $f \in \mathbb{L}^2$, $\text{Proj}_{-i}(I - \text{Proj}_{-i})f = 0$. So

$$\begin{aligned}\int_{E_i} G_A(\mathbf{x}_A)\text{P}_{X_i}(dx_i) &= \text{Proj}_{-i}G_A(x)\\ &= \text{Proj}_{-i}(I - \text{Proj}_{-i})\prod_{j\in A, j\neq i}(I - \text{Proj}_{-j})\text{Proj}_A G(x),\\ &= 0 \text{ (a.s.)}.\end{aligned}$$

Thus, we have proved that there exists a decomposition of $G$ in the form of Eq. (3.6) satisfying properties 1 and 2 Now, considering a decomposition of $G$ in the form of Eq. (3.6),

$$G(\mathbf{x}) = \sum_{A\in\mathcal{P}_d} G_A(\mathbf{x}_A) \quad \text{a.s.},$$

satisfying properties 1 and 2, we will prove that it also satisfies Eq. (3.7), leading to a.s. uniqueness. First, note that $\mathbb{E}[G(\mathbf{X})] = \int G(\mathbf{x})\text{P}_\mathbf{X}(d\mathbf{x}) = G_\emptyset$. Indeed, as $G(\mathbf{x}) = G_\emptyset + \sum_{A\in\mathcal{P}_d, A\neq\emptyset} G_A(\mathbf{x}_A)$ and for all $A \in \mathcal{P}_d$, $A \neq \emptyset$, for any $i \in A$,

$$\int_E G_A(\mathbf{x}_A)\text{P}_\mathbf{X}(d\mathbf{x}) = \int_{\prod_{j\neq i}E_j}\underbrace{\int_{E_i} G_A(\mathbf{x}_A)\text{P}_{X_i}(dx_i)\prod_{j\neq i}}_{0}\text{P}_{X_j}(dx_j) = 0.$$

Now let $i \in [1:d]$. We have $\int_{\prod_{j\neq i}E_j} G(\mathbf{x})\prod_{j\neq i}\text{P}_{X_j}(dx_j) = G_\emptyset + G_{\{i\}}(X_i)$ as by property 2.

$$\sum_{A\neq i, A\neq\emptyset}\underbrace{\int_{\prod_{j\neq i}E_j} G_A(\mathbf{x}_A)\prod_{j\neq i}\text{P}_{X_j}(dx_j)}_{0} = 0.$$

Thus, by Eq. (3.4), for any $i \in [1:d]$,

$$\begin{aligned}G_{\{i\}}(x_i) &= \int_{\prod_{j\neq i}E_j} G(\mathbf{x})\prod_{j\neq i}\text{P}_{X_j}(dx_j) - G_\emptyset\\ &= \mathbb{E}[G(\mathbf{X})|X_i = x_i] - \mathbb{E}\,G(\mathbf{X}).\end{aligned}$$

More generally, let $A \in \mathcal{P}_d$. We can write

$$G(\mathbf{x}) = \sum_{B\in\mathcal{P}_d, \bar{A}\cap B\neq\emptyset} G_B(\mathbf{x}_B) + \sum_{B\subset A} G_B(\mathbf{x}_B).$$

Let $B \in \mathcal{P}_d$, $\bar{A} \cap B \neq \emptyset$. For any $i \in \bar{A} \cap B$, we have, using one more time property 2,

$$\begin{aligned}\mathbb{E}[G_B(\mathbf{X}_B)|\mathbf{X}_A = \mathbf{x}_A] &= \int_{\prod_{j\notin A}E_j} G_B(\mathbf{x}_B)\prod_{j\notin A}\text{P}_{X_j}(dx_j)\\ &= \int_{\prod_{j\notin A, j\neq i}E_j}\underbrace{\int_{E_i} G_B(\mathbf{x}_B)\text{P}_{X_i}(dx_i)\prod_{j\notin A, j\neq i}}_{0}\text{P}_{X_j}(dx_j) = 0.\end{aligned}$$

Using the assumed expansion of $G$ and the last equation, we thus have, for $\mathrm{P_X}$ a.s. all $\mathbf{x} \in \prod_{i=1}^d E_i$,

$$\mathbb{E}[G(\mathbf{X})|\mathbf{X}_A = \mathbf{x}_A] = \sum_{B\subset A} \mathbb{E}[G_B(\mathbf{X}_B)|\mathbf{X}_A = \mathbf{x}_A] = \sum_{B\subset A} G_B(\mathbf{x}_B). \tag{3.9}$$

Defining for $\mathrm{P_X}$ a.s. all $\mathbf{x} \in \prod_{i=1}^d E_i$, $g(\mathbf{x}, A) = \mathbb{E}[G(\mathbf{X})|\mathbf{X}_A = \mathbf{x}_A]$, and $f(\mathbf{x}, B) = G_B(\mathbf{x}_B)$, we get $g(\mathbf{x}, A) = \sum_{B\subset A} f(\mathbf{x}, B)$. Applying the Rota inversion formula [400], we get

$$f(\mathbf{x}, A) = \sum_{B\subset A} (-1)^{|A|-|B|} g(\mathbf{x}, B).$$

Thus, we may conclude that

$$G_A(\mathbf{x}_A) = \sum_{B\subset A} (-1)^{|A|-|B|} \mathbb{E}[G(\mathbf{X})|\mathbf{X}_B = \mathbf{x}_B]. \quad \square$$

**Corollary 3.4.** *Let $G \in \mathbb{L}^2(\mathrm{P_X})$. Then under Assumption $\mathcal{A}_1$, the two following properties hold:*

1. *Let $A \in \mathcal{P}_d$ be such that $A \neq \emptyset$. Then $\mathbb{E}[G_A(\mathbf{X}_A)] = 0$.*

2. *Let $A \neq A' \in \mathcal{P}_d$. Then $G_A$ and $G_{A'}$ are orthogonal in $\mathbb{L}^2(\mathrm{P_X})$. That is, $\langle G_A, G_B\rangle = \mathbb{E}[G_A(\mathbf{X}_A)G_{A'}(\mathbf{X}_{A'})] = 0$.*

***Proof.*** Let $A \in \mathcal{P}_d$ be such that $A \neq \emptyset$. Let $i \in A$. Using point 2 of the previous theorem, we have

$$\mathbb{E}[G_A(\mathbf{X}_A)] = \int_{\prod_{j\in A, j\neq i} E_j} \underbrace{\int_{E_i} G_A(\mathbf{x}_A)\mathrm{P}_{X_i}(dx_i) \prod_{j\in A, j\neq i}}_{0} \mathrm{P}_{X_j}(dx_j).$$

Let $A \neq A' \in \mathcal{P}_d$. As $A \neq A'$, either there exists $i \in A \cap \bar{A}'$ or there exists $i \in A' \cap \bar{A}$. Assume without loss of generality that there exists $i \in A' \cap \bar{A}$, as the other case can be handled similarly. We have

$$\mathbb{E}[G_A(\mathbf{X}_A)G_{A'}(\mathbf{X}_{A'})] = \int_{\prod_{j\neq i} E_j} \underbrace{\int_{E_i} G_{A'}(\mathbf{x}_{A'})\mathrm{P}_{X_i}(dx_i)}_{0} G_A(\mathbf{x}_A) \prod_{j\neq i} \mathrm{P}_{X_j}(dx_j) = 0. \quad \square$$

The first term in the Hoeffding decomposition (Eq. (3.6)) corresponds to the mean behavior of the model $G$. Then any higher-order term in this decomposition can be interpreted as the contribution on the initial function $G$ of a group of variables $\mathbf{X}_A$ when removing all the lower-order interaction contributions related to $\mathbf{X}_B, B \subset A$. For example, the contribution $G_{\{i\}}$ ($i = 1, \dots, d$) provides the centered effect of the variable $X_i$ alone, while $G_{\{i,j\}}$ ($1 \le i < j \le d$) describes the centered interaction effect of the variables $X_i$ and $X_j$ removing the single effects.

From Corollary 3.4, we know that the summand in the Hoeffding decomposition is centered (except for $A = \emptyset$) and uncorrelated. Thus, the variance of the function $G(\mathbf{X})$ can be computed as the sum of the variances of the nonconstant terms of the decomposition. This is the foundation stone for second-order methods in SA. Indeed, as for classical variance analysis [427], this variance decomposition allows one to quantify, for any $A \in \mathcal{P}_d$, $A \neq \emptyset$, the part of variance of $G(\mathbf{X})$ explained by interactions between components in $\mathbf{X}_A$. Actually, the classical analysis of variance appears to be a Euclidean special case where the probability distribution $P_{\mathbf{X}}$ is supported by a finite number of points. The following corollary formally states the variance decomposition.

**Corollary 3.5 (FANOVA).** *Let $G \in \mathbb{L}^2(P_{\mathbf{X}})$. For any $A \in \mathcal{P}_d$, $A \neq \emptyset$, let $V_A = \operatorname{Var} G_A(\mathbf{X}_A)$. Then, under Assumption $\mathcal{A}_1$, we get*

$$V = \operatorname{Var} G(\mathbf{X}) = \sum_{A \in \mathcal{P}_d, A \neq \emptyset} V_A.$$

*Furthermore, for any $A \in \mathcal{P}_d$, $A \neq \emptyset$,*

$$V_A = \sum_{B \subset A} (-1)^{|A|-|B|} \operatorname{Var} \mathbb{E}[G(\mathbf{X})|\mathbf{X}_B].$$

***Proof.*** The first part of the corollary is straightforward from Corollary 3.4. Indeed, as $G(\mathbf{X}) - G_\emptyset = \sum_{A \in \mathcal{P}_d, A \neq \emptyset} G_A(\mathbf{X}_A)$ and from the results in Corollary 3.4, we get

$$\begin{aligned} V &= \mathbb{E}[G(\mathbf{X}) - G_\emptyset]^2 \\ &= \sum_{A \in \mathcal{P}_d, A \neq \emptyset} \mathbb{E}\, G_A^2(\mathbf{X}_A) + \sum_{\emptyset \subsetneq A \neq A' \in \mathcal{P}_d} \mathbb{E}[G_A(\mathbf{X}_A) G_{A'}(\mathbf{X}_{A'})] \\ &= \sum_{A \in \mathcal{P}_d, A \neq \emptyset} \operatorname{Var} G_A(\mathbf{X}_A) = \sum_{A \in \mathcal{P}_d, A \neq \emptyset} V_A. \end{aligned}$$

Now let $A \in \mathcal{P}_d$, $A \neq \emptyset$. From Eq. (3.9) in the proof of Theorem 3.3, we have $P_{\mathbf{X}_A}$ a.s. for all $\mathbf{x}_A$, $\mathbb{E}[G(\mathbf{X})|\mathbf{X}_A = \mathbf{x}_A] = \sum_{B \subset A} G_B(\mathbf{x}_B)$. Then, from Corollary 3.4, we get

$$\operatorname{Var} \mathbb{E}[G(\mathbf{X})|\mathbf{X}_A] = \operatorname{Var}\left(\sum_{B \subset A} G_B(\mathbf{X}_B)\right) = \sum_{B \subset A} \operatorname{Var} G_B(\mathbf{X}_B).$$

Once more, applying the Rota inversion formula [400], we get

$$\operatorname{Var} G_A(\mathbf{X}_A) = \sum_{B \subset A} (-1)^{|A|-|B|} \operatorname{Var} \mathbb{E}[G(\mathbf{X})|\mathbf{X}_B]. \quad \square$$

**Remark 6.** *In the decompositions of Theorem* 3.3 *and Corollary* 3.5*, the main assumption is that the components of the input vector are independent. Note that the possible nature of these components is very general and can be composite. For example, some of the components may be real or vectorial random variables, while others may be random functions or fields having values in some functional space. Furthermore, for the sake of simplicity, the output function $G$ is assumed to be scalar valued, but extensions to more general frameworks are possible. For example, the case of vector-valued output is considered in Lamboni et al. [260] and in [147, 148]. We refer the reader to Section* 3.3 *for further developments in the framework of nonscalar outputs.*

### 3.1.2 ▪ Sobol' indices

The FANOVA decomposition stated in Corollary 3.5 is the key relation for variance-based SA. It is possible to prove that $G_A$ is the best approximation of $G$ in the $\mathbb{L}^2(P_{\mathbf{X}})$ sense that belongs to $\mathbb{L}_A^2$ and is orthogonal (with respect to the Hilbertian scalar product in $\mathbb{L}^2(P_{\mathbf{X}})$) to any function in

$\mathbb{L}^2_{A'}$, $A' \subsetneq A$ (see, e.g., [198]). Therefore, $V_A$ is a quantitative evaluation of the *randomness* contained in this best approximation and so is well suited to measure the sensitivity of the function $G$ with respect to the set of variables $\mathbf{X}_A$. More often, the ratio between $V_A$ and the variance of $G(X)$ is considered. The following definition, coming from Sobol' [442], gives several useful families of sensitivity indices based on the FANOVA decomposition stated in Corollary 3.5.

**Definition 3.6.** *Let $G \in \mathbb{L}^2(\mathrm{P}_\mathbf{X})$. Assume that $\mathcal{A}_1$ is satisfied. Let $A \in \mathcal{P}_d$.*

- *The Sobol' sensitivity index (in short, the Sobol' index) associated to $A$ is defined as*

$$S_A = \frac{V_A}{V} = \frac{\sum_{B\subset A}(-1)^{|A|-|B|}\operatorname{Var}\mathbb{E}[G(\mathbf{X})|\mathbf{X}_B]}{\operatorname{Var} G(\mathbf{X})}.$$

- *$S_j = S_{\{j\}}$ is the Sobol' index associated to the singleton $\{j\}$. It is called the first-order index for the input variable $X_j$. More generally, if $p = |A|$, then $S_A$ is called the Sobol' index of order $p$ associated to $\mathbf{X}_A$.*

- *The closed Sobol' index associated to $A$ is defined as*

$$S_A^{\text{clos}} = \sum_{A'\subset A} S_{A'} = \frac{\operatorname{Var}\mathbb{E}[G(\mathbf{X})|\mathbf{X}_A]}{\operatorname{Var} G(\mathbf{X})}.$$

  *This index is also called the first-order Sobol' index associated to the input vector $\mathbf{X}_A$.*

- *The total Sobol' index associated to $\mathbf{X}_A$ is defined as*

$$S_A^{\mathrm{T}} = 1 - S_{\bar{A}}^{\text{clos}}.$$

**Remark 7.** *The Sobol' index is equivalent to the square of the correlation ratio (measure of the relationship between the statistical dispersion within individual categories and the dispersion across a whole population or a sample), well known in statistics and introduced by Karl Pearson as part of analysis of variance (see, e.g., [233]). In the SA domain, it has been introduced by Sobol' [441, 442] in a general context but has appeared before in some applicative papers for the case where $A$ reduces to one element, as in Cukier et al. [95] and McKay [319]. The Sobol' index has often been called the variance-based importance measure.*

The so-called *total variance theorem* is stated below, as it is required in many proofs in this book.

**Lemma 3.7.** *Let $G \in \mathbb{L}^2(\mathrm{P}_\mathbf{X})$. For any $A \in \mathcal{P}_d$, we have*

$$\operatorname{Var}(G(\mathbf{X})) = \mathbb{E}\left[\operatorname{Var}(G(\mathbf{X})|\mathbf{X}_{\bar{A}})\right] + \operatorname{Var}\left(\mathbb{E}[G(\mathbf{X})|\mathbf{X}_{\bar{A}}]\right).$$

***Proof.*** The proof of Lemma 3.7 is straightforward and does not require the independence of the components of $\mathbf{X}$. □

In the next proposition, we give some direct consequences of the FANOVA decomposition (Corollary 3.5) and of the total variance theorem (Lemma 3.7).

**Proposition 3.8.** *Let $G \in \mathbb{L}^2(\mathrm{P}_\mathbf{X})$. Assume that $\mathcal{A}_1$ is satisfied. We have*

1. $$\sum_{A \in \mathcal{P}_d} S_A = 1.$$

2. *If $\sum_{j=1}^d S_j = 1$, then the model is additive (that is, $G$ is nothing more than the sum of functions involving only one variable).*

3. $$\begin{aligned} S_A^{\mathrm{T}} &= \frac{\mathbb{E}\left[\mathrm{Var}\left(G(\mathbf{X})|\mathbf{X}_{\bar{A}}\right)\right]}{\mathrm{Var}\, G(\mathbf{X})} \\ &= \sum_{A' \in \mathcal{P}_d, A' \cap A \neq \emptyset} S_{A'}. \end{aligned} \tag{3.10}$$

   *For $A = \{j\}$, the last equality is $S_j^{\mathrm{T}} = \sum_{A' \in \mathcal{P}_d, j \in A'} S_{A'}$.*

**Remark 8.** *Notice that the first equality in Eq.* (3.10) *always holds even for dependent components of* $\mathbf{X}$. *This is not true for the second one.*

We now turn to two examples to illustrate both Theorem 3.3 and Corollary 3.5 and to show the relevance of Definition 3.6.

**Example 3.9 (Sobol' $g$-function).** Let us consider for $G$ the Sobol' $g$-function described in Appendix C.7. Recall that $g(\mathbf{x}) = \prod_{j=1}^d g_j(x_j)$ with $g_j(x_j) = \frac{|4x_j - 2| + a_j}{1 + a_j}$, $a_j > 0$. Assume that $X_1, \ldots, X_d$ are independent random variables uniformly distributed on $[0,1]$. Then straightforward computations yield, for $A \in \mathcal{P}_d$, $S_A = V_A/V$, with $V_A = \prod_{j \in A} V_j$, $V = \prod_{j=1}^d (V_j + 1) - 1$, and $V_j = (3(1 + a_j)^2)^{-1}$. Thus, inputs $X_j$ corresponding to values $a_j$ close to zero have a strong impact on the output. Moreover, main effects are higher than pure interaction effects for that model. More precisely, for any $A \in \mathcal{P}_d$ such that $|A| < d$, $S_A > S_{A \cup \{j\}}$ for all $j \notin A$. ∎

**Example 3.10 (Ishigami).** Let us consider for $G$ the Ishigami function described in Appendix C.4. Recall that $Y = G(X_1, X_2, X_3) = \sin(X_1) + 7\sin^2(X_2) + 0.1 X_3^4 \sin(X_1)$. Assume that $X_1, \ldots, X_3$ are independent random variables uniformly distributed on $[-\pi, \pi]$. Straightforward computations lead to $S_1 = 0.3139$, $S_2 = 0.4424$, $S_3 = 0$, $S_{\{1,2\}}^{\mathrm{clos}} = 0.7563$, $S_{\{2,3\}}^{\mathrm{clos}} = 0.4424$, $S_{\{1,3\}}^{\mathrm{clos}} = 0.5575$, and $S_{\{1,2,3\}}^{\mathrm{clos}} = 1$. The model is nonadditive, as $S_1 + S_2 + S_3 = 0.7563 < 1$. The third input has no influence by itself. It only acts on the output through its interaction with input $X_1$. ∎

### 3.1.3 ▪ Effective dimension

Many functions $G$, although defined on $(E, \mathcal{E}, \mathrm{P}_\mathbf{X})$, may be thought to have an effective dimension smaller than $d$ in the sense that it can be closely approximated by certain sums of functions that involve fewer than $d$ components of $\mathbf{X}$. More precisely, Caflisch et al. [62] introduce two types of effective dimension:

- The function $G$ is said to have effective dimension in the superposition sense equal to $d_s \leq d$ if $\sum_{0 < |A| \leq d_s} V_A \geq 0.99\, V$, which means that $99\%$ of the variance of the model is explained by the interactions up to order $d_s$.

- The function $G$ is said to have effective dimension in the truncation sense equal to $d_t$ if $\sum_{\max\{j\,,\,j\in A\}\leq d_t} V_A \geq 0.99\,V$, which means that 99% of the variance of the model is explained by only the $d_t$ most influential inputs.

The choice of 99th percentile is arbitrary but common in the context of numerical quadratures based on quasi-Monte Carlo (QMC) sampling (see, e.g., [62] and [194] for more details).

Later, Owen [348] introduced the notion of dimension distribution. He suggests choosing a subset $U \in \mathcal{P}_d$ at random with

$$\mathbb{P}(U = A) = \mu(A) = \frac{V_A}{V}$$

and then measuring the size of this random $U$ either through its cardinality $|U|$ or through the maximum of its elements' indices $\max\{j\,,\,j \in U\}$. Then he defines

- the dimension distribution of $G$ in the superposition sense as the probability distribution of $|U|$ when $\mathbb{P}(U = A) = V_A/V$, which has probability mass $\nu_s(m) = \sum_{|A|=m} V_A/V$, $m = 1, \ldots, d$;
- the dimension distribution of $G$ in the truncation sense as the probability distribution of $\max\{j\,,\,j \in U\}$ when $\mathbb{P}(U = A) = V_A/V$, which has probability mass $\nu_t(m) = \sum_{\max\{j\,,\,j\in A\}=m} V_A/V$, $m = 1, \ldots, d$.

Then the effective dimension of $G$ in the superposition sense can be interpreted as the 99th percentile of the dimension distribution in the superposition sense. In the same vein, the effective dimension of $G$ in the truncation sense can be interpreted as the 99th percentile of the dimension distribution in the truncation sense.

It is also possible to consider the mean of dimension distribution (in the superposition or truncation sense) or any of its moments. More precisely,

- the mean dimension of $G$ in the superposition sense is defined as

$$\bar{\nu}_s = \sum_{A\in\mathcal{P}_d, A\neq\emptyset} \frac{V_A}{V}|A|,$$

- while the mean dimension of $G$ in the truncation sense is defined as

$$\bar{\nu}_t = \sum_{A\in\mathcal{P}_d, A\neq\emptyset} \frac{V_A}{V} \max\{j\,,\,j \in A\}.$$

From Definition 3.6 in Section 3.1.2, the links between the mean dimensions and the Sobol' indices are straightforward as

- $\bar{\nu}_s = \sum_{A\in\mathcal{P}_d, A\neq\emptyset} S_A\,|A|$,
- $\bar{\nu}_t = \sum_{A\in\mathcal{P}_d, A\neq\emptyset} S_A \max\{j\,,\,j \in A\}$.

We refer the reader to Liu and Owen [287] for further links between dimension distribution and variance-based global SA.

Analyzing the dimension structure of functions comes from the theory of QMC integration, which consists in approximating integrals of the form $I(h) = \int_{[0,1]^d} h(\mathbf{x})d\mathbf{x}$ by numerical integration $\hat{I}(h) = \frac{1}{n}\sum_{i=1}^{n} h(\mathbf{x}^i)$, with $(\mathbf{x}^1, \ldots, \mathbf{x}^n)$ a QMC sample on the $d$-dimensional unit hypercube $[0,1]^d$. If $h$ can be decomposed as a sum $\sum_{A\in\mathcal{P}_d} h_A(\mathbf{x}_A)$, then it is possible to derive a QMC error bound of the form

$$\left|I(h)-\hat{I}(h)\right| \leq \sum_{A\in\mathcal{P}_d} D(\mathbf{x}_A^1,\ldots,\mathbf{x}_A^n)\|h_A\|_A$$

with suitable definitions of discrepancy $D$ and variation $\|\cdot\|_A$. We refer the reader to Section A.4.1 in Appendix A.4 for a brief description of and references on QMC samples. We also refer the reader to the book in progress by Owen [351] available online. As summarized in Hoyt and Owen [199], the error bound stated above shows that a good result can be expected from (randomized) QMC for numerical integration of a function $h$ if all the conditions listed below hold:

(i) the (randomized) QMC points have highly uniform low-dimensional projections;

(ii) the integrand $h$ is nearly a sum of low-dimensional part;

(iii) those parts are regular enough to benefit from (randomized) QMC.

Seeking a low effective dimension of $h$ defined on $([0,1]^d, \mathcal{B}([0,1]^d, \mathcal{U}([0,1]^d)$ is a way of asserting point (iii) above. However, quantiles can be hard to estimate directly, while moments are often easier to estimate; thus, it may be interesting to assert point (ii) by estimating the mean dimension of $h$.

### 3.1.4 ▪ General links between Sobol' indices and DGSM

In this section, the input variables $X_1,\ldots,X_d$ have been assumed to be real. Derivative-based global sensitivity measures (DGSM) have been introduced in Section 2.3.3. Sobol' and Kucherenko [444] (in some particular cases), Lamboni et al. [259] (in a more general setting), and Roustant et al. [401] (giving optimal results) have proposed a bound of the total Sobol' index $S_j^{\mathrm{T}}$ (see Definition 3.6) by the corresponding DGSM $\nu_j$ (Eq. (2.7), page 41) up to a constant related to the probability measure of $X_j$, namely, $\mu_{X_j}$, with probability distribution $\mathrm{P}_{X_j}$. The main ingredient to derive such a bound is the existence of a Poincaré inequality for $\mu_{X_j}$. Concerning Poincaré inequality, we refer the reader to the paper of Bakry et al. [19].

To begin with, let us recall the classical $L^2$-Poincaré inequality for a measure $\mu$ on $\mathbb{R}$. We say that the measure $\mu$ satisfies a Poincaré inequality whenever the following inequality holds for any centered function $h\in L^2(\mu)$ with $|h'|\in L^2(\mu)$:

$$\int h(x)^2 d\mu(x) \leq C(\mu)\int |h'(x)|^2 d\mu(x). \tag{3.11}$$

Here, $C(\mu)$ is a constant. The optimal constant $C_{\text{opt}}(\mu)$, called the Poincaré constant, is the best possible constant (that is, the smallest possible constant in Eq. (3.11); see [401]).

Then a connection between total indices and DGSM has been established by Lamboni et al. [259] for $\mu_{X_j}$ satisfying a Poincaré inequality.

**Theorem 3.11 (Lamboni et al. [259]).** *Assume that $\mu_{X_j}$ satisfies a Poincaré inequality and that $G$ is differentiable. Then the following inequality holds:*

$$S_j^{\mathrm{T}} \leq \frac{C(\mathrm{P}_{X_j})}{V}\nu_j, \tag{3.12}$$

*with $\nu_j = \mathbb{E}\left(\frac{\partial G(\mathbf{X})}{\partial x_j}\right)^2$ the DGSM defined in Eq. (2.7) (page 41).*

***Proof.*** First, note that if $\frac{\partial G}{\partial x_j} \notin \mathbb{L}^2(\mathrm{P}_{X_j})$, then Eq. (3.12) obviously holds true. Let us now assume that $\frac{\partial G}{\partial x_j} \in \mathbb{L}^2(\mathrm{P}_{X_j})$. Using Eq. (3.6), we may write, for any $j \in [1:d]$,

$$G(\mathbf{x}) = \sum_{A \in \mathcal{P}_d, j \in A} G_A(\mathbf{x}_A) + \sum_{A \in \mathcal{P}_d, j \notin A} G_A(\mathbf{x}_A) = G_j^{\text{tot}}(\mathbf{x}) + (G(\mathbf{x}) - G_j^{\text{tot}}(\mathbf{x})).$$

Due to orthogonality constraints in the Hoeffding decomposition, we have

$$\int_{E_j} G_j^{\text{tot}}(\mathbf{x}) \mathrm{P}_{X_j}(dx_j) = 0$$

for any $j = 1, \ldots, d$. By applying the Poincaré inequality to $G_j^{\text{tot}}(\mathbf{x})$ and then integrating with respect to all the variables except $x_j$, we obtain

$$\operatorname{Var} G_j^{\text{tot}}(\mathbf{X}) \leq C(\mathrm{P}_{X_j}) \nu_j.$$

Indeed, the Fubini–Tonelli theorem applies to the left-hand term, as all the integrands are non-negative. Moreover, $\frac{\partial (G - G_j^{\text{tot}})}{\partial x_j} = 0$, as $G - G_j^{\text{tot}}$ does not depend on $x_j$. Dividing by $V$ concludes the proof. □

Theorem 3.11 is useful for performing a fast screening on the input variables in $\mathbf{X}$ whenever a fast evaluation of the derivatives of $G$ is available. Indeed, a single sample of the model output derivative with respect to $X_j$ is required to evaluate $\nu_j$ (see Section 2.3.3 and references therein). In particular, when $\nu_j$ is close to zero, certifying that $S_j^{\mathrm{T}}$ is close to zero (which allows to screen out $X_j$), a low sample size is sufficient (contrarily to the sample size required to estimate $S_j^{\mathrm{T}}$). However, in the case of nonnegligible $\nu_j$, it is desirable to get the best constant (i.e., the Poincaré one) in Eq. (3.12).

Finding the values of Poincaré constants in 1D Poincaré inequalities for any continuous probability measures has been tackled in Roustant et al. [401]. Using the spectral theorem related to Poincaré inequalities [20], the Poincaré constant is equal to the inverse of the first positive eigenvalue of an equivalent spectral problem. A general numerical scheme (using a finite element method) is then proposed to solve the spectral (elliptic) problem, allowing the effective computation of the Poincaré constant.

**Remark 9.** *The following rough bounds for the optimal constant $C_{opt}(\mu)$ can also be obtained [401]:*

$$\begin{aligned}\frac{1}{2}\int_{\mathbb{R}} (x - x')^2 \mu(dx)\mu(dx') \leq C_{opt}(\mu) &\leq 4\left(\sup_{\mathbb{R}} \frac{F_X(1 - F_X)}{f_X}\right)^2 \\ &\leq 4\left(\sup_{\mathbb{R}} \frac{\min(F_X, 1 - F_X)}{f_X}\right)^2.\end{aligned}$$

*Here, $F_X$ (resp., $f_X$) is the probability distribution (resp., density function) of $X \sim \mu$. Technically, the first (resp., second) upper bound is obtained by transporting the probability measure $\mu$ to the logistic (resp., double-exponential) probability distribution; see [401] for more details. Note that the constant obtained by the logistic transport is always more accurate than the one obtained by the double-exponential transport (that has been given in Lamboni et al. [259]).*

In the sensitivity R package, all of these constants can be directly obtained by using the functions PoincareConstant() and PoincareOptimal().

## 3.2 ▪ Sampling-based estimators

Once the Sobol' indices have been defined, the question of their effective computation or estimation remains open. In practice, one has to estimate (in a statistical sense) these indices using a finite sample (of minimal size typically on the order of several hundreds) of model evaluations. In the present chapter, the inputs have been assumed to be independent so that the probability distribution $P_{\mathbf{X}}$ of $\mathbf{X}$ is a product probability distribution. We will mainly focus on the case of scalar inputs ($\mathbf{X}$ is a vector of $\mathbb{R}^d$). Nevertheless, most of the results presented here still hold for more general inputs.

In the next section, we focus on the estimation of first-order and total Sobol' indices with the so-called pick-freeze scheme first introduced in Sobol' [442]. In Section 3.2.2, we discuss different sampling strategies that can be used along with the pick-freeze scheme. The generalization of the pick-freeze scheme to the estimation of higher-order Sobol' indices typically requires many more model evaluations to produce accurate estimates. In Section 3.2.3, we present a tricky sampling strategy first introduced in Saltelli [409] to estimate at a low cost (in terms of model evaluations) the full set of first-order, second-order, and total Sobol' indices accurately. Then, in Sections 3.2.4 and 3.2.5, we present both the asymptotic and the finite horizon properties of the pick-freeze estimators of Sobol' indices. Note that one of the main advantages of the pick-freeze estimation scheme is that the only assumption it requires on $G$ is the square integrability with respect to $P_{\mathbf{X}}$. In Section 3.2.6, numerical illustrations with R are presented.

### 3.2.1 ▪ Pick-freeze scheme for the estimation of first-order and total Sobol' indices

Let $\mathbf{X}$ and $\mathbf{X}'$ be two independent copies of the input vectors. Let $A$ be in $\mathcal{P}_d$. The quantities of interest in this section are the first-order and total Sobol' indices associated to $\mathbf{X}_A$, namely, $S_A^{\text{clos}}$ and $S_A^{\text{T}}$. We define $\mathbf{X}^{\sim A} = (\mathbf{X}_A, \mathbf{X}'_{\bar{A}})$, $\mathbf{Y} = G(\mathbf{X})$, and $\mathbf{Y}^{\sim A} = G(\mathbf{X}^{\sim A})$.

As noticed in [217], Lemma 3.12 is probably the key ingredient motivating the so-called pick-freeze scheme for the estimation of $S_A^{\text{clos}}$ and $S_A^{\text{T}}$.

**Lemma 3.12 (Lemma 2.2 in [217]).** *We have* $\operatorname{Var}\mathbb{E}[G(\mathbf{X})|\mathbf{X}_A] = \operatorname{Cov}(\mathbf{Y}, \mathbf{Y}^{\sim A})$, *which implies*

$$S_A^{\text{clos}} = \frac{\operatorname{Cov}(\mathbf{Y}, \mathbf{Y}^{\sim A})}{V} \tag{3.13}$$

*and*

$$S_A^{\text{T}} = 1 - \frac{\operatorname{Cov}(\mathbf{Y}, \mathbf{Y}^{\sim \bar{A}})}{V}. \tag{3.14}$$

***Proof.*** We focus on the proof of Eq. (3.13), the proof being similar for Eq. (3.14). The objective is to prove that

$$\operatorname{Cov}(\mathbf{Y}, \mathbf{Y}^{\sim A}) = \operatorname{Var}\mathbb{E}(\mathbf{Y}|\mathbf{X}_A).$$

Since $\mathbf{Y}$ and $\mathbf{Y}^{\sim A}$ have the same distribution, we can write

$$\operatorname{Cov}(\mathbf{Y}, \mathbf{Y}^{\sim A}) = \mathbb{E}(\mathbf{Y}\mathbf{Y}^{\sim A}) - \mathbb{E}\mathbf{Y}^2.$$

Moreover, $\mathbf{Y}$ and $\mathbf{Y}^{\sim A}$ are conditionally independent on $\mathbf{X}_A$. Then

$$\mathbb{E}(\mathbf{Y}\mathbf{Y}^{\sim A}) = \mathbb{E}[\mathbb{E}(\mathbf{Y}, \mathbf{Y}^{\sim A}|\mathbf{X}_A)] = \mathbb{E}[\mathbb{E}(\mathbf{Y}|\mathbf{X}_A)\mathbb{E}(\mathbf{Y}^{\sim A}|\mathbf{X}_A)] = \mathbb{E}[\mathbb{E}(\mathbf{Y}|\mathbf{X}_A)^2]. \qquad \square$$

Based on Eqs. (3.13) and (3.14), we present the pick-freeze scheme to estimate first-order and total Sobol' indices associated to $\mathbf{X}_A$ from a sample of the input vector and the corresponding outputs. For the sake of clarity in the presentation, we consider the case $A = \{j\}$,

$j=1,\dots,d$, in the following. Then the term $\operatorname{Var}\mathbb{E}(\mathbf{Y}|\mathbf{X}_A)$ coincides with $V_A = \operatorname{Var} G_A(\mathbf{X}_A)$ and $S_A^{\text{clos}} = S_{\{j\}}^{\text{clos}} = S_j$. Moreover, due to Lemma 3.12, $V_j = \operatorname{Cov}(\mathbf{Y},\mathbf{Y}^{\sim j})$, and it seems intuitive to estimate $S_j$ by $\widehat{V}_j/\widehat{V}$ with $\widehat{V}_j$ an empirical estimator of $\operatorname{Cov}(\mathbf{Y},\mathbf{Y}^{\sim j})$ and $\widehat{V}$ an empirical estimator of $V$. In the same way, an intuitive estimator for $S_j^{\text{T}}$ is defined as $(\widehat{V}-\widehat{V}_{\bar{j}})/\widehat{V}$ with $\widehat{V}_{\bar{j}}$ an empirical estimator for $\operatorname{Cov}(\mathbf{Y},\mathbf{Y}^{\sim\bar{j}})$. Various choices are possible for the estimates of the variance $V$ and of the parts of variance $V_j$ and $V_{\bar{j}}$.

In Sobol' [442] and Saltelli et al. [411], the authors suggest independently a simple sampling plan that supports estimation of $S_j$ and $S_j^{\text{T}}$, $j=1,\dots,d$. The construction of the plan begins by drawing two $n\times d$ arrays, $\mathbf{X}^n$ and $\mathbf{X}'^n$, whose $n$ lines are sampled from the random input vector. In the following, the superscript $n$ is omitted. Let

$$\mathbf{X} = \begin{pmatrix} x_1^{(1)} & \dots & x_d^{(1)} \\ x_1^{(2)} & \dots & x_d^{(2)} \\ \dots & \dots & \dots \\ x_1^{(n)} & \dots & x_d^{(n)} \end{pmatrix} = (\mathbf{x}_1,\mathbf{x}_2,\mathbf{x}_3,\dots,\mathbf{x}_d)$$

and

$$\mathbf{X}' = \begin{pmatrix} \mathbf{x}'^{(1)}_1 & \dots & \mathbf{x}'^{(1)}_d \\ \mathbf{x}'^{(2)}_1 & \dots & \mathbf{x}'^{(2)}_d \\ \dots & \dots & \dots \\ \mathbf{x}'^{(n)}_1 & \dots & \mathbf{x}'^{(n)}_d \end{pmatrix} = (\mathbf{x}'_1,\mathbf{x}'_2,\mathbf{x}'_3,\dots,\mathbf{x}'_d).$$

A sampling plan supporting estimation of $S_j$ and $S_j^{\text{T}}$, $j=1,\dots,d$, is comprised of the locations specified by the rows of $\mathbf{X}$, together with those prescribed by $d$ new arrays, the $j$th of which is constructed by replacing the $j$th column of $\mathbf{X}'$ with the $j$th column of $\mathbf{X}$:

$$\mathbf{X}^{\sim j} = (\mathbf{x}'_1,\dots,\mathbf{x}'_{j-1},\mathbf{x}_j,\mathbf{x}'_{j+1},\dots,\mathbf{x}'_d),\ j=1,2,3,\dots,d.$$

The lines in the $n\times d$ array $\mathbf{X}^{\sim j}$ are denoted by $\mathbf{x}^{\sim j,(i)}$, $1\le i\le n$. This sampling plan is known as the *substituted columns* sampling in Morris et al. [331] or as the *pick-freeze* sampling procedure in [218].

The estimators proposed in Saltelli [409] (see also [197]) are then defined by

$$\widehat{V}_j = \frac{1}{n}\sum_{i=1}^{n} G(\mathbf{x}^{(i)})\left\{G(\mathbf{x}^{\sim j,(i)}) - G(\mathbf{x}'^{(i)})\right\}, \tag{3.15}$$

$$\widehat{V}_{\bar{j}} = \frac{1}{n}\sum_{i=1}^{n} G(\mathbf{x}'^{(i)})\left\{G(\mathbf{x}^{\sim j,(i)}) - G(\mathbf{x}^{(i)})\right\},$$

$$\hat{V} = \frac{1}{n}\sum_{i=1}^{n} G(\mathbf{x}^{(i)})^2 - \left(\frac{1}{n}\sum_{i=1}^{n} G(\mathbf{x}^{(i)})\right)^2. \tag{3.16}$$

This sampling plan calls for a total of $N = n\times(d+2)$ model runs for estimating all first-order and total Sobol' indices $S_j$ and $S_j^{\text{T}}$, $j=1,\dots,d$.

Among the numerous choices for the estimators based on the substituted columns sampling plan, several numerical tests performed in Saltelli et al. [412] allow one to provide some practical recommendations, some of them being reminded in Owen [349]. For the estimation of total Sobol' indices, it is recommended to estimate $V_{\bar{j}}$ by

$$\widehat{V}_{\bar{j}} = \frac{1}{2n}\sum_{i=1}^{n}\left\{G(\mathbf{x}'^{(i)}) - G(\mathbf{x}^{\sim j,(i)})\right\}^2$$

as was proposed in Jansen [220]. In Janon et al. [218], it is proved that considering

$$\widehat{V}_j = \frac{1}{n}\sum_{i=1}^{n} G(\mathbf{x}^{(i)})G(\mathbf{x}^{\sim j,(i)}) - \left(\frac{1}{n}\sum_{i=1}^{n}\frac{G(\mathbf{x}^{(i)})+G(\mathbf{x}^{\sim j,(i)})}{2}\right)^2, \tag{3.17}$$

$$\widehat{V} = \frac{1}{n}\sum_{i=1}^{n}\frac{\left(G(\mathbf{x}^{(i)})\right)^2+\left(G(\mathbf{x}^{\sim j,(i)})\right)^2}{2} - \left(\frac{1}{n}\sum_{i=1}^{n}\frac{G(\mathbf{x}^{(i)})+G(\mathbf{x}^{\sim j,(i)})}{2}\right)^2 \tag{3.18}$$

leads to an efficient estimation of $S_j$ in a sense which is made precise in Remark 11 in Section 3.2.4. It is also recommended to estimate total Sobol' indices by $(\widehat{V} - \widehat{V}_{\bar{j}})/\widehat{V}$ with

$$\widehat{V} - \widehat{V}_{\bar{j}} = \widehat{V} - \left(\frac{1}{n}\sum_{i=1}^{n} G(\mathbf{x}^{(i)})G(\mathbf{x}^{\sim \bar{j},(i)}) - \left(\frac{1}{n}\sum_{i=1}^{n}\frac{G(\mathbf{x}^{(i)})+G(\mathbf{x}^{\sim \bar{j},(i)})}{2}\right)^2\right) \tag{3.19}$$

and

$$\widehat{V} = \frac{1}{n}\sum_{i=1}^{n}\frac{\left(G(\mathbf{x}^{(i)})\right)^2+\left(G(\mathbf{x}^{\sim \bar{j},(i)})\right)^2}{2} - \left(\frac{1}{n}\sum_{i=1}^{n}\frac{G(\mathbf{x}^{(i)})+G(\mathbf{x}^{\sim \bar{j},(i)})}{2}\right)^2. \tag{3.20}$$

One can enrich the substituted columns sampling plan with

$$\mathbf{X}" = \begin{pmatrix} x"^{(1)}_1 & \dots & x"^{(1)}_d \\ x"^{(2)}_1 & \dots & x"^{(2)}_d \\ \dots & \dots & \dots \\ x"^{(n)}_1 & \dots & x"^{(n)}_d \end{pmatrix} = (\mathbf{x}"_1, \mathbf{x}"_2, \mathbf{x}"_3, \dots, \mathbf{x}"_d)$$

and $d$ new arrays, the $j$th of which is constructed by replacing the $j$th column of $\mathbf{X}$ with the $j$th column of $\mathbf{X}"$:

$$\mathbf{X}"^{\sim j} = (\mathbf{x}_1, \dots, \mathbf{x}_{j-1}, \mathbf{x}"_j, \mathbf{x}_{j+1}, \dots, \mathbf{x}_d),\ j = 1, 2, 3, \dots, d.$$

It is then recommended in Owen [349] to consider for small first-order Sobol' indices the following estimate (called *Correlation 2* in Owen [349]):

$$\widehat{V}_j = \frac{1}{n}\sum_{i=1}^{n}\left(G(\mathbf{x}^{(i)}) - G(\mathbf{x}"^{\sim j,(i)})\right)\left(G(\mathbf{x}^{\sim j,(i)}) - G(\mathbf{x}'^{(i)})\right).$$

We also refer the reader to Glen and Isaacs [164], where a correction of the aforementioned estimators is proposed to remove some of the spurious correlations. Indeed, it happens that two independent and identically distributed (i.i.d.) $n$-samples from two independent random variables have in practice a small empirical correlation that is qualified as spurious correlation.

## 3.2.2 ▪ Sampling ways

### Different sampling strategies for arrays $\mathbf{X}$, $\mathbf{X}'$, and $\mathbf{X}^{\sim j}$

A naive sampling strategy is to sample independently lines in $\mathbf{X}$ and $\mathbf{X}'$ and to sample $\mathbf{X}$ independently from $\mathbf{X}'$. However, independent sampling is not always the best strategy, especially for small sample sizes $n$.

To simplify the notation in this section, we now assume that $P_{\mathbf{X}}$ is the uniform distribution on the $d$-dimensional unit-hypercube. Thanks to the inversion method (see, e.g., [113]), a model $Y = G(\mathbf{X})$, where the $X_i$'s are independent and arbitrarily distributed, can always be transformed into a model $Y = \widetilde{G}(\mathbf{U})$, where the $U_i$'s are independent random variables uniformly distributed on $[0, 1]$, defined by $U_i = F_i(X_i)$ with $F_i$ the cumulative distribution function of $X_i$. In the substituted columns sampling described in Section 3.2.1, it is then also possible to use independent low-discrepancy samples of size $n$ based, e.g., on independent Sobol' LP$_\tau$-sequences for each column of $\mathbf{X}$ and $\mathbf{X}'$. QMC methods are deterministic versions of Monte Carlo methods. Their use in estimating high-dimensional integrals (and thus for the estimation of Sobol' indices) is motivated in Appendix A. One could also choose to sample via a space-filling design, as a Latin hypercube sample (LHS), in order to fill the space more regularly (see Appendix A).

### Permuted Column Sampling

In this section, the construction of a *permuted column sample* is recalled, as described in McKay [319]. We start, as previously, by drawing an $n \times d$ array $\mathbf{X}$:

$$\mathbf{X} = \begin{pmatrix} x_1^{(1)} & \dots & x_d^{(1)} \\ x_1^{(2)} & \dots & x_d^{(2)} \\ \dots & \dots & \dots \\ x_1^{(n)} & \dots & x_d^{(n)} \end{pmatrix} = (\mathbf{x}_1, \mathbf{x}_2, \mathbf{x}_3, \dots, \mathbf{x}_d).$$

$\kappa \times d$ permutation operators are selected randomly and independently from the $n!$ permutations of $[1 : d]$: $\pi_j^k$, $k = 1, \dots, \kappa$, $j = 1, \dots, d$. Then, for $k = 1, \dots, \kappa$, we define $\mathbf{X}^{r_k}$ as the following replicated designs:

$$\mathbf{X}^{r_k} = \begin{pmatrix} x_1^{(\pi_1^k(1))} & \dots & x_d^{(\pi_d^k(1))} \\ x_1^{(\pi_1^k(2))} & \dots & x_d^{(\pi_d^k(2))} \\ \dots & \dots & \dots \\ x_1^{(\pi_1^k(n))} & \dots & x_d^{(\pi_d^k(n))} \end{pmatrix} = (\mathbf{x}_1^{r_k}, \mathbf{x}_2^{r_k}, \mathbf{x}_3^{r_k}, \dots, \mathbf{x}_d^{r_k}).$$

An important property of replicated designs is given in the following. By applying permutation $(\pi_j^k)^{-1}$ to the lines of $\mathbf{X}^{r_k}$, we obtain a design whose $j$th column is similar to the one in $\mathbf{X}$ and whose other $d-1$ columns are resampled. This new design is denoted as $\mathbf{X}^{r_k,\sim j}$. Note that the set of points in $\mathbf{X}^{r_k,\sim j}$ is similar to the set of points in $\mathbf{X}^{r_k}$ except that they are sampled in a different order. Thus, evaluating $G$ on $\mathbf{X}^{r_k,\sim j}$ does not require additional evaluations of $G$ if we have already evaluated $G$ on $\mathbf{X}^{r_k}$.

We define

$$\widehat{V}^{r_k} = \frac{1}{n}\sum_{i=1}^{n} G(\mathbf{x}^{r_k,(i)})^2 - \left(\frac{1}{n}\sum_{i=1}^{n} G(\mathbf{x}^{r_k,(i)})\right)^2,$$

with $\mathbf{x}^{r_k,(i)}$ the $i$th line of $\mathbf{x}^{r_k}$ and

$$\widehat{V}_j^{r_1,\dots,r_\kappa,(i)} = \frac{1}{\kappa}\sum_{i=1}^{\kappa} G(\mathbf{x}^{r_k,\sim j,(i)})^2 - \left(\frac{1}{\kappa}\sum_{i=1}^{\kappa} G(\mathbf{x}^{r_k,\sim j,(i)})\right)^2.$$

We then define the estimator of $S_j$ introduced in Morris et al. [332] by $\widehat{V}_j/\widehat{V}$ with

$$\widehat{V}_j = \frac{1}{\kappa}\sum_{k=1}^{\kappa}\widehat{V}^{r_k} - \frac{1}{n}\sum_{i=1}^{n}\widehat{V}_j^{r_1,\dots,r_\kappa,(i)},$$

$$\widehat{V} = \frac{1}{\kappa}\sum_{k=1}^{\kappa}\widehat{V}^{r_k}.$$

This approach requires the computation of $n \times \kappa$ model outputs, namely, the evaluation of $G$ on $\mathbf{X}, \mathbf{X}^{r_1}, \dots, \mathbf{X}^{r_\kappa}$. To estimate accurately the first-order Sobol' indices with this approach, we have to choose $n$ and $\kappa$ large enough; thus, this approach is not that efficient. However, the idea of replicated samples can be reused to propose a more efficient estimation procedure in terms of number of model evaluations.

### Random balance design–based procedure

We start by constructing a permuted column matrix $\mathbf{X}^{r_1}$ from an initial sampling plan $\mathbf{X}$. We then define $\mathbf{X}^{r_1,\sim j}$ as in the previous section. It has already been noticed that, applying permutation $(\pi_j^k)^{-1}$ on the lines of $\mathbf{X}^{r_k}$, we obtain a design whose $j$th column is similar to the one in $\mathbf{X}$ and whose other $d-1$ columns are resampled and that evaluating $G$ on $\mathbf{X}^{r_1,\sim j}$ does not require additional evaluations of $G$ if we have already evaluated $G$ on $\mathbf{X}^{r_1}$.

We then go back to the pick-freeze estimator and define, as in Mara and Joseph [302] and Tissot and Prieur [484], the random balance design–based pick-freeze estimator of $S_j$ with

$$\widehat{V}_j = \frac{1}{n}\sum_{i=1}^{n} G(\mathbf{x}^{(i)})G(\mathbf{x}^{r_1,\sim j,(i)}) - \left(\frac{1}{n}\sum_{i=1}^{n}\frac{G(\mathbf{x}^{(i)}) + G(\mathbf{x}^{r_1,\sim j,(i)})}{2}\right)^2, \tag{3.21}$$

$$\widehat{V} = \frac{1}{n}\sum_{i=1}^{n}\frac{\left(G(\mathbf{x}^{(i)})\right)^2 + \left(G(\mathbf{x}^{r_1,\sim j,(i)})\right)^2}{2} - \left(\frac{1}{n}\sum_{i=1}^{n}\frac{G(\mathbf{x}^{(i)}) + G(\mathbf{x}^{r_1,\sim j,(i)})}{2}\right)^2. \tag{3.22}$$

With this random balance design–based approach, we only use $2 \times n$ model evaluations to estimate all the first-order Sobol' indices, namely, the evaluations of $G$ on $\mathbf{X}$ and $\mathbf{X}^{r_1}$.

### Replicated Latin hypercube sampling

In both previous sections, a possible refinement of the permuted column sample is to generate a *replicated Latin hypercube sample* (replicated LHS) as was first proposed in McKay [319].

**Definition 3.13.** *Let $d$, $\kappa$ and $n$ in $\mathbb{N}^*$, and consider $\Pi_n$ the set of all the permutations of $[1:n]$. We say that $(\mathbf{LH}_k)_{k=1,\dots,\kappa}$ are $\kappa$ replicated LHS of size $n$ in $[0,1]^d$ if, for all $k \in [1:\kappa]$, $i \in [1:n]$,*

$$\mathbf{LH}_k^{(i)} = \left(\frac{\pi_1^k(i) - U_{\pi_1^k(i),1}}{n}, \dots, \frac{\pi_d^k(i) - U_{\pi_d^k(i),d}}{n}\right),$$

*with the $\pi_j^k$'s and the $U_{i,j}$'s independent random variables uniformly distributed on $\Pi_n$ and $[0,1]$, respectively.*

The matrices $\mathbf{X}^{r_k}$, $k = 1, \dots, \kappa$, can now be replaced, in the procedure of Morris et al. [332], by the replicated LHS $\mathbf{LH}_k$, $k = 1, \dots, \kappa$. In the same way, in the efficient strategy defined by Eqs. (3.21) and (3.22), $\mathbf{X}$ and $\mathbf{X}^{r_1}$ can be replaced by two replicated LHS, $\mathbf{LH}_1$ and $\mathbf{LH}_2$. The result is that the sample used to estimate the first-order Sobol' indices is more space-filling. The asymptotic properties of this last estimate can also be derived; see Section 3.2.4 and [484].

### Example in R

We illustrate the designs produced by the random balance design–based procedure and the replicated LHS with $d = 3$ inputs (each on $[0,1]$) and an initial sample of $n = 5$ points. We use the Sobol' indices estimation functions of the sensitivity R package (see Section 3.2.6). The following R commands give such designs that are shown in Fig. 3.1:

```
> set.seed(12345) # To reproduce the same results
> library(sensitivity)
> d <- 3
> n <- 5
> # Random balance design-based procedure (order 1)
> X1 <- data.frame(matrix(runif(d * n), nrow = n))
> sa1 <- sobolmara(model = NULL, X1 = X1)
> # Replicated latin hypercube sampling (order 1)
> sa2 <- sobolroalhs(model = NULL, factors = d, N = n, order = 1)

> par(mfcol = c(2,3))
> mat <- matrix(c(1,2,1,3,2,3), nrow = 2)
> for (i in 1:3){
+    plot(sa1$X[1:n,c(mat[1,i],mat[2,i])],xlim=c(0,1),ylim=c(0,1))
+    points(sa1$X[(n+1):(2*n),c(mat[1,i],mat[2,i])],col='blue',pch=4)
+    plot(sa2$X[1:n,c(mat[1,i],mat[2,i])],xlim=c(0,1),ylim=c(0,1))
+    points(sa2$X[(n+1):(2*n),c(mat[1,i],mat[2,i])],col='blue',pch=4)}
```

In each scatterplot (2D projection of each pair of variables) of Fig. 3.1, the 5 black points constitute the initial design, while the 5 blue points constitute the additional design that will

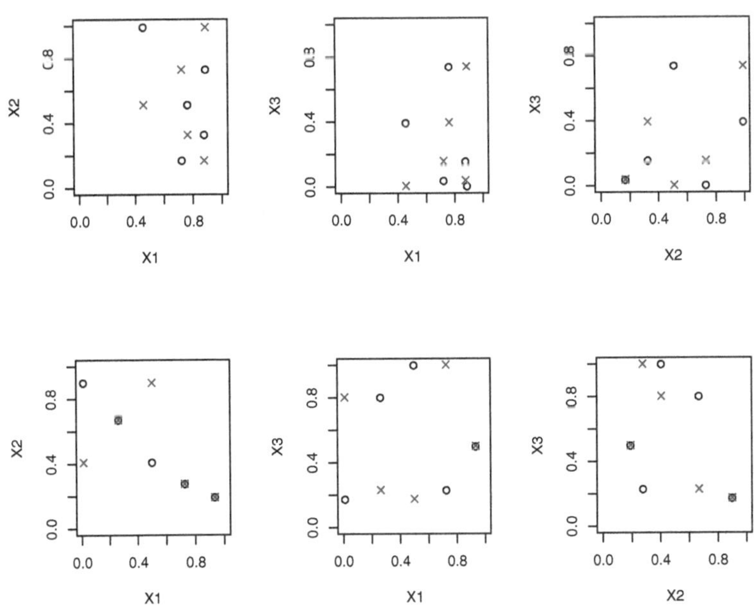

**Figure 3.1.** *2D projections of a random balance design (up) and a replicated LHS (bottom) with $d = 3$ and $n = 5$. Points of the initial design are shown by black circles, and points of the additional sample are shown by blue crosses.*

allow to estimate first-order Sobol' indices. Each axis and ordinate coordinate is replicated (with one point from the initial design and one point from the additional design). Clearly, the replicate LHS (lower plots) offers a better coverage of the space than the random balance design (upper plots).

### 3.2.3 ▪ Estimation of second-order Sobol' indices

In this section, the estimation procedures presented above are generalized to the estimation of closed second-order sensitivity indices, $S^{\text{clos}}_{\{j_1,j_2\}}$ with $1 \leq j_1 \neq j_2 \leq d$. In many applications, interactions up to order two contain most of the information on the variability of the output. If it is not the case, that is, if $\sum_{j=1}^d S_j + \sum_{1\leq j<k\leq d} S_{\{j,k\}}$ is far from 1, then we can estimate total Sobol' indices $S_1^{\text{T}}, \dots, S_d^{\text{T}}$ to complete the information. Now recall that there holds the relation

$$S_{\{j_1,j_2\}} = S^{\text{clos}}_{\{j_1,j_2\}} - S_{j_1} - S_{j_2}$$

such that it is possible to construct an estimate of the second-order Sobol' index $S_{\{j_1,j_2\}}$ from estimates of the closed second-order index $S^{\text{clos}}_{\{j_1,j_2\}}$ and of the first-order indices $S_{j_1}$ and $S_{j_2}$. For the sake of conciseness in the following, we denote $S^{\text{clos}}_{\{j_1,j_2\}}$ by $S^{\text{clos}}_{j_1 j_2}$. Remark that $S^{\text{clos}}_{-j_1 j_2}$ stands for the closed index associated to the complementary $\overline{j_1 j_2}$ of $\{j_1, j_2\}$.

#### Saltelli's approach

This section presents the sampling strategy proposed in Saltelli [409].

**Theorem 3.14 (Theorem 2 in Saltelli [409]).** *Allowing for $N = n \times (2d+2)$ model evaluations, we obtain*

(1) *double estimates for each $S_j^{\text{T}}$ and $S_j$ indices;*

(2) *double estimates for each $\begin{pmatrix} k \\ 2 \end{pmatrix}$ for each $S^{\text{clos}}_{j_1 j_2}$ index;*

(3) *double estimates for each $\begin{pmatrix} k \\ 2 \end{pmatrix}$ for each $S^{\text{clos}}_{-j_1 j_2}$ index.*

Let us generalize the definition of the pick-freeze sampling plan $\mathbf{X}^{\sim j}$ to the sampling plan $\mathbf{X}^{\sim A}$, $A \in \mathcal{P}_d$, which is comprised of the runs specified by a new array, which is constructed by replacing the columns of $\mathbf{X}'$ indexed by $j \in A$ with the columns of $\mathbf{A}$ indexed by $j \in A$:

$$\mathbf{X}^{\sim A} = (\mathbf{x}_A, \mathbf{x}'_{\bar{A}}), \ A \in \mathcal{P}_d.$$

Now, to explain the procedure of Theorem 3.14 above, we give the illustration in dimension $d = 4$. We use the notation

$$U_A = \text{Var}\,(\mathbb{E}(Y|\mathbf{X}_A)) + \mathbb{E}^2(Y).$$

We then have

$$\begin{aligned} S_j &= \frac{U_j - \mathbb{E}^2(Y)}{V}, \\ S_j^{\text{T}} &= \frac{U_{-j} - \mathbb{E}^2(Y)}{V}, \\ S^{\text{clos}}_{j_1 j_2} &= \frac{U_{j_1 j_2} - U_{j_1} - U_{j_2} + \mathbb{E}^2(Y)}{V}. \end{aligned}$$

In Table 3.1, we see, e.g., that $U_1$ can be estimated from $\mathbf{x}^{\sim\{2,3,4\}}$ and $\mathbf{x}'$ but also from $\mathbf{x}^{\sim\{1,2,3,4\}}$ and $\mathbf{x}^{\sim 1}$. More generally, we read in this table that we only need $N = n \times (2d+2) = 10n$ model evaluations to estimate twice all the following indices: $S_j$, $S_j^{\mathrm{T}}$, $j = 1, 2, 3, 4$, $S_{j_1 j_2}^{\mathrm{clos}}$, $S_{-j_1 j_2}^{\mathrm{clos}}$, $1 \leq j_1 \neq j_2 \leq 4$.

**Table 3.1.** *The table gives for each cell the term that can be estimated by evaluating the model on the corresponding input vectors, $d = 4$. For example, $U_{-12}$ can be estimated from the evaluation of the model on two $n$-samples $\mathbf{x}^{\sim 2,(i)}$ and $\mathbf{x}^{\sim 1,(i)}$, $i = 1, \ldots, n$.*

| | $\mathbf{x}'$ | $\mathbf{x}^{\sim 1}$ | $\mathbf{x}^{\sim 2}$ | $\mathbf{x}^{\sim 3}$ | $\mathbf{x}^{\sim 4}$ | $\mathbf{x}^{\sim\{2,3,4\}}$ | $\mathbf{x}^{\sim\{1,3,4\}}$ | $\mathbf{x}^{\sim\{1,2,4\}}$ | $\mathbf{x}^{\sim\{1,2,3\}}$ | $\mathbf{x}^{\sim\{1,2,3,4\}}$ |
|---|---|---|---|---|---|---|---|---|---|---|
| $\mathbf{x}'$ | $V$ | | | | | | | | | |
| $\mathbf{x}^{\sim 1}$ | $U_{-1}$ | $V$ | | | | | | | | |
| $\mathbf{x}^{\sim 2}$ | $U_{-2}$ | $U_{-12}$ | $V$ | | | | | | | |
| $\mathbf{x}^{\sim 3}$ | $U_{-3}$ | $U_{-13}$ | $U_{-23}$ | $V$ | | | | | | |
| $\mathbf{x}^{\sim 4}$ | $U_{-4}$ | $U_{-14}$ | $U_{-24}$ | $U_{-34}$ | $V$ | | | | | |
| $\mathbf{x}^{\sim\{2,3,4\}}$ | $U_1$ | | $U_{12}$ | $U_{13}$ | $U_{14}$ | $V$ | | | | |
| $\mathbf{x}^{\sim\{1,3,4\}}$ | $U_2$ | $U_{12}$ | | $U_{23}$ | $U_{24}$ | $U_{-12}$ | $V$ | | | |
| $\mathbf{x}^{\sim\{1,2,4\}}$ | $U_3$ | $U_{13}$ | $U_{23}$ | | $U_{34}$ | $U_{-13}$ | $U_{-23}$ | $V$ | | |
| $\mathbf{x}^{\sim\{1,2,3\}}$ | $U_4$ | $U_{14}$ | $U_{24}$ | $U_{34}$ | | $U_{-14}$ | $U_{-24}$ | $U_{-34}$ | $V$ | |
| $\mathbf{x}^{\sim\{1,2,3,4\}}$ | $\mathbb{E}^2 Y$ | $U_1$ | $U_2$ | $U_3$ | $U_4$ | $U_{-1}$ | $U_{-2}$ | $U_{-3}$ | $U_{-4}$ | $V$ |

### Replicated sampling plan based on randomized orthogonal array

In the procedure described in Section 3.2.3, all first-order, total, and closed second-order indices could be estimated with a cost linear in the dimension. In the present section, we present a procedure introduced in Tissot and Prieur [484] which allows the estimation of all closed second-order indices with a cost $N = 2n$. We have a constraint on $n$ to apply this procedure, which is due to the construction of orthogonal arrays of strength 2: $n \geq (d-1)^2$. Note that the choice of $n$ is related to the available budget (in terms of model evaluations) but also to the accuracy of the estimation of Sobol' indices. As the convergence (see Section 3.2.4) of the proposed estimators is "slow" ($\sqrt{n}$), high value for $n$ is chosen most of the times. It explains why in many practical situations, $n \geq (d-1)^2$ is not really constraining.

In order to describe the procedure, we need now to define a randomized orthogonal array; see [273, 345]. Recall that the definition of orthogonal array has been given in Appendix A.

**Definition 3.15.** *Let $(\mathbf{A}_j^{(i)})_{\substack{i=1,\ldots,n\\ j=1,\ldots,d}}$ be an orthogonal array in dimension $d$, with $n$ points and $q$ levels in $[1:q]$, and consider $\Pi_q$ the set of all the permutations of $[1:q]$. We say that $\mathbf{X}$ and $\mathbf{X}'$ are two replicated designs based on the randomized orthogonal array $(\mathbf{A}_j^{(i)})_{\substack{i=1,\ldots,n\\ j=1,\ldots,d}}$ if, for all $i \in [1:n]$,*

$$\mathbf{X}^{(i)} = \left( \frac{\pi_1(A_1^{(i)}) - U_{\pi_1(A_1^{(i)}),1}}{q}, \ldots, \frac{\pi_d(A_d^{(i)}) - U_{\pi_d(A_d^{(i)}),d}}{q} \right),$$

$$\mathbf{X}'^{(i)} = \left( \frac{\pi_1'(A_1^{(i)}) - U_{\pi_1'(A_1^{(i)}),1}}{q}, \ldots, \frac{\pi_d'(A_d^{(i)}) - U_{\pi_d'(A_d^{(i)}),d}}{q} \right),$$

*where the $\pi_j$'s, the $\pi_i j$'s, and the $U_{i,j}$'s are independent random variables uniformly distributed on $\Pi_q$, $\Pi_q$, and $[0,1]$, respectively.*

The procedure presented in the subsection *Replicated Latin hypercube sampling* of Section 3.2.2 allows to estimate all first-order Sobol' indices with only two replicated LHS.

Similarly, the procedure presented in the subsection *Replicated sampling plan based on randomized orthogonal array* of Section 3.2.3 allows us to estimate all second-order Sobol' indices with only two replicated designs of strength two. We start by constructing two replicated designs based on an orthogonal array of strength two, $\mathbf{X}$ and $\mathbf{X}'$. The lines in $\mathbf{X}'$ are then reordered in such a way that the $j$th and the $k$th columns in this new array $\mathbf{X}^{\sim jk}$ coincide with the $j$th and the $k$th columns of $\mathbf{X}$. We have defined $d(d-1)/2$ new arrays $\mathbf{X}^{\sim jk}$ corresponding to different permutations of the lines of the array $\mathbf{X}'$. Thus, the number of distinct inputs contained in the concatenation of the $\mathbf{X}^{\sim jk}$, $1 \leq j < k \leq d$, is equal to $n$ and not to $n \times d(d-1)/2$. We then go back to the pick-freeze estimator. Let us define, as in Tissot and Prieur [484], the random balance design–based pick-freeze estimator of $S_{jk}$ with

$$\widehat{V_{jk}^{\text{clos}}} = \frac{1}{n}\sum_{i=1}^{n} G(\mathbf{x}^{(i)})G(\mathbf{x}^{\sim j,k,(i)}) - \left(\frac{1}{n}\sum_{i=1}^{n}\frac{G(\mathbf{x}^{(i)}) + G(\mathbf{x}'^{(i)})}{2}\right)^2, \tag{3.23}$$

$$\widehat{V} = \frac{1}{n}\sum_{i=1}^{n}\frac{\left(G(\mathbf{x}^{(i)})\right)^2 + \left(G(\mathbf{x}^{\sim j,k,(i)})\right)^2}{2} - \left(\frac{1}{n}\sum_{i=1}^{n}\frac{G(\mathbf{x}^{(i)}) + G(\mathbf{x}^{\sim j,k,(i)})}{2}\right)^2, \tag{3.24}$$

with $V_{jk}^{\text{clos}} = \text{Var}\left(\mathbb{E}(Y|X_j, X_k)\right)$.

This procedure unfortunately does not allow us to estimate total indices. However, we are not always interested in total indices, as for many models most of the information is contained in the first- and second-order interactions. If one wants to estimate first-order, second-order, and total Sobol' indices, it is possible to choose Saltelli's strategy [409] recalled earlier in Theorem 3.14. More recently, Gilquin et al. [160] proposed two new strategies: the first one, called REP18A in their paper, to estimate the full set of first- and second-order Sobol' indices at a cost of $2n$ model evaluations, and the second one, called REP18B in their paper, to estimate the full set of first-order, second-order, and total Sobol' indices at a cost of $n(d+2)$ model evaluations. Strategy REP18A is based on two replicated OAs of strength two. As already mentioned, there exist constraints on the construction of strength two OAs, leading to $n = q^2 \geq d-1$ with $q$ a prime number. This last assumption is the one required if the so-called method of differences [49] is used to construct strength OAs. It can be relaxed by using other constructions (see, e.g., [186]). Strategy REP18B is an adaptation of REP18A so as to incorporate the estimation of total-effect indices: First-order and second-order indices are estimated with two replicated OAs, then all total Sobol' indices are estimated using one of the two replicated designs plus $d$ additional designs obtained following Saltelli's scheme [409]. Strategy REP18B is preferred to the one in Saltelli [409] (see Theorem 3.14 above) for models up to dimension $d \approx 100$. For higher dimensions, the constraint due to the construction of OAs is too restrictive, and the approach in Saltelli [409] is preferred. In the following, the approach in Saltelli [409] is called SAL02*.

The procedures REP18A, REP18B, and SAL02* are summarized in Table 3.2, and the corresponding formulas for the estimators are given in Table 3.3.

Recall that $q$ is a prime number with $q^2 \geq d-1$. REP18A is recommended for estimating first-order and second-order Sobol' indices, REP18B for estimating first-order, second-order, and total Sobol' indices for small to moderate dimension (up to $d \approx 100$), and SAL02* for estimating

**Table 3.2.** *Summary of procedures REP18A, REP18B, and SAL02*.*

| name | 1st | 2nd | total | cost | sampling |
|---|---|---|---|---|---|
| REP18A | yes | yes | no | $2q^2$ | OA |
| REP18B | yes | yes | yes | $q^2(d+2)$ | OA + MC/QMC |
| SAL02* | yes | yes | yes | $n(2d+2)$ | MC/QMC |

**Table 3.3.** *Estimator formulas of procedures REP18A, REP18B, and SAL02*.*

| name | 1st order | 2nd order | total |
|---|---|---|---|
| REP18A | (3.17)/(3.18) | (3.23)/(3.24) | |
| REP18B | (3.17)/(3.18) | (3.23)/(3.24) | (3.19)/(3.20) |
| SAL02* | (3.17)/(3.18) | (3.23)/(3.24) | (3.19)/(3.20) |

first-order, second-order, and total Sobol' indices for high dimension. The corresponding functions in the package sensitivity are given in Table 3.4. The functions sobolrep() and sobolSalt() also provide bootstrap confidence intervals.

**Table 3.4.** *Estimator formulas of procedures REP18A, REP18B, and SAL02*.*

| name | ref. | corresp. function in sensitivity |
|---|---|---|
| REP18A | [160] | sobolrep() with total=FALSE |
| REP18B | [160] | sobolrep() with total=TRUE |
| SAL02* | [409] | sobolSalt() with scheme="B" |

### 3.2.4 ▪ Asymptotic results

In this section, the asymptotic properties of pick-freeze estimators of Sobol' indices when the sampling scheme is purely Monte Carlo are studied. Most of these results, although intuitive, are quite recent. In this context and not for a pick-freeze sampling, a second-order efficient semiparametric method of estimation is proposed and studied in Da Veiga and Gamboa [101].

#### Pick-freeze estimators based on substituted columns sampling

Pick-freeze estimators based on substituted columns sampling were introduced in Section 3.2.1. We present in this section the asymptotic results for the estimation of first-order and total Sobol' indices only, although they can be generalized to any-order Sobol' indices; see [149] for the generalization.

It is possible to derive asymptotic properties of any-order Sobol' index pick-freeze estimators based on substituted columns sampling; however, we focus our attention on the pick-freeze estimator of first-order Sobol' indices defined, for any $j = 1, \dots, d$, by $\widehat{S}_j = \frac{\widehat{V}_j}{\widehat{V}}$ with $\widehat{V}_j$ defined by Eq. (3.17) and $\widehat{V}$ defined by Eq. (3.18) when an i.i.d. sampling scheme is used. More precisely, we consider in that case two independent i.i.d. samples, of size $n \in \mathbb{N}_*$, distributed as the input vector $\mathbf{X}$

$$\mathbf{X}^{(i)} = (X_1^{(i)}, \dots, X_{j-1}^{(i)}, X_j^{(i)}, X_{j+1}^{(i)}, \dots, X_d^{(i)}),\ i = 1, \dots, n,$$

$$\mathbf{X}'^{(i)} = (X_1'^{(i)}, \dots, X_{j-1}'^{(i)}, X_j'^{(i)}, X_{j+1}'^{(i)}, \dots, X_d'^{(i)}),\ i = 1, \dots, n.$$

Further, for any $j = 1, \dots, d$ and $i = 1, \dots, n$, we set

$$\mathbf{X}^{\sim j,(i)} = (X_1'^{(i)}, \dots, X_{j-1}'^{(i)}, X_j^{(i)}, X_{j+1}'^{(i)}, \dots, X_d'^{(i)}),$$

and we define $Y^{\sim j,(i)} = G(\mathbf{X}^{\sim j,(i)})$ and $Y^{(i)} = G(\mathbf{X}^{(i)})$. We then define the pick-freeze estimator, for $j = 1, \ldots, d$, as $\widehat{S}_j = \frac{\widehat{V}_j}{\widehat{V}}$, with $\widehat{V}_j$ defined by Eq. (3.17) and $\widehat{V}$ defined by Eq. (3.18). We thus have

$$\widehat{V}_j = \frac{1}{n}\sum_{i=1}^{n} Y^{(i)}Y^{\sim j,(i)} - \left(\frac{1}{n}\sum_{i=1}^{n}\frac{Y^{(i)} + Y^{\sim j,(i)}}{2}\right)^2,$$

$$\widehat{V} = \frac{1}{n}\sum_{i=1}^{n}\frac{\left(Y^{(i)}\right)^2 + \left(Y^{\sim j,(i)}\right)^2}{2} - \left(\frac{1}{n}\sum_{i=1}^{n}\frac{Y^{(i)} + Y^{\sim j,(i)}}{2}\right)^2.$$

This estimator was first introduced in Monod et al. [324], and later its asymptotic optimality was studied in [218] for the 1D case and in [149] for the multidimensional one.

The following central limit theorem (CLT) holds.

**Theorem 3.16.** *Assume that $\mathbb{E}[Y] = 0$, and let $\widehat{S} = (\widehat{S}_1, \ldots, \widehat{S}_d)$ be the vector comprising the estimators for the $d$ first-order Sobol' indices $S_1, \ldots, S_d$ and $S = (S_1, \ldots, S_d)$. Assume further that $\mathbb{E}Y^4 < +\infty$. We then have*

$$\sqrt{n}\left(\widehat{S} - S\right) \xrightarrow[n\to+\infty]{\mathcal{D}} \mathcal{N}_d\left(0, \Sigma_S\right),$$

*where $\Sigma_S = ((\Sigma_S)_{l,j})_{1\le j,k\le d}$ with*

$$(\Sigma_S)_{j,k} = \frac{\mathrm{Cov}(YY^{\sim j}, YY^{\sim k}) - S_j\mathrm{Cov}(YY^{\sim k}, M) - S_k\mathrm{Cov}(YY^{\sim j}, M) + S_jS_k\mathrm{Var}\,M}{(\mathrm{Var}\,Y)^2}$$

*and*

$$M = \frac{1}{d+1}\left(Y^2 + \sum_{j=1}^{d}(Y^{\sim j})^2\right).$$

**Remark 10.** *Let us make two remarks concerning the previous theorem. First, note that the statement still holds for a noncentered output $Y$. Obviously, Sobol' indices are invariant by a constant translation of the output. Further, one can easily check that this is also the case both for $\widehat{V}_j$ ($j \in [\![1:d]\!]$) and $\widehat{V}$ and so for $\widehat{S}$. Hence, if $\mathbb{E}[Y] \neq 0$, one may apply the last theorem to $Y - \mathbb{E}[Y]$, so that the asymptotic normality for $\sqrt{n}(\widehat{S} - S)$ built on $Y$ still holds. Nevertheless, the asymptotic covariance matrix has to be updated, replacing the output of the code by its centered version. For example, for $i, j \in [\![1:d]\!]$, the term $\mathrm{Cov}(YY^{\sim j}, YY^{\sim k})$ becomes*

$$\mathrm{Cov}((Y - \mathbb{E}[Y])(Y^{\sim j} - \mathbb{E}[Y]), (Y - \mathbb{E}[Y])(Y^{\sim k} - \mathbb{E}[Y])).$$

*To alleviate the statement, we would rather consider only the centered case in the last theorem. The second comment is that in an applied context, one should of course estimate the asymptotic covariance matrix $\Sigma_S$ to perform statistical inference like the computation of trust regions or statistical hypothesis testing.*

**Remark 11.** *In Janon et al. [218], the authors prove the optimality of the estimator $\widehat{S}$ (in terms of asymptotic variance) among regular estimators which are functions of the designs $\{\mathbf{X}^{(i)}\}_{i=1}^n$ and $\{\mathbf{X}^{\sim j,(i)}\}_{i=1}^n$, $1 \le j \le d$. This feature is called asymptotic efficiency and is a generalization of the notion of minimum variance unbiased estimator (see Chapters 8 and 25 in [488] or [201] for more details).*

***Proof.*** We define $W_i = (Y^{(i)}Y^{\sim j,(i)}, j = 1,\ldots,d, Y^{(i)}, Y^{\sim j,(i)}, j = 1\ldots,d, M^{(i)})^t$, $i = 1,\ldots,n$. Let $\Sigma$ denote the covariance matrix of $W_i$. The vectorial CLT implies that

$$\sqrt{n}\left(\frac{1}{n}\sum_{i=1}^{n} W_i - \mathbb{E}W\right) \xrightarrow[n\to+\infty]{\mathcal{D}} \mathcal{N}_{2d+2}\left(0,\Sigma\right).$$

Define $g$ from $\mathbb{R}^{2d+2}$ into $\mathbb{R}^d$ as

$$g(x_1,\ldots,x_d,y,y^{\sim 1},\ldots,y^{\sim d},z) = \left(\frac{x_1 - \left(\frac{y+y^{\sim 1}}{2}\right)^2}{z - \left(\frac{y+y^{\sim 1}+\ldots+y^{\sim d}}{d+1}\right)^2},\ldots,\frac{x_k - \left(\frac{y+y^{\sim d}}{2}\right)^2}{z - \left(\frac{y+y^{\sim 1}+\ldots+y^{\sim d}}{d+1}\right)^2}\right).$$

We then apply the so-called *Delta method* [488] to $W$ and $g$, so that

$$\sqrt{n}\left(g\left(\frac{1}{n}\sum_{i=1}^{n} W^{(i)}\right) - g(\mathbb{E}W)\right) \xrightarrow[n\to+\infty]{\mathcal{D}} \mathcal{N}\left(0, J_g(\mathbb{E}W)\Sigma J_g(\mathbb{E}W)^t\right)$$

with $J_g(\mathbb{E}W)$ the Jacobian of $g$ at point $\mathbb{E}W$.

For $j = 1,\ldots,d$, $k = 1,\ldots,d$,

$$\begin{cases} \dfrac{\partial g_k}{\partial x_j}u(\mathbb{E}W) = \dfrac{1}{V}\delta_{j,k}, \\ \dfrac{\partial g_k}{\partial y}(\mathbb{E}W) = 0, \\ \dfrac{\partial g_k}{\partial y^{\sim j}}(\mathbb{E}W) = 0, \\ \dfrac{\partial g_k}{\partial z}(\mathbb{E}W) = -\dfrac{S_k}{V}. \end{cases}$$

Thus $\Sigma_S = J_g(\mathbb{E}W)\Sigma J_g(\mathbb{E}W)^t$ is as stated in Theorem 3.16. □

It is also possible to prove asymptotic results for the procedure presented in the subsection *Replicated Latin hypercube sampling* in Section 3.2.2 using two replicated LHS to estimate first-order Sobol' indices. We refer the interested reader to Tissot and Prieur [484].

## 3.2.5 ▪ Nonasymptotic properties

In this section, some nonasymptotic results for the pick-freeze estimators of the Sobol' indices are discussed. The main sources for this section come from the paper [149]. These results are of two kinds. We first give concentration inequalities and then Berry–Esseen-like results. Roughly speaking, concentration inequalities give bounds on the probability that the estimator deviates the target value, while the Berry–Esseen theorem bounds the error made when one uses the approximation by the asymptotic Gaussian distribution given in Theorem 3.16. On the one hand, these results are interesting, as they are true for finite size samples, which is always the case in practice. On the other hand, as they are quite general, the bounds given are very rough and so quite conservative. For concentration inequalities, we refer the reader to the two excellent books of Ledoux [268] and of Boucheron et al. [50]. For the sake of simplicity, the nonasymptotic results are set only for one component of the vector $\widehat{S}$, so that, without loss of generality, we only consider $\widehat{S}_1$. For the concentration inequality, the multidimensional case could be tackled with a substantial extra work by using general arguments on Lipschitz functions (see the first chapter of the book of Ledoux [268]). By contrast, for Berry–Esseen-like results, this also could be done but at the cost of paying a high price in technical investment on Gaussian approximation; see [36]. To begin with, let us define, for $x > -1$,

$$h(x) = (1+x)\ln(1+x) - x.$$

Let us introduce the random variables $K^{(i)\pm} = Y^{(i)}Y^{\sim 1,(i)} - (S_1 \pm y)\frac{(Y^{(i)})^2+(Y^{\sim 1,(i)})^2}{2}$. We denote by $V_K^+$ (resp., $V_K^-$) the second moment of the i.i.d. random variable $K^{(i)+}$ (resp., $K^{(i)-}$).

**Theorem 3.17.** *Let $b > 0$ and $y > 0$. Assume that $Y \in [-b, b]$. Then*

$$\mathbb{P}\left(\widehat{S}_1 \geq S_1 + y\right) \leq m_1 + 2m_2 \mathbb{1}_{\{S_1+y-1\geq 0\}}, \tag{3.25}$$

$$\mathbb{P}\left(\widehat{S}_1 \leq S_1 - y\right) \leq m_3 + 2m_4 \mathbb{1}_{\{S_1+y-1\geq 0\}}, \tag{3.26}$$

*where*

$$m_1 = \exp\left\{-\frac{nV_K^+}{b_U^2} h\left(\frac{b_U}{V_K^+}\frac{yV}{2}\right)\right\},\ m_2 = \exp\left\{-\frac{n(V+V_1)}{2b^2} h\left(\frac{b}{V+V_1}\sqrt{\frac{2yV}{S_1+y-1}}\right)\right\},$$

$$m_3 = \exp\left\{-\frac{nV_K^-}{b_U^2} h\left(\frac{b_U}{V_K^-}\frac{yV}{2}\right)\right\},\ m_4 = \exp\left\{-\frac{n(V+V_1)}{2b^2} h\left(\frac{b}{V+V_1}\sqrt{\frac{2yV}{y+1-S_1}}\right)\right\},$$

*and $b_U = b^2(1 + S_1 + y)$.*

***Proof.*** One may assume without loss of generality that $Y$ is centered. Obviously, $K^{(i)+}$ and $K^{(i)-}$ are upper bounded by $b_U$, $\mathbb{E}\, K^{(i)+} = -yV$, $\mathbb{E}\, K^{(i)-} = yV$.

Set, for $i = 1, \dots, n$, $Z^{(i)} = (Y^{(i)} + Y^{\sim 1,(i)})/2$. $Z^{(i)}$ is also supported by $[-b, b]$. Let $\overline{Z}_n$ denote the empirical mean of the variables $Z^{(i)}$ ($i = 1, \dots, n$). We have $\mathbb{E}\, Z^{(i)} = 0$ and $\mathbb{E}\,(Z^{(i)})^2 = \frac{V+V_1}{2}$. We first prove Eq. (3.25). One gets, if $S_1 + y - 1 \geq 0$,

$$\begin{aligned}
\mathbb{P}\left(\widehat{S}_1 \geq S_1 + y\right) &= \mathbb{P}\left(\frac{\frac{1}{n}\sum_{i=1}^n Y^{(i)}Y^{\sim 1,(i)} - \left(\overline{Z}_n\right)^2}{\frac{1}{n}\sum_{i=1}^n \frac{(Y^{(i)})^2+(Y^{\sim 1,(i)})^2}{2} - \left(\overline{Z}_n\right)^2} \geq S_1 + y\right) \\
&= \mathbb{P}\left(\frac{1}{n}\sum_{i=1}^n \left(K^{(i)+} - \mathbb{E}\, K^{(1)+}\right) + \left(\overline{Z}_n\right)^2 (S_1 + y - 1) \geq yV\right) \\
&\leq \mathbb{P}\left(\sum_{i=1}^n \left(K^{(i)+} - \mathbb{E}\, K^{(1)+}\right) \geq n\frac{yV}{2}\right) \\
&\quad + \mathbb{P}\left(\sum_{i=1}^n Z^{(i)} \geq n\sqrt{\frac{yV}{2(S_1+y-1)}}\right) \\
&\quad + \mathbb{P}\left(\sum_{i=1}^n (-Z^{(i)}) \geq n\sqrt{\frac{yV}{2(S_1+y-1)}}\right).
\end{aligned}$$

Inequality (3.25) comes directly by applying the Bennett inequality (see, e.g., [50]) to the random variables $K^{(i)+}$, $Z^{(i)}$, and $-Z^{(i)}$.

Let us turn to the proof of (3.26). One gets, since $y+1-S_1>0$,

$$\begin{aligned}\mathbb{P}\left(\widehat{S}_1 \le S_1 - y\right) &= \mathbb{P}\left(\frac{1}{n}\sum_{i=1}^{n}\left(-K^{(i)-} + \mathbb{E}\,K^{(1)-}\right) + \left(\overline{Z}_n\right)^2(y+1-S_1) \ge yV\right)\\ &\le \mathbb{P}\left(\sum_{i=1}^{n}\left(-K^{(i)-} + \mathbb{E}\,K^{(1)-}\right) \ge n\frac{yV}{2}\right)\\ &+\mathbb{P}\left(\sum_{i=1}^{n} Z^{(i)} \ge n\sqrt{\frac{yV}{2(y+1-S_1)}}\right)\\ &+\mathbb{P}\left(\sum_{i=1}^{n}(-Z^{(i)}) \ge n\sqrt{\frac{yV}{2(y+1-S_1)}}\right).\end{aligned}$$

Inequality (3.26) comes from the Bennett inequality applied to the random variables $K^{(i)-}$, $Z^{(i)}$, and $-Z^{(i)}$. □

To put in action the previous theorem, one needs to know many parameters on the model $(V, V_1, V_K^+, \ldots)$. This is somehow quite unrealistic, as, for example, knowing $V$ and $V_1$ allows us to compute directly $S_1$. A classical way in asymptotic statistics to avoid this drawback is to replace the unknown parameters by some estimates. If the sample size $n$ is large enough, this strategy leads to plausible bounds. Another way, when the variance $V$ is known, is to use the more conservative bounds given in the following theorem.

**Theorem 3.18.** *Let $b>0$ and $y>0$. Assume that $Y \in [-b,b]$. Then*

$$\mathbb{P}\left(\widehat{S}_1 \ge S_1 + y\right) \le \exp\left(-\frac{nV^2}{128}\left(1-\frac{1}{n}\right)^2\left(\frac{y}{1+y}\right)^2\right).$$

*Assume further that $\frac{(1-S_1)(1+S_1)}{2n-(1+S_1)} \le y \le 1$. Then*

$$\mathbb{P}\left(\widehat{S}_1 \le S_1 - y\right) \le \exp\left(-\frac{nV^2}{128}\left(y-\frac{9}{8n}\right)^2\right).$$

***Proof.*** First, one can use Corollary 1.17 of the book of Ledoux [268] to obtain

$$\mathbb{P}\left(\widehat{S}_1 \ge S_1 + y\right) \le \exp\left(-\frac{nV^2}{32}\left(\frac{y-(1/2n)(S_1+y-1)(S_1+1)}{1+S_1+y+|S_1+y-1|}\right)^2\right).$$

Assume further that $\frac{(1-S_1)(1+S_1)}{2n-(1+S_1)} \le y \le 1$. Then

$$\mathbb{P}\left(\widehat{S}_1 \le S_1 - y\right) \le \exp\left(-\frac{nV^2}{32}\left(\frac{y+(1/2n)(S_1-y-1)(S_1+1)}{1+S_1-y+|S_1-y-1|}\right)^2\right).$$

In order to obtain the results of the previous theorem, just bound $S_1$ by 1. □

We now turn to the Berry–Esseen-like results. For the sake of simplicity, we restrict ourselves to the centered case, where one knows the mean value of $Y$. The more general case, where this mean is unknown, is much more intricate. We refer the reader for this case to [149], where some

Berry–Esseen results based on a general theorem of Pinelis and Molzon [369] are stated. In this section, a Berry–Esseen theorem is given for the estimator

$$\widehat{\widehat{S}}_1 = \frac{\frac{1}{n}\sum_{i=1}^n Y^{(i)}Y^{\sim 1,(i)}}{\frac{1}{n}\sum_{i=1}^n (Y^{(i)})^2}$$

in the centered case. Further, let $\kappa \approx 0.42$ be the last best constant known in the classical Berry–Esseen theorem; see [246]. Recall that $\Phi$ denotes the distribution function of the standard Gaussian distribution. We then have the following.

**Theorem 3.19.** *Assume that the random variable $Y$ has finite moment up to order 6. Then, for all $t \in \mathbb{R}$,*

$$\left|\mathbb{P}\left(\frac{\sqrt{n}}{\sigma}(\widehat{\widehat{S}}_1 - S_1) \leq t\right) - \Phi(t)\right| \leq \frac{\kappa\mu_{3,n}}{\sqrt{n}} + \left|\Phi(t) - \Phi\left(\frac{t}{\sqrt{1 + \frac{t\nu_n}{\sigma\sqrt{n}V^2}}}\right)\right|. \tag{3.27}$$

*Here*

$$\sigma^2 = \mathrm{Var}\left(\frac{1}{V}\left(YY^{\sim 1} - S_1 Y^2\right)\right) \tag{3.28}$$

*is the asymptotic variance of $\sqrt{n}\widehat{\widehat{S}}_1$ and*

$$\begin{aligned}
\mu_{3,n} &= \mathbb{E}\left[\left|\frac{\Delta_n - \mathbb{E}\Delta_n}{\sqrt{\mathrm{Var}\,\Delta_n}}\right|^3\right],\\
\Delta_n &= \sigma^{-1}V\left[YY^{\sim 1} - \left(S_1 + \frac{t\sigma}{\sqrt{n}}\right)Y^2\right],\\
\nu_n &= \left(\frac{t\sigma}{\sqrt{n}} + 2S_1\right)\mathrm{Var}\,Y^2 - 2\mathrm{Cov}(YY^{\sim 1}, Y^2).
\end{aligned}$$

***Proof.*** To begin with, we compute the asymptotic variance $\sigma^2$ of $\sqrt{n}\widehat{\widehat{S}}_1$. The Delta method is applied (see [488]) to $(YY^{\sim 1}, (Y')^2)$ and $\Psi(u,v) = uv^{-1}$ $(u \in \mathbb{R}, v > 0)$. Then $\sigma^2 = J_\Psi(\mathbb{E}W)\Sigma J_\Psi(\mathbb{E}W)^t$ ($J_\Psi$ the Jacobian of $\Psi$), and the expression given in (3.28) follows obviously.

Now, for $t \in \mathbb{R}$, set

$$A_t = \left\{\frac{\sqrt{n}}{\sigma}\left(\widehat{\widehat{S}}_1 - S_1\right) \leq t\right\}.$$

Obvious algebraic manipulations lead to

$$A_t = \left\{\sqrt{n^{-1}}\sum_{i=1}^n \Delta_n^{(i)} \leq 0\right\},$$

where, for $i = 1, \ldots, n$,

$$\Delta_n^{(i)} = \sigma^{-1}\left[Y^{(i)}Y^{\sim 1,(i)}V - \left(V_1 + \frac{t\sigma}{\sqrt{n}}V\right)(Y^{(i)})^2\right].$$

Now, we have

$$\mathbb{E}\Delta_n = \frac{-tV^2}{\sqrt{n}} \quad \text{and} \quad \mathrm{Var}\,\Delta_n = V^4\left[1 + \frac{t\nu_n}{\sigma\sqrt{n}V^2}\right].$$

Denoting by $\Delta_{n,\cdot}$ the empirical mean of $(\Delta_n^{(i)})_{i=1,\ldots,n}$, we therefore obtain

$$A_t = \{\sqrt{n}\Delta_{n,\cdot} \leq 0\} = \left\{ \sqrt{n}\left(\frac{\Delta_{n,\cdot} - \mathbb{E}\Delta_n}{\sqrt{\operatorname{Var}\Delta_n}}\right) \leq \frac{t}{\sqrt{1+\frac{t\nu_n}{\sigma\sqrt{n}V^2}}} \right\}.$$

Now, to conclude, the classical Berry–Esseen theorem [246] and the triangular inequality are applied to obtain Eq. (3.27). □

**Remark 12.** *We denote by $B(t)$ the right-hand side of the Berry–Esseen inequality Eq.* (3.27). *It is clear that, for any $y > 0$, we have*

$$\begin{aligned}\mathbb{P}(-y \leq \widehat{\widetilde{S}}_1 - S_1 \leq y) \geq &\left[\Phi\left(\frac{\sqrt{n}}{\sigma}y\right) - \Phi\left(-\frac{\sqrt{n}}{\sigma}y\right)\right] \\ &- \left[B\left(\frac{\sqrt{n}}{\sigma}y\right) + B\left(-\frac{\sqrt{n}}{\sigma}y\right)\right]\end{aligned}$$

*and*

$$\begin{aligned}\mathbb{P}(-y \leq \widehat{\widetilde{S}}_1 - S_1 \leq y) \leq &\left[\Phi\left(\frac{\sqrt{n}}{\sigma}y\right) - \Phi\left(-\frac{\sqrt{n}}{\sigma}y\right)\right] \\ &+ \left[B\left(\frac{\sqrt{n}}{\sigma}y\right) + B\left(-\frac{\sqrt{n}}{\sigma}y\right)\right].\end{aligned} \tag{3.29}$$

*Hence, the actual confidence level of the asymptotic confidence interval for $S_1$ using $\widetilde{S}_1$ is greater than the theoretical level (first term of the sum above) minus a correction term given by the Berry–Esseen theorem (second term). The upper bound given by Eq.* (3.29) *may also be of practical interest: An overly conservative (overconfident) interval is not always desirable, as a more precise interval with an accurate level may exist.*

**Remark 13.** *The results in Sections 3.2.4 and 3.2.5 are given for estimators of first-order Sobol' indices. They are still true for the corresponding estimators of closed Sobol' indices of any order as established in Janon et al. [218] and Gamboa et al. [149].*

### 3.2.6 ▪ Examples in R

The following pick-freeze–type estimators of Sobol' indices are integrated into the sensitivity package of R:

- Sobol' scheme [442] (see Section 3.2.1) to compute the indices given by the variance decomposition up to a specified order (sobol() function);
- Saltelli's scheme [409] (see Section 3.2.3) to compute first-order, second-order, and total indices (sobolSalt() function);
- Saltelli's scheme [409] (see Section 3.2.1) to compute first-order and total indices (sobol2002() function);
- Mauntz and Kucherenko's scheme [446] to compute first-order and total indices using improved formulas for small indices (sobol2007() function);
- Jansen and Sobol's scheme [220] (see Section 3.2.1) to compute first-order and total indices using improved formulas (soboljansen() function);

- Martinez's scheme using correlation coefficient–based formulas [22] to compute first-order and total indices, associated with theoretical confidence intervals (sobolmartinez() and soboltouati() functions);
- Janon and Monod's scheme [324, 218] (see Section 3.2.1) to compute first-order indices with optimal asymptotic variance (sobolEff() function);
- Mara's scheme [302] (see Section 3.2.2) to compute first-order indices with a cost independent of the dimension via permutations of a single matrix (sobolmara() function);
- Owen's scheme [349] to compute first-order and total indices using improved formulas (via 3 input independent matrices) for small indices (sobolowen() function);
- total interaction indices using Liu and Owen's scheme [287] (sobolTIIlo() function) and pick-freeze scheme [144] (sobolTIIpf() function);
- estimation of the Sobol' first-order and closed second-order indices using replicated orthogonal array–based LHS [484] (see Section 3.2.3, sobolroalhs() and sobolrec() functions);
- estimation of the Sobol' first-order, second-order, and total indices using the generalized method with replicated orthogonal array–based Latin hypecube sample (see Section 3.2.3, sobolrep() function);
- Sobol' indices estimation under inequality constraints [161] by extension of the replication procedure [484] (sobolroauc() function).

On the Sobol' $g$-function (see Section C.7 with independent $X_j \sim \mathcal{U}(0,1)$), the pick-freeze scheme of sobolEff() is illustrated for computing first-order Sobol' indices with their asymptotic confidence intervals:

```
> set.seed(12345) # To reproduce the same results
> library(sensitivity)
> d <- 8
> n <- 1000
> X1 <- data.frame(matrix(runif(d * n), nrow = n))
> X2 <- data.frame(matrix(runif(d * n), nrow = n))
> sa <- sobolEff(model = sobol.fun, X1 = X1, X2 = X2, nboot = 0)
> print(sa)

Call:
sobolEff(model = sobol.fun, X1 = X1, X2 = X2, nboot = 0)

Model runs: 9000

Model variance: 0.4491169


Sobol indices
   original std. error min. c.i. max. c.i.
X1 0.714442   0.015658  0.683752  0.745132
X2 0.189149   0.031528  0.127356  0.250942
```

```
X3 0.041871   0.031577 -0.020018  0.103760
X4 0.026143   0.031106 -0.034823  0.087109
X5 0.021640   0.030402 -0.037947  0.081227
X6 0.020967   0.030405 -0.038627  0.080561
X7 0.021261   0.030391 -0.038304  0.080826
X8 0.021250   0.030426 -0.038383  0.080883
```

The argument `original` gives the Sobol' indices values, `std.`error is the standard deviation, and `min.`c.i. (resp., `max.`c.i.) is the minimal (resp., maximal) bound of the confidence interval (of level 95% by default). Figure 3.2, which gives the Sobol' indices estimates and their 95% confidence intervals for each input, is obtained with the following command:

```
> library(ggplot2) ; ggplot(sa)
```

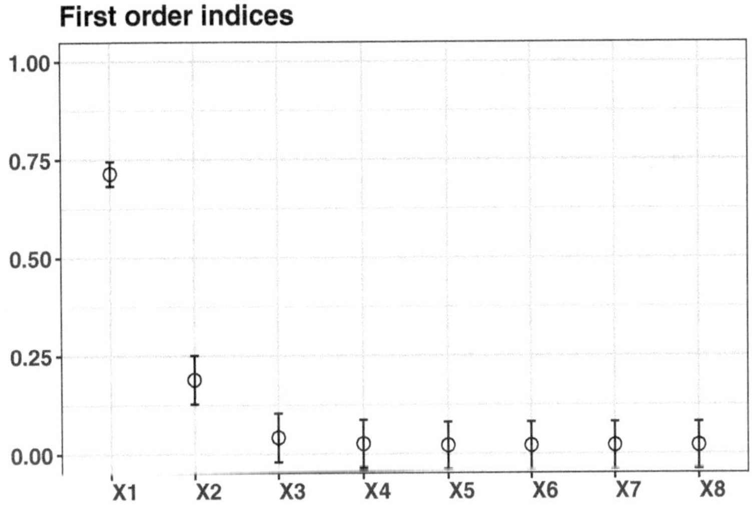


**Figure 3.2.** *Plotted results of the* sobolEff() *function on the Sobol' g-function.*

The pick-freeze scheme of the sobolSalt() function allows to obtain the first-order, second-order, and total Sobol' indices with their bootstrap-based confidence intervals. On the Ishigami function (see Section C.4 with independent $X_j \sim \mathcal{U}(-\pi, \pi)$), it gives

```
> set.seed(12345) # To reproduce the same results
> d <- 3
> n <- 1000
> X1 <- data.frame(matrix(runif(d*n, min = -pi, max = pi), nrow=n))
> X2 <- data.frame(matrix(runif(d*n, min = -pi, max = pi), nrow=n))
> sa <- sobolSalt(model=ishigami.fun, X1, X2, scheme="B", nboot=100)
```

Figure 3.3, which gives the estimates with their 95% confidence intervals, is obtained with the following commands:

```
> library(ggplot2)
> ggplot(sa, choice = 1) ; ggplot(sa, choice = 2)
```

Finally, we illustrate that the first-order (as well the second-order) Sobol' indices can be obtained with a lower computational cost (independent of $d$) by a direct sampling scheme, for

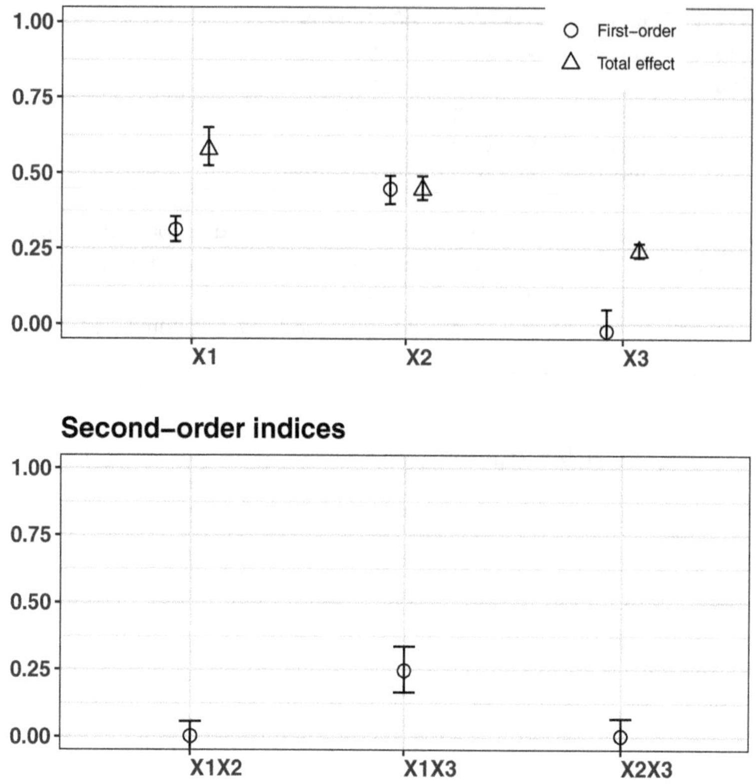


**Figure 3.3.** *Plotted results of the* sobolSalt() *function on the Ishigami model.*

example, the one of the sobolroalhs() function. On the same Sobol' $g$-function as above, it gives

```
> set.seed(12345) # To reproduce the same results
> library(sensitivity)
> d <- 8
> n <- 1000
> sa <- sobolroalhs(sobol.fun, factors=d, N=n, order=1, nboot=100)
```

Figure 3.4, which gives the estimates with their 95% confidence intervals, is obtained with the following commands:

```
> library(ggplot2)
> ggplot(sa)
```

One thousand model runs were required here, while Fig. 3.2 required 9000 model runs (and shows similar confidence intervals on the estimates). This function also gives access to second-order Sobol' indices at a computational cost independent of $d$.

## 3.3 ▪ Nonscalar outputs

The algorithms of global SA have mainly been developed for a scalar output. However, numerical models often simulate complex spatial and temporal evolutions [61, 438, 312]. Such studies can

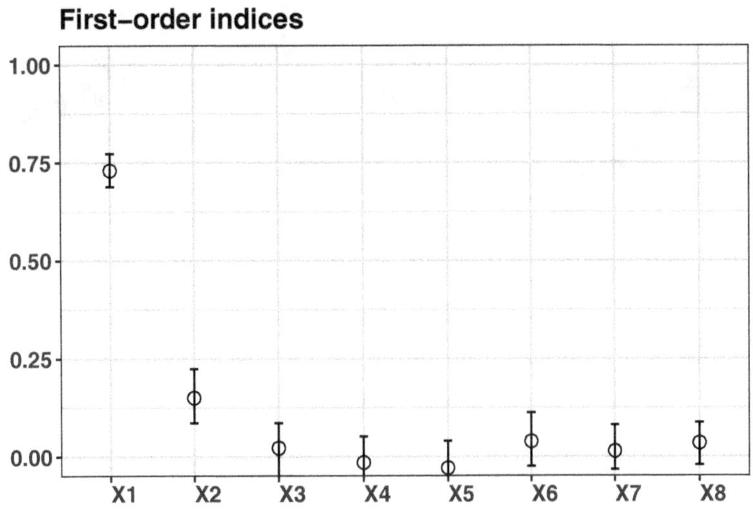


**Figure 3.4.** *Plotted results of the* sobolroalhs() *function on the Sobol' g-function.*

be rather complicated as soon as they rely on numerical models computing the evolution in time with tens of variables on thousands or millions of grid points, systems coupling several models, or prediction systems using not only one model but also observations through, for instance, data assimilation techniques.

### 3.3.1 ▪ A direct, naive, and informational approach: Ubiquitous indices

The *ubiquitous* SA (term proposed in Terraz et al. [479]) is the generalization of SA for outputs defined by a scalar spatiotemporal field $(Y(\mathbf{u},t))_{\mathbf{u}\in\mathbb{R}^k, t\geq 0}$ with $k = 2, 3$. In this case, the model becomes

$$Y(\mathbf{u},t) = G(\mathbf{u}, t, X_1, \ldots, X_d),$$

with $\mathbf{u}$ the spatial location in 2D or 3D, depending on the application and $t > 0$ the time. In most of the implementations, the spatiotemporal field is discretized on a spatiotemporal grid with $n_{\text{space}}$ spatial nodes and $n_{\text{time}}$ time steps. Therefore, ubiquitous SA provides a matrix of Sobol' indices $M_{\text{Sob}} = (S_A(\mathbf{u}_j, t_k))_{1\leq j\leq n_{\text{space}}, 1\leq k\leq n_{\text{time}}}$.

The usual way to proceed is then to focus the analysis on a limited number of probes in the spatiotemporal domain; see [491] for an example of this standard practice. It is also possible to consider the estimation of the whole set of Sobol' indices in $M_{\text{Sob}}$; see [312] for an overview on this subject. In usual implementations, simulation results are stored to disk, and statistics are computed afterward. Even if supercomputers are capable of running large studies, scientists are then constrained to run low-resolution simulations with a limited number of probes (low numbers $n_{\text{space}}$ and $n_{\text{time}}$) to keep the amount of intermediate storage manageable. In a recent paper, Terraz et al. [479] propose a file-avoiding, adaptive, fault-tolerant, and elastic framework that enables high-resolution global SA at large scale. Their approach combines iterative statistics and in-transit processing to compute Sobol' indices without any intermediate storage. Statistics are updated on the fly as soon as the in-transit parallel server receives results from one of the running simulations.

As an illustration of global SA for an output defined by a scalar spatial field, Fig. 3.5 gives the spatial maps of first-order Sobol' indices obtained on the flooding simulation model (see Section 1.3.2 and [166, 167]). It shows that the Strickler coefficients $\text{CF}_1$, $\text{CF}_2$, $\text{CF}_3$, and $\text{CF}_5$

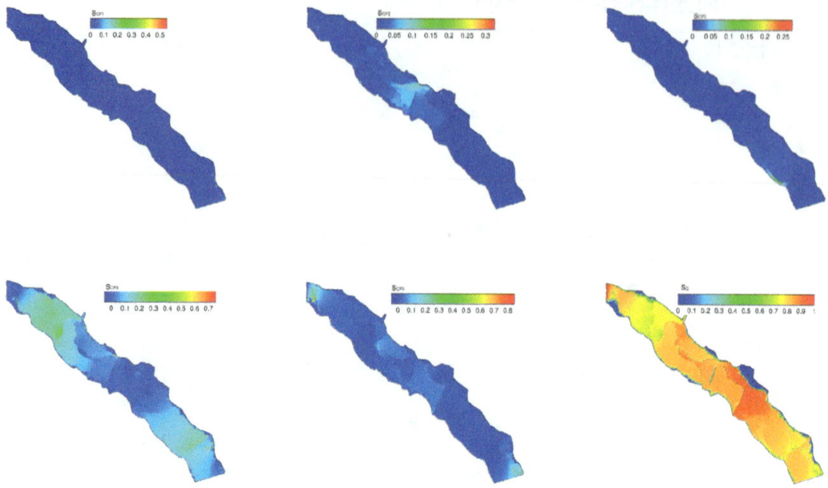

**Figure 3.5.** *Ubiquitous first-order Sobol' indices maps of the flooding model applied on the Garonne Valley. From left to right and top to bottom, the inputs are $CF_1$, $CF_2$, $CF_3$, $CF_4$, $CF_5$, and $Q$. Courtesy of Cédric Goeury (EDF R&D).*

are not influential at any point in the geographical area of interest, $\mathrm{CF}_4$ is slightly influential in the north-west and south-east domains, and the flow rate $Q$ is largely influential everywhere and predominant in the central zone.

Applications of these techniques on environmental assessment can be found, for example, in Higdon et al. [195], Marrel et al. [308], and Perrin et al. [364] for spatial outputs and in Marrel et al. [311] and De Lozzo and Marrel [107] for spatiotemporal outputs. All these works have shown that obtaining temporal/spatial/spatiotemporal sensitivity maps leads to powerful information for the analysts. In general, the parameter effects are localized in time or space and can be easily put in relation with the studied physical phenomena. For alternative works, see Ribés et al. [396], who proposed a visual sensitivity tool in order to jointly analyze a functional output and its corresponding inputs. It allows one to apply the cobweb plot principle (see Section 2.2) when the output is a function (and not only a vector of scalars).

However, it is sometimes useful to synthesize this rich information in one single index per input in order to have a sole indicator of the influence of the input on the whole variability of the model output (see, e.g., [12]). The following section develops such an approach.

## 3.3.2 ▪ Sobol' indices for vectorial outputs

### Introductory example

Let us begin with one simple example that sheds light on the need of a proper definition of sensitivity indices in the case of a vectorial output. Consider the following nonlinear model:

$$Y = G^{a,b}(X_1, X_2) = \begin{pmatrix} G_1^{a,b}(X_1, X_2) \\ G_2^{a,b}(X_1, X_2) \end{pmatrix} = \begin{pmatrix} X_1 + X_1X_2 + X_2 \\ aX_1 + bX_1X_2 + X_2 \end{pmatrix}.$$

Here, $X_1$ and $X_2$ are assumed to be i.i.d. standard Gaussian random variables. Let us begin with the computation of first-order Sobol' indices of both components of $G^{a,b}$ with respect to

$X_j$ ($j = 1, 2$). We get

$$\begin{cases} (S_1(G_1^{a,b}), S_1(G_2^{a,b})) = (1/3, a^2/(1+a^2+b^2)), \\ (S_2(G_1^{a,b}), S_2(G_2^{a,b})) = (1/3, 1/(1+a^2+b^2)). \end{cases}$$

Note that the ratios

$$\frac{S_1(G_i^{a,b})}{S_2(G_i^{a,b})}, \quad i = 1, 2,$$

do not depend on $b$. Furthermore, for $|a| > 1$, this ratio is greater than 1, so that we may think that $X_1$ is the most influential input. Now, compute the Sobol' indices associated to $\|Y\|^2 = \sqrt{(G_1^{a,b}(X_1, X_2))^2 + (G_2^{a,b}(X_1, X_2))^2}$. We find

$$S_1(\|Y\|^2) \geq S_2(\|Y\|^2) \iff (a-1)(a^3 + a^2 + 5a + 5 - 4b) \geq 0,$$

which is represented in Fig. 3.6.

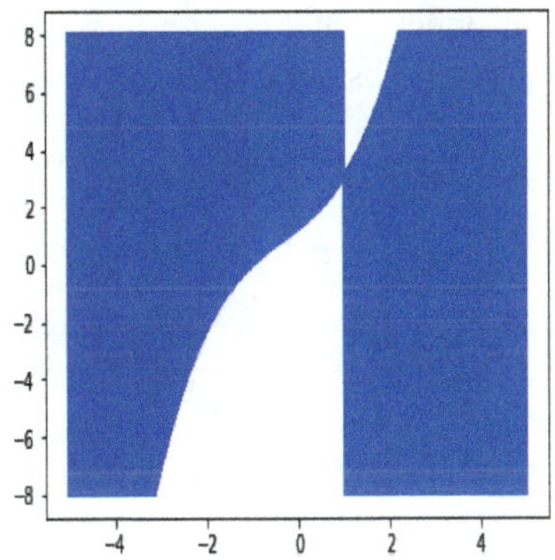


**Figure 3.6.** *In blue,* $(a-1)(a^3 + a^2 + 5a + 5 - 4b) \geq 0$ $(S_1(\|Y\|^2) \geq S_2(\|Y\|^2))$.

We may conclude that for $\|Y\|^2$, the region where $X_1$ is the most influential depends on $b$. Moreover, this region is not obvious and cannot be deduced by our first component-by-component SA. Another motivation to introduce a suitable Sobol' index is related to estimation cost. Indeed, as the dimension of the output increases, the statistical estimation of the whole set of Sobol' indices, corresponding to the Sobol' indices of each component of the vectorial output, becomes more and more expensive. Moreover, the interpretation of the vector of Sobol' indices is not necessarily easy. For all these reasons, it may be reasonable to propose an interpretable scalar summary of the information contained in this vector.

### Aggregated Sobol' indices

In the following, we define aggregated indices, first introduced by Lamboni et al. [261] and further studied in Gamboa et al. [147]. As in the previous chapters, we are dealing with an input-output relationship

$$Y = G(X_1, \ldots, X_d),$$

except that $G$ is no longer valued in $\mathbb{R}$ but in $\mathbb{R}^k$ ($k \in \mathbb{Z}_{>0}$). As before, $X_1, \ldots, X_d$ are assumed to be independent and $Y$ square integrable (i.e., $\mathbb{E}\,\|Y\|^2 < \infty$). The covariance matrix of $Y$

is further assumed to be positive definite. Note that if this is not the case, one may reduce the dimension in order to fulfill this assumption. As in Theorem 3.3 (page 49), for $A \in \mathcal{P}_d$, we may write

$$G(\mathbf{X}) = G_\emptyset + G_A(\mathbf{X}_A) + G_{\bar{A}}(\mathbf{X}_{\bar{A}}) + G_{A,\bar{A}}(\mathbf{X}_A, \mathbf{X}_{\bar{A}}), \tag{3.30}$$

where, vectorially,

$$\begin{cases} G_\emptyset = \mathbb{E}Y, \\ G_A(\mathbf{X}_A) = \mathbb{E}(Y|\mathbf{X}_A) - G_\emptyset, \\ G_{\bar{A}}(\mathbf{X}_{\bar{A}}) = \mathbb{E}(Y|\mathbf{X}_{\bar{A}}) - G_\emptyset, \\ G_{A,\bar{A}}(\mathbf{X}_A, \mathbf{X}_{\bar{A}}) = Y - G_A(\mathbf{X}_A) - G_{\bar{A}}(\mathbf{X}_{\bar{A}}) - G_\emptyset. \end{cases}$$

The decomposition in (3.30) is orthogonal in $\mathbb{L}^2(\mathrm{P}_\mathbf{X})$ so that we may write

$$\Sigma = C_A + C_{\bar{A}} + C_{A,\bar{A}},$$

where $\Sigma$, $C_A$, $C_{\bar{A}}$ and $C_{A,\bar{A}}$ denote, respectively, the $k \times k$ covariance matrices of $Y$, $G_A(\mathbf{X}_A)$, $G_{\bar{A}}(\mathbf{X}_{\bar{A}})$, and $G_{A,\bar{A}}(\mathbf{X}_A, \mathbf{X}_{\bar{A}})$. Because of the noncommutativity of the matrix product, a direct generalization of the usual Sobol' index is not straightforward. Nevertheless, for any square matrix $M$ of size $k$, we may *scalarize* the last equation by projecting it onto the direction spanned by $M$,

$$\mathrm{Tr}(M\Sigma) = \mathrm{Tr}(MC_A) + \mathrm{Tr}(MC_{\bar{A}}) + \mathrm{Tr}(MC_{A,\bar{A}}).$$

Thus, as soon as $\mathrm{Tr}(M\Sigma) \neq 0$, we may define the $M$-sensitivity indices of $Y$ associated to $\mathbf{X}_A$ as

$$S_A^{\mathrm{clos}}(M) = \frac{\mathrm{Tr}(MC_A)}{\mathrm{Tr}(M\Sigma)}.$$

We define analogously $S_{\bar{A}}^{\mathrm{clos}}(M)$ and $S_{A,\bar{A}}(M)$. The following lemma is then easily obtained.

**Lemma 3.20 (Lemma 2.1 in [147]).**

1. *The aggregated sensitivity measures sum up to* 1*:*

$$S_A^{\mathrm{clos}}(M) + S_{\bar{A}}^{\mathrm{clos}}(M) + S_{A,\bar{A}}(M) = 1. \tag{3.31}$$

2. *Left-composing $G$ by a linear operator $O$ of $\mathbb{R}^k$ changes the sensitivity measure accordingly to*

$$S_A^{\mathrm{clos}}(M; OG) = S_A^{\mathrm{clos}}(O^T M O; G). \tag{3.32}$$

3. *For $k = 1$ and for any $M \neq 0$, we have $S_A^{\mathrm{clos}}(M) = S_A^{\mathrm{clos}}$.*

We now turn to the important special case $M = \mathrm{Id}_k$ (the identity matrix of dimension $k$). To alleviate the notation, we set in the following $S_A^{\mathrm{clos}} = S_A^{\mathrm{clos}}(\mathrm{Id}_k)$. It is then possible to prove Proposition 3.21 below.

**Proposition 3.21 (Proposition 2.2 in [147]; see also Proposition 3.1 in [148]).** *We have the following properties:*

- $0 \leq S_A^{\mathrm{clos}} \leq 1$;
- *$S_A^{\mathrm{clos}}$ is invariant by left composition of $G$ by any isometry on $\mathbb{R}^k$ or any nonzero homothety.*

***Proof.*** The nonnegativity of $S_A^{\text{clos}}$ is a straightforward consequence of the positivity of $C_A$ and $\Sigma$. From (3.31) in Lemma 3.20, we deduce that $S_A^{\text{clos}} \leq 1$. The proof of the invariance by isometry is obtained by applying (3.32) in Lemma 3.20 with $O^T O = \text{Id}_k$ and $M = \text{Id}_k$. The invariance by nonzero homothety is straightforward. □

The invariance by any nonzero homothety is true even when choosing $M \neq \text{Id}_k$. It is a natural property that implies that the sensitivity index $S_A^{\text{clos}}(M)$ is not modified when changing the measurement units of the output. The isometry invariance property ensures that the index $S_A^{\text{clos}}(\text{Id}_k)$ is *intrinsic* in the sense that it is invariant by an isometric change of parametrization of the output.

In Proposition 3.22 below, we state why, within all possible choices for the matrix $M$, the identity matrix $\text{Id}_k$ is a good candidate.

**Proposition 3.22.** *Let $M$ be a square matrix of size $k$ having full rank. Assume that, for any $G$ such that* $\text{Tr}(M\Sigma) \neq 0$, *the index $S_A^{\text{clos}}(M)$ is invariant by left composition of $G$ by any isometry of $\mathbb{R}^k$. Then $S_A^{\text{clos}}(M) = S_A^{\text{clos}}$.*

***Proof.*** The trace of an antisymmetric matrix vanishes, so that, decomposing $M$ as a sum of symmetric and antisymmetric matrices, we may assume without loss of generality that $M$ is symmetric. Now, as $M$ is symmetric, we may write $M = PDP^T$, where $P^T P = \text{Id}_k$ and $D$ is diagonal. We have

$$S_A^{\text{clos}}(M) = \frac{\text{Tr}(PDP^T C_A)}{\text{Tr}(PDP^T \Sigma)} = \frac{\text{Tr}(DP^T C_A P)}{\text{Tr}(DP^T \Sigma P)} = S_A^{\text{clos}}(D; P^T G) = S_A^{\text{clos}}(D).$$

Hence, $M$ may be assumed to be diagonal. We now show that $M = \lambda \text{Id}_k$ for some $\lambda \in \mathbb{R}^*$. Suppose, by contradiction, that $M$ has two different diagonal coefficients $\lambda_1 \neq \lambda_2$. We may restrict to the case $k = 2$, $A = \{1\}$ and take $G(X_1, X_2) = \left(\frac{X_1 - \mathbb{E}X_1}{\text{Var}\, X_1}, \frac{X_2 - \mathbb{E}X_2}{\text{Var}\, X_2}\right)^T$. We then have $\Sigma = \text{Id}_2$, $C_A = \left(\begin{smallmatrix} 1 & 0 \\ 0 & 0 \end{smallmatrix}\right)$. On the one hand, $S_A^{\text{clos}}(M) = \frac{\lambda_1}{\lambda_1 + \lambda_2}$. On the other hand, choosing $O$ as the isometry exchanging the two vectors of the canonical basis of $\mathbb{R}^2$, we get $S_A^{\text{clos}}(M; OG) = \frac{\lambda_2}{\lambda_1 + \lambda_2}$. Thus $\lambda_1 = \lambda_2$, as the invariance by any isometry of $\mathbb{R}^k$ has been assumed. Thus $\exists\, \lambda \in \mathbb{R}^*$ such that $M = \lambda \text{Id}_k$. Now, the invariance by nonzero homothety yields $S_A^{\text{clos}}(\lambda \text{Id}_k) = S_A^{\text{clos}}(\text{Id}_k)$. □

It is also possible to write the Sobol' index $S_A^{\text{clos},Y}$ of the vectorial output $Y \in \mathbb{R}^k$ associated to $A$ as a weighted mean of the corresponding Sobol' indices $S_A^{\text{clos},Y_j}$ of the components of $Y_j$, $j = 1, \dots, k$, of $Y$. The precise statement is written in Lemma 3.23 below.

**Lemma 3.23.** *For any $A \subset [1:d]$, for any $Y = (Y_1, \dots, Y_k)^T \in \mathbb{R}^k$ such that* $\text{Tr}(\Sigma) \neq 0$, *with $\Sigma$ the covariance matrix of $Y$, we have*

$$S_A^{\text{clos},Y} = \sum_{j=1}^{k} \frac{\text{Var}\, Y_j}{\sum_{l=1}^{k} \text{Var}\, Y_l} S_A^{\text{clos},Y_j}.$$

***Proof.*** Straightforward computations lead to

$$\text{Tr}(\Sigma) = \sum_{l=1}^{k} \text{Var}\, Y_l \quad \text{and} \quad \text{Tr}(C_A) = \sum_{j=1}^{k} \text{Var}\, \mathbb{E}(Y_j | X_A) = \sum_{j=1}^{k} \text{Var}\, Y_j\, S_A^{\text{clos},Y_j}.$$ □

**Remark 14.** *Let us define the total inertia of $Y$ as $\mathrm{Tr}(\Sigma)$. Then $\mathrm{Var}\, Y_j / \sum_{l=1}^{k} \mathrm{Var}\, Y_k$ measures the relative contribution of $Y_j$ to the total inertia of $Y$. Then the aggregated sensitivity index $S_A^{\mathrm{clos},Y}$ is a normalized aggregate of the sensitivity indices $S_A^{\mathrm{clos},Y_j}$ weighted by the relative contribution of $Y_j$ to the total inertia of $Y$.*

**Remark 15.** *Following Definition 3.6, the total Sobol' index associated to $A$ for a vectorial output $Y$ can be defined as $S_A^{\mathrm{T}} = 1 - S_{\bar{A}}^{\mathrm{clos}}$.*

In practice, the covariance matrices $C_A$ and $\Sigma$ are unknown. In the scalar case ($k = 1$), the previous chapters have shown that it is customary to estimate $S_A^{\mathrm{clos}}$ by using a pick-freeze method (see Section 3.2). We now propose and study the statistical properties of such a type of estimator in the multidimensional framework.

We set $Y^{\sim A} = G(\mathbf{X}_A, \mathbf{X}'_{\bar{A}})$ and $Y = G(\mathbf{X})$, where $\mathbf{X}'$ is an independent copy of $\mathbf{X}$. Let $n$ be a positive integer. We take $n$ independent copies $Y^{(1)}, \dots, Y^{(n)}$ (resp., $Y^{\sim A(1)}, \dots, Y^{\sim A(n)}$) of $Y$ (resp., $Y^{\sim A}$). Finally, for $l = 1, \dots, k$ and $i = 1, \dots, n$, we denote by $Y_l^{(i)}$ (resp., $Y_l^{\sim A(i)}$) the $l$th component of $Y^{(i)}$ (resp., $Y^{\sim A(i)}$). The following estimator of $S_A^{\mathrm{clos}}$ is then defined as

$$S_{A,n}^{\mathrm{clos}} = \frac{\sum_{l=1}^{k} \left( \frac{1}{n} \sum_{i=1}^{n} Y_l^{(i)} Y_l^{\sim A(i)} - \left( \frac{1}{n} \sum_{i=1}^{n} \frac{Y_l^{(i)} + Y_l^{\sim A(i)}}{2} \right)^2 \right)}{\sum_{l=1}^{k} \left( \frac{1}{n} \sum_{i=1}^{n} \frac{(Y_l^{(i)})^2 + (Y_l^{\sim A(i)})^2}{2} - \left( \frac{1}{n} \sum_{i=1}^{n} \frac{Y_l^{(i)} + Y_l^{\sim A(i)}}{2} \right)^2 \right)}. \tag{3.33}$$

Note that this estimator can be rewritten in the more condensed form

$$S_{A,n}^{\mathrm{clos}} = \frac{\mathrm{Tr}\,(C_{A,n})}{\mathrm{Tr}\,(\Sigma_n)},$$

where $C_{A,n}$ and $\Sigma_n$ are the empirical estimators defined by

$$C_{A,n} = \frac{1}{n} \sum_{i=1}^{n} Y^{\sim A(i)} Y^{(i)} - \left( \frac{1}{n} \sum_{i=1}^{n} \frac{Y^{(i)} + Y^{\sim A(i)}}{2} \right) \left( \frac{1}{n} \sum_{i=1}^{n} \frac{Y^{(i)} + Y^{\sim A(i)}}{2} \right)^T$$

and

$$\begin{aligned} \Sigma_n = &\frac{1}{n} \sum_{i=1}^{n} \frac{Y^{(i)} Y^{(i)T} + Y^{\sim A(i)} (Y^{\sim A(i)})^T}{2} \\ &- \left( \frac{1}{n} \sum_{i=1}^{n} \frac{Y^{(i)} + Y^{\sim A(i)}}{2} \right) \left( \frac{1}{n} \sum_{i=1}^{n} \frac{Y^{(i)} + Y^{\sim A(i)}}{2} \right)^T. \end{aligned}$$

The asymptotic properties of this estimator are now studied. First, note that a straightforward application of the strong law of large numbers leads to the following proposition.

**Proposition 3.24.** $S_{A,n}^{\mathrm{clos}} \xrightarrow[n \to +\infty]{\text{a.s.}} S_A^{\mathrm{clos}}$.

The asymptotic normality of $(S_{A,n}^{\mathrm{clos}})_{n \geq 1}$ is stated in Proposition 3.25 below.

**Proposition 3.25.** *Assume that for all $l = 1, \dots, k$, $\mathbb{E} Y_l^4 < \infty$. For $l = 1, \dots, k$, we set*

$$U_l = (Y_l - \mathbb{E} Y_l)(Y_l^{\sim A} - \mathbb{E} Y_l), \qquad V_l = (Y_l - \mathbb{E} Y_l)^2 + (Y_l^{\sim A} - \mathbb{E} Y_l)^2.$$

*Then*

$$\sqrt{n}\left(S_{A,n}^{\text{clos}} - S_A^{\text{clos}}\right) \xrightarrow[n\to+\infty]{\mathcal{D}} \mathcal{N}(0,\sigma^2)$$

*and*

$$\sigma^2 = a^2 \sum_{l,l'\in[1:k]} \text{Cov}(U_l, U_{l'}) + b^2 \sum_{l,l'\in[1:k]} \text{Cov}(V_l, V_{l'}) + 2ab \sum_{l,l'\in[1:k]} \text{Cov}(U_l, V_{l'}), \quad (3.34)$$

*with*

$$a = \frac{1}{\sum_{l=1}^k \text{Var}\, Y_l}, \qquad b = -\frac{a}{2} S_A^{\text{clos}}, \qquad [1:k] = \{1,\dots,k\}.$$

***Proof.*** Since $S_{A,n}^{\text{clos}}$ remains invariant when $Y$ is changed to $Y - \mathbb{E}Y$, we may write

$$S_{A,n}^{\text{clos}} = \Phi\left(\frac{1}{n}\sum_{i=1}^n W_i\right),$$

where

$$W_i = \begin{pmatrix} (Y_1^{(i)} - \mathbb{E}Y_1)(Y_1^{\sim A(i)} - \mathbb{E}Y_1) \\ \vdots \\ (Y_k^{(i)} - \mathbb{E}Y_k)(Y_k^{\sim A(i)} - \mathbb{E}Y_k) \\ (Y_1^{(i)} - \mathbb{E}Y_1) + (Y_1^{\sim A(i)} - \mathbb{E}Y_1) \\ \vdots \\ (Y_k^{(i)} - \mathbb{E}Y_k) + (Y_k^{\sim A(i)} - \mathbb{E}Y_k) \\ (Y_1^{(i)} - \mathbb{E}Y_1)^2 + (Y_1^{\sim A(i)} - \mathbb{E}Y_1)^2 \\ \vdots \\ (Y_k^{(i)} - \mathbb{E}Y_k)^2 + (Y_k^{\sim A(i)} - \mathbb{E}Y_k)^2 \end{pmatrix}$$

and

$$\Phi(x_1,\dots,x_k,y_1,\dots,y_k,z_1,\dots,z_k) = \frac{\sum_{l=1}^k \left(x_l - (y_l/2)^2\right)}{\sum_{l=1}^k \left(z_l/2 - (y_l/2)^2\right)}.$$

The Delta method [488, Theorem 3.1] gives

$$\sqrt{n}\left(S_{A,n}^{\text{clos}} - S_A^{\text{clos}}\right) \xrightarrow[n\to+\infty]{\mathcal{D}} \mathcal{N}(0,\sigma^2),$$

where $\sigma^2 = g^T \Gamma g$, $\Gamma$ is the covariance matrix of $W_1$ and

$$g = \nabla\Phi(\mathbb{E}W_1).$$

We have

$$\mathbb{E}W_1 = \left(\text{Cov}(Y_1, Y_1^{\sim A}), \dots, \text{Cov}(Y_k, Y_k^{\sim A}), 0, \dots, 0, 2\text{Var}\, Y_1, \dots, 2\text{Var}\, Y_k\right)^T,$$

and by differentiation of $\Phi$, $g = (a,\dots,a,0,\dots,0,b,\dots,b)^T$. A simple matrix computation leads to (3.34). □

**Remark 16.** *In* (3.33)*, the numerator (resp., the denominator) is defined by using Eq.* (3.17) *(page* 60*) (resp., Eq.* (3.18)*) inside the sum over $k$ on the output components. Note that asymptotic normality still holds true when, for instance, replacing Eqs.* (3.17) *and* (3.18) *(page* 60*) by* (3.15) *and* (3.16) *or, more generally, by using any pick-freeze procedure presented in Section* 3.2.1.

**Remark 17.** *In practice, the cost of the estimation procedure defined by* (3.33) *increases with both the input and output space dimension. One way to tackle this issue can be to perform a preprocessing path consisting in a* principal component analysis *(PCA) reduction of $Y$. This preprocessing had been early proposed by Lamboni et al. [261]. A PCA reduction tool is well tailored here, as the index is invariant by left composition by an isometry. Furthermore, PCA generally leads to an orthonormal change of coordinates such that the variance is almost spanned by very few coordinates.*

### 3.3.3 ▪ Extension to functional outputs

In this section, we consider an output $Y$ valued in a functional space $\mathcal{H}$. More precisely, let $\mathcal{H}$ be a separable Hilbert space endowed with the scalar product $\langle\cdot,\cdot\rangle$ and the associated norm $||\cdot||$. Let $H$ be an $\mathcal{H}$-valued function. Note that this means that $Y$ is valued in the abstract Hilbert space $\mathcal{H}$. Hence, we are working with a quite general space for the output of the studied input-output relationship. We assume that $\mathbb{E}||Y||^2 < \infty$. Recall that $\mathbb{E}Y$ is defined by duality as the unique element in $\mathcal{H}$ satisfying

$$\mathbb{E}\langle h, Y\rangle = \langle h, \mathbb{E}Y\rangle \quad \forall\, h \in \mathcal{H}.$$

Recall also that the covariance operator associated with $Y$ is the endomorphism $\Gamma$ on $\mathcal{H}$ defined, for $h \in \mathcal{H}$, by $\Gamma(h) = \mathbb{E}(\langle Y-\mathbb{E}Y, h\rangle(Y-\mathbb{E}Y))$. Then a classical result states that $\mathbb{E}||Y||^2 < \infty$ implies that $\Gamma$ is a trace class operator. Consequently, its trace is well defined. In this framework, we may generalize the definition of $S_A^{\text{clos}}$ introduced previously for functional outputs.

**Definition 3.26.** *Recall that $Y^{\sim A} = G(\mathbf{X}_A, \mathbf{X}'_{\bar{A}})$ and $Y = G(\mathbf{X})$, where $\mathbf{X}'$ is an independent copy of $\mathbf{X}$. We then define $S_A^{\text{clos}}$ as*

$$S_A^{\text{clos}} = \frac{\text{Tr}(\Gamma_A)}{\text{Tr}(\Gamma)},$$

*where, for any $h \in \mathcal{H}$, $\Gamma_A$ is the endomorphism on $\mathcal{H}$ defined by*

$$\Gamma_A(h) = \mathbb{E}(\langle Y^{\sim A} - \mathbb{E}Y, h\rangle(Y - \mathbb{E}Y)).$$

The estimator proposed in Fort et al. [138] to estimate $S_A^{\text{clos}}$ introduced in Definition 3.26 above is based on the so-called polar decomposition of the traces of $\Gamma$ and $\Gamma_A$ recalled in Lemma 3.27 below.

**Lemma 3.27.** *We have*

$$\begin{aligned}
\text{Tr}(\Gamma) &= \mathbb{E}||Y||^2 - ||\mathbb{E}Y||^2,\\
\text{Tr}(\Gamma_A) &= \frac{1}{4}\left[\mathbb{E}||Y + Y^{\sim A}||^2 - \mathbb{E}||Y - Y^{\sim A}||^2 - 4||\mathbb{E}Y||^2\right].
\end{aligned}$$

Now, let $(\varphi_l)_{l\geq 1}$ be an orthonormal basis of $\mathcal{H}$. Then

$$||Y||^2 = \sum_{l=1}^{\infty}\langle Y, \varphi_l\rangle^2.$$

To estimate $S_A^{\text{clos}}$ (thus, to estimate first $\text{Tr}(\Gamma)$ and $\text{Tr}(\Gamma_A)$), the previous sum is truncated by setting

$$\|Y\|_m^2 = \sum_{l=1}^{m} \langle Y, \varphi_l \rangle^2 = \|Y_m\|^2$$

with $Y_m$ the projection of $Y$ onto $\text{Span}(\varphi_1, \dots, \varphi_m)$:

$$Y_m = \sum_{l=1}^{m} \langle Y, \varphi_l \rangle \varphi_l.$$

Based on the polar decomposition (see Lemma 3.27), Fort et al. [138] introduce the following estimator of $S_A^{\text{clos}}$:

$$S_{A,m,n}^{\text{clos}} = \frac{\frac{1}{4n}\sum_{i=1}^{n}\left(\|Y^{(i)} + Y^{\sim A(i)}\|_m^2 - \|Y^{(i)} - Y^{\sim A(i)}\|_m^2 - \|\overline{Y} + \overline{Y^{\sim A}}\|_m^2\right)}{\frac{1}{n}\sum_{i=1}^{n}\left(\frac{\|Y^{(i)}\|_m^2 + \|Y^{\sim A(i)}\|_m^2}{2} - \left\|\frac{\overline{Y} + \overline{Y^{\sim A}}}{2}\right\|_m^2\right)}.$$

In order to state the asymptotic normality of this estimator, we introduce some notation and Lemma 3.28 below. Let $T$ be an $\mathcal{H}$-valued random variable and $(T_i)_{i \geq 1}$ a sequence of i.i.d. copies of $T$. We define

$$D_{m,n}(T) = \frac{1}{n}\sum_{i=1}^{n}\left(\|T_i\|_m^2 - \|\overline{T}\|_m^2\right)$$

and

$$e_l = \mathbb{E}\langle T_i, \varphi_l \rangle, \; v_l = \mathbb{E}\langle T_i, \varphi_l \rangle^2, \; Z_{i,l} = \langle T_i, \varphi_l \rangle - e_l, \; W_{i,l} = \langle T_i, \varphi_l \rangle^2 - v_l.$$

The key tool for the study of the asymptotic behavior of $S_{A,m,n}^{\text{clos}}$ in Fort et al. [138] is a decomposition of $D_{m,n}(T)$ stated in Lemma 3.28 below, whose proof is omitted here. The interested reader should refer to Gamboa et al. [148] for more details.

**Lemma 3.28 (Proposition 1 and proof of Theorem 1 in [138]).**

1. *$D_{m,n}(T)$ can be rewritten as the sum of a totally degenerated U-statistic of order* 2, *a centered linear term, and a deterministic term as*

$$D_{m,n}(T) - \mathbb{E}\|T\|^2 + \|\mathbb{E}T\|^2 = -U_n K(T) + P_n L(T) - B_m(T),$$

*where*

$$U_n K(T) = \sum_{l=1}^{m} \frac{1}{n^2} \sum_{1 \leqslant i \neq j \leqslant n} Z_{i,l} Z_{j,l},$$

$$P_n L(T) = \frac{1}{n}\left(1 - \frac{1}{n}\right)\sum_{l=1}^{m}\sum_{i=1}^{n}\left(W_{i,l} - 2e_l Z_{i,l}\right),$$

$$B_m(T) = \sum_{l>m}\left(v_l - e_l^2\right) + \frac{1}{n}\sum_{l=1}^{m}\left(v_l - e_l^2\right).$$

2. *Assume that there exists $\delta > 1$ such that*

$$v_l = \mathbb{E}\left(\langle T, \varphi_l \rangle^2\right) = O\left(l^{-(\delta+1)}\right) \tag{3.35}$$

*and there exists* $\delta' > 1$ *such that*

$$\mathbb{E}\left(\langle T, \phi_l \rangle^4\right) = O\left(l^{-\delta'}\right). \tag{3.36}$$

*Then, for any* $m = m(n)$ *such that*

$$\frac{m(n)}{n^{\frac{1}{2\delta}}} \to +\infty, \quad \frac{m(n)}{\sqrt{n}} \to 0, \tag{3.37}$$

*we have*

(a) $B_m^2(T) = o(1/n)$,

(b) $\mathbb{E}\left((U_n K(T))^2\right) = o(1/n)$,

(c) $\sqrt{n}(P_n L(T) - P_n L'(T))$ *converges to* 0 *in probability,*

*where* $P_n L'(T) = \frac{1}{n}\left(1 - \frac{1}{n}\right) \sum_{l=1}^{\infty} \sum_{i=1}^{n} \left[W_{i,l} - 2e_l Z_{i,l}\right]$.

Our main theorem follows.

**Theorem 3.29 (Theorem 2 in [138]).** *Assume that assumptions* (3.35), (3.36), *and* (3.37) *of Lemma* 3.28 *are fulfilled for* $Y'$. *Then we have*

$$\sqrt{n}(S^{\text{clos}}_{A,m,n} - S^{\text{clos}}_A) \xrightarrow[n \to +\infty]{\mathcal{D}} \mathcal{N}(0, \sigma^2)$$

*with* $\sigma^2$ *depending on the moments of* $\left(D_{n,m}(Y), D_{n,m}(Y + Y^{\sim A}), D_{n,m}(Y - Y^{\sim A})\right)$.

Before starting the proof of Theorem 3.29, we state an auxiliary lemma.

**Lemma 3.30.**

$$\sqrt{n}\left\|\overline{Y} - \overline{Y^{\sim A}}\right\|_m^2 \xrightarrow[n \to +\infty]{\mathbb{P}} 0.$$

***Proof.*** We start by upper bounding the term $\sqrt{n}\|\overline{Y} - \overline{Y^{\sim A}}\|_m^2$:

$$\sqrt{n}\left\|\overline{Y} - \overline{Y^{\sim A}}\right\|_m^2 \leq \sqrt{n}\left\|\overline{Y} - \overline{Y^{\sim A}}\right\|^2 = \frac{1}{\sqrt{n}}\left\|\sqrt{n}\left[\frac{1}{n}\sum_{i=1}^{n}\left(Y^{(i)} - Y^{\sim A(i)}\right)\right]\right\|^2.$$

We conclude by using the CLT for random variables valued in a Hilbert space [269]. □

***Proof of Theorem 3.29.*** First, we note that

$$S^{\text{clos}}_{A,m,n} = \frac{1}{2}\frac{D_{m,n}(Y + Y^{\sim A}) - D_{m,n}(Y - Y^{\sim A}) - \left\|\overline{Y} - \overline{Y^{\sim A}}\right\|_m^2}{D_{m,n}(Y) + D_{m,n}(Y^{\sim A}) + 1/2\left\|\overline{Y} - \overline{Y^{\sim A}}\right\|_m^2}.$$

The proof is decomposed into three steps:

- Step 1: We prove a vectorial CLT for

$$\mathbb{V}_n = \left(D_{m,n}(Y), D_{m,n}(Y^{\sim A}), D_{m,n}(Y + Y^{\sim A}), D_{m,n}(Y - Y^{\sim A})\right).$$

$\mathbb{V}_n$ can be decomposed as in Lemma 3.28 in the following way:

$$\mathbb{V}_n - \mathbb{E}\mathbb{V}_n = -U_n\mathbb{K} + P_n\mathbb{L}' + (P_n\mathbb{L} - P_n\mathbb{L}') - \mathbb{B}_m,$$

where

$$\begin{aligned} U_n\mathbb{K} &= \left(U_nK(Y), U_nK(Y^{\sim A}), U_nK(Y + Y^{\sim A}), U_nK(Y - Y^{\sim A})\right), \\ P_n\mathbb{L} &= \left(P_nL(Y), P_nL(Y^{\sim A}), P_nL(Y + Y^{\sim A}), P_nL(Y - Y^{\sim A})\right), \\ P_n\mathbb{L}' &= \left(P_nL'(Y), P_nL'(Y^{\sim A}), P_nL'(Y + Y^{\sim A}), P_nL'(Y - Y^{\sim A})\right), \\ \mathbb{B}_m &= \left(B_m(Y)B_m(Y^{\sim A}), B_m(Y + Y^{\sim A}), B_m(Y - Y^{\sim A})\right). \end{aligned}$$

By Lemma 3.28 2., it is enough to prove a CLT for $P_n\mathbb{L}'$. This CLT holds true since we consider an empirical sum of i.i.d. centered random vectors.

- Step 2: Using Lemma 3.30 and the Delta method, we derive a CLT for

$$\left(D_{m,n}(Y + Y^{\sim A}) - D_{m,n}(Y - Y^{\sim A}) - \left\|\overline{Y} - \overline{Y^{\sim A}}\right\|_m^2, \right. \\ \left. D_{m,n}(Y) + D_{m,n}(Y^{\sim A}) + 1/2\left\|\overline{Y} - \overline{Y^{\sim A}}\right\|_m^2\right).$$

- Step 3: We conclude using the Delta method (see, e.g., [488]). □

**Remark 18.** *We also refer the reader to Alexanderian et al. [3] for the study of efficient estimates for $S_A^{\text{clos}}$ and $S_A^{\text{T}} = 1 - S_{\overline{A}}^{\text{clos}}$ in the framework where the output is a time-dependent process.*

### Hessian-based bounds

Let $A \in \mathcal{P}_d$. Assume that the probability distribution of input vector $\mathbf{X}$ is $\mu = \mathcal{N}(m, \Sigma)$, where $\Sigma \in \mathbb{R}^{d\times d}$ is a nonsingular covariance matrix. We denote by

$$\mathcal{H} = L^2(\mathbb{R}^d, \mathcal{B}(\mathbb{R}^d), \mu; \mathbb{R}^k)$$

the Hilbert space which contains all the measurable functions $v : \mathbb{R}^d \to \mathbb{R}^k$ such that $\|v\| < \infty$. In this section, $\|\cdot\|$ is the norm associated with the scalar product $\langle\cdot,\cdot\rangle$ defined by

$$\langle u, v\rangle = \int \langle u(\mathbf{x}), v(\mathbf{x})\rangle_{\mathbb{R}^k}\, \mathrm{d}\mu(\mathbf{x})$$

for any $u, v \in \mathcal{H}$, with $\langle\cdot,\cdot\rangle_{\mathbb{R}^k}$ denoting the usual scalar product on $\mathbb{R}^k$. Let us assume that $G \in \mathcal{H}$ and that $G$ is continuously differentiable. Let $\nabla G(\mathbf{x}) \in \mathbb{R}^{k\times d}$ denote the Jacobian matrix of $G(\mathbf{x}) = (G_1(\mathbf{x}), \ldots, G_k(\mathbf{x}))$ at point $\mathbf{x}$ given by

$$\nabla G(\mathbf{x}) = \begin{pmatrix} \frac{\partial G_1}{\partial x_1}(\mathbf{x}) & \cdots & \frac{\partial G_1}{\partial x_d}(\mathbf{x}) \\ \vdots & \ddots & \vdots \\ \frac{\partial G_k}{\partial x_1}(\mathbf{x}) & \cdots & \frac{\partial G_k}{\partial x_d}(\mathbf{x}) \end{pmatrix}.$$

Let $H \in \mathbb{R}^{d\times d}$ be the matrix defined by

$$H = \int_{\mathbb{R}^d} (\nabla G(\mathbf{x}))^T (\nabla G(\mathbf{x})) \, \mathrm{d}\mu(\mathbf{x}).$$

Then Zahm et al. [512, Section 4] have proved the following bound:

$$S_A^{\text{clos}} \geq 1 - \frac{\sum_{j\notin A} \operatorname{Var}(X_j) H_{j,j}}{\|G - \mathbb{E}_\mu G\|^2} .$$

Let $S_A^{\text{T}} = 1 - S_{\overline{A}}^{\text{clos}}$. They have also proved

$$S_A^{\text{T}} \leq \frac{\sum_{j\in A} \operatorname{Var}(X_j) H_{j,j}}{\|G - \mathbb{E}_\mu G\|^2} . \tag{3.38}$$

The proof relies on a so-called subspace Poincaré inequality (see [512, Proposition 2.4]).

The fact that the DGSM can bound the total Sobol' index $S_A^{\text{T}}$ has already been noted for scalar-valued functions (see Section 3.1.4). In the scalar-valued case ($k = 1$) indeed, $H_{j,j} = \int (\partial_j G)^2 \, \mathrm{d}\mu$ coincides with the $j$th DGSM defined by Eq. (2.7) (page 41) in Chapter 2. Thus, choosing $A = \{j\}$, we recover inequality (3.12) (page 56) stated in Theorem 3.11.

Here, under the assumption of Gaussian probability measure $\mu$, inequality (3.38) provides a generalization of these bounds to the case of vector-valued functions, where the $i$th DGSM ought to be defined as $H_{i,i} = \int \|\partial_i f\|_V^2 \mathrm{d}\mu$. These results can be extended to more general probability distribution of the input vector $\mathbf{X}$ provided that one can prove a subspace Poincaré inequality.

Derivative-based bounds for $S_A^{\text{T}}$ were also derived in Cleaves et al. [83] for models with functional output in the framework where the probability distribution on the input space is a product probability distribution by extending the definition of DGSM (see Section 2.3.3) to functional outputs.

### 3.3.4 ▪ Examples in R

Ubiquitous and aggregated Sobol' indices are illustrated on the 1D-functional output Arctangent temporal function (see Section C.1 with independent $X_j \sim \mathcal{U}(-7, 7)$), which contains two random inputs. We first provide several realizations of this function (see Fig. 3.7) with the following command lines:

```
> set.seed(12345)
> library(sensitivity)
> n <- 100
> X <- matrix(c(runif(2 * n, -7, 7)), ncol = 2)
> y <- atantemp.fun(X)

> plot(y[1,], ylim = c(-2,2), type = "l", xlab = "t", ylab = "Y")
> apply(y, 1, lines)
```

We then apply on this model the sobolMultOut() function of the sensitivity package:

```
> d <- 2
> n <- 1000
> X1 <- data.frame(matrix(runif(d * n), nrow = n))
> X2 <- data.frame(matrix(runif(d * n), nrow = n))
> sa <- sobolMultOut(model = atantemp.fun, q = 100, X1, X2,
+                    MCmethod = "soboljansen", ubiquitous = TRUE)
> plotMultOut(sa)
```

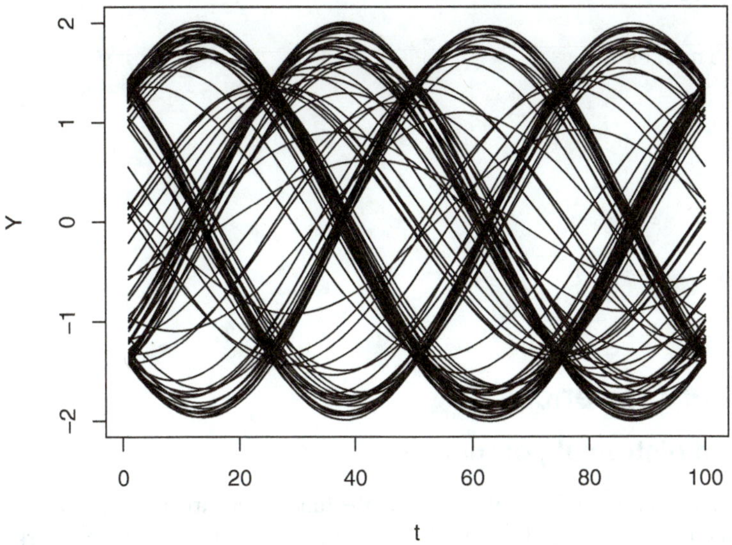


**Figure 3.7.** 100 *realizations of the Arctangent temporal function.*

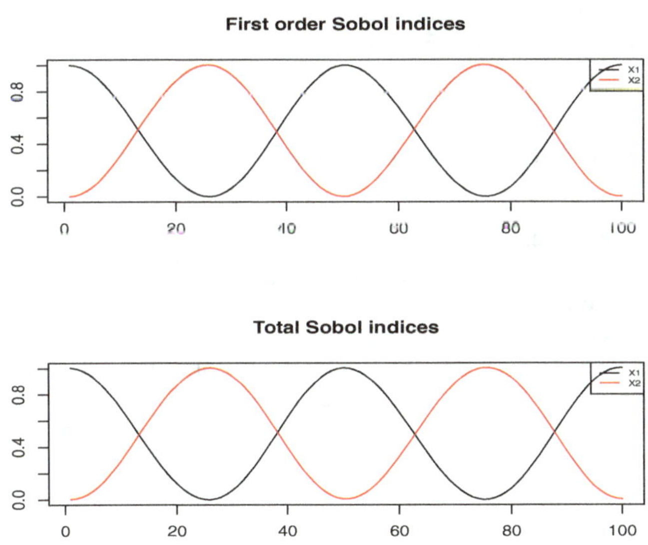


**Figure 3.8.** *Ubiquitous Sobol' indices for the Arctangent temporal function.*

Figure 3.8 is directly obtained from the plotMultOut() call. It shows the ubiquitous (i.e., all along the temporal discretization) first-order and total Sobol' indices of the two inputs. Finally, the following command provides the aggregated first-order and total Sobol' indices which are trivial in this case:

```
> print(sa)

Call:
soboljansen(model = NULL, X1 = X1, X2 = X2)
```

```
Model runs: 4000

First order indices:
    original
X1 0.5023336
X2 0.4936351

Total indices:
    original
X1 0.5063649
X2 0.4976664
```

## 3.4 ▪ Further references

### 3.4.1 ▪ A brief historical perspective

As pointed out and widely studied in this chapter, the functional variance decomposition method (FANOVA) is the keystone for global SA. The story begins with the pioneering works of Cukier et al. [95, 97, 96] followed by the trailblazing mathematical ones of Efron and Stein [122] and Antoniadis [7] and in the early 1990s the work of Sobol' [441, 442] and Owen [346, 344]. The techniques have then been highly popularized by the European Commission Joint Research Centre around Saltelli (see [414, 410] and earlier references therein). The variance decomposition method remains a very active scientific field at the crossroads of applied mathematics and engineering.

### 3.4.2 ▪ Other estimators and expansions

For the estimation issues, numerous pick-freeze estimators of Sobol' indices have been covered in this chapter. Complementary elements can be found in Saltelli et al. [412, 290], Glen and Isaacs [164], Prieur and Tarantola [373], Sun et al. [469], and Damblin and Ghione [104]. For the small sample cases, note that a new promising statistical estimation framework, based on rank statistics, has been proposed in Gamboa et al. [146]. In fact, this approach is closely related to the nearest-neighbor approach developed in Section 6.1 in Broto et al. [56] as an intermediate brick for the estimation of Shapley effects (see also Section 5.5.4 of Chapter 5 of the present book). In the sensitivity R package, the rank statistics and the nearest-neighbor approaches are available in the sobolshap_knn() function (see also the sobolrank() and shapleySubsetMc() functions).

### 3.4.3 ▪ Stochastic computer codes

As stated in Section 1.1 and by Eq. (1.2) of Chapter 1, only deterministic models have been considered in this chapter (and in the rest of the book). A deterministic model is characterized by the fact that the same set of input variables always gives the same output value. In comparison, for a stochastic computer code, the result of each run is random. The literature on global sensitivity analysis for stochastic models is more recent and still in expansion. Examples of stochastic models are, e.g., agent-based models, stochastic differential equations, or Monte Carlo–based numerical models.

In case we are only interested in the mean value relative to the intrinsic randomness of the model, a quite common procedure consists in replacing the mean quantity by the empirical mean and then performing usual global SA. Le Gratiet [263] and Iooss et al. [207] have also extended results on Sobol' indices of deterministic simulator to this noisy (or Monte Carlo) simulator case. If one is interested in the full probability distribution of the output, both papers [212] and [307] propose a strategy based on a joint modeling of the mean and dispersion of stochastic model outputs. From these joint models, it is possible to compute, for each uncertain parameter, the first-order Sobol' index and any Sobol' index associated to an interaction with a subset of uncertain parameters. Finally, the recent work of Fort et al. [137] proposes a general framework based on Wasserstein metrics to deal with stochastic computer code outputs.

It is also possible to compute a total Sobol' index which contains the part of the output variance explained by the intrinsic noise of the model, by itself or in interaction with the uncertain parameters. However, such an approach does not allow one to separate the effects enclosed in this total index. In Hart et al. [181] and Mazo [318], the authors propose another point of view. They define Sobol' indices as random variables and study their statistical properties. Azzi et al. [16] have proposed alternative sensitivity indices.

Global sensitivity analysis for parametrized stochastic differential equations has been considered in Le Maître and Knio [267] and Jimenez et al. [222] and more recently in Etoré et al. [125]. Finally, note that this stochastic code topic remains an active field of research [18], especially for metamodel fitting issues (see, e.g., [6, 334, 59, 39, 15, 515]).

## Chapter 4

# Spectral and metamodel-based estimation

In Chapter 3, we studied estimation techniques of the Sobol' indices based on sampling methods. In the present one we develop a quite different point of view. Roughly speaking, the method studied here relies on some special regression models whose Sobol' indices either have a very special form or are cheap to simulate. On the one hand, the archetype of the regression model in which the Sobol' indices have very simple forms is the linear regression model with orthonormal regressors. On the other hand, the Gaussian regression is a well-tailored Bayesian method having both good approximation properties and a low computation cost.

Section 4.1 presents the so-called spectral approaches, which are based on Fourier expansion and thus require regularity assumptions on the model. However, these methods are no longer applicable when the required number of model evaluations is too large with respect to the cost of the computer model in terms of CPU time. In such a case, Section 4.2 explains how to use a mathematical approximation of the computer model, also called a metamodel, in order to estimate the Sobol' indices and control the bias induced by the approximation.

## 4.1 ▪ Spectral approaches

In this section, we focus on spectral approaches for the estimation of Sobol' indices. In case of a model $G$ with some regularity properties, one might opt for spectral rather than Monte Carlo approaches for estimating Sobol' indices. Such approaches can indeed reach convergence rates of order $n^{-1}$ with $n$ being the size of the experimental design. We first introduce the setting for spectral estimation of Sobol' indices. We then discuss two classical approaches, namely, the *Fourier amplitude sensitivity test* (FAST) and *random balance design* (RBD), and we relate these estimation procedures to the more general context. Finally, a bias correction for RBD is provided.

### 4.1.1 ▪ Notation and definitions

For this purpose, consider, for all $j = 1, \ldots, d$, an orthonormal Hilbert basis of $\mathbb{L}^2_{\mathbb{R}}(\mathrm{P}_{X_j})$, denoted by $(\Phi_{j,k}\,,\, k \in \mathbb{Z})$ such that $\Phi_{j,0} = 1$.

**Remark 19.** *Given that* $\mathbb{L}^2_{\mathbb{R}}(\mathrm{P}_{X_j}) \subset \mathbb{L}^2_{\mathbb{C}}(\mathrm{P}_{X_j})$, *one can also replace the basis* $(\Phi_{j,k}, k \in \mathbb{Z})$ *by a Hilbert basis of* $\mathbb{L}^2_{\mathbb{C}}(\mathrm{P}_{X_j})$. *It has a practical interest when considering, e.g., the Fourier basis.*

Consider the tensorized basis of $\mathbb{L}^2_{\mathbb{K}}(\mathrm{P}_{\mathbf{X}})$, $\mathbb{K} = \mathbb{R}$ or $\mathbb{C}$, whose elements are indexed by $\mathbf{k} = (k_1, \ldots, k_d) \in \mathbb{Z}^d$ and defined as the product $\Phi_{1,k_1}(\cdot) \times \cdots \times \Phi_{d,k_d}(\cdot)$. Then $G(\mathbf{X})$ can be decomposed as follows:

$$Y = G(\mathbf{X}) = \sum_{\mathbf{k} \in \mathbb{Z}^d} c_{\mathbf{k}}(G) \Phi_{1,k_1}(X_1) \ldots \Phi_{d,k_d}(X_d).$$

Now define, for $A \in \mathcal{P}_d$,

$$\left(\mathbb{Z}^d\right)_{\neq 0} = \{\mathbf{k} \in \mathbb{Z}^d \,|\, \mathbf{k} \neq (0, \ldots, 0)\},$$

$$\mathbb{Z}_{A,\neq 0} = \{\mathbf{k} \in \mathbb{Z}^d \,|\, \forall j \in A,\ k_j \in \mathbb{Z}_{\neq 0} \forall j \notin A,\ k_j = 0\}.$$

Thanks to the orthonormality of the bases $(\Phi_{j,k}\ ,\ k \in \mathbb{Z})$, it is possible to identify the components in the FANOVA decomposition as

$$G_A(\mathbf{X}_A) = \sum_{\mathbf{k} \in \mathbb{Z}_{A,\neq 0}} c_{\mathbf{k}}(G) \Phi_{1,k_1}(X_1) \ldots \Phi_{d,k_d}(X_d). \tag{4.1}$$

**Proposition 4.1.** *Assume that $G \in \mathbb{L}^2_{\mathbb{K}}(\mathrm{P}_{\mathbf{X}})$ and that $\operatorname{Var} G(\mathbf{X}) \neq 0$. Then, for any nonempty subset $A \in \mathcal{P}_d$,*

$$S_A = \frac{\sum_{\mathbf{k} \in \mathbb{Z}_{A,\neq 0}} |c_{\mathbf{k}}(G)|^2}{\operatorname{Var} G(\mathbf{X})}. \tag{4.2}$$

***Proof.*** Recall that we endow the space $\mathbb{L}^2_{\mathbb{K}}(\mathrm{P}_{\mathbf{X}})$ with the scalar product associated to $\mathrm{P}_{\mathbf{X}}$: $\langle f, g\rangle = \int f(\mathbf{x})\overline{g(\mathbf{x})}\mathrm{P}_{\mathbf{X}}(d\mathbf{x})$. Equality (4.2) derives from a straightforward consequence of the identification equality (4.1) and of Parseval's identity. □

Recall that, for all $j = 1, \ldots, d$, $I_j$ is the support of the probability distribution of $X_j$. For $n \in \mathbb{N}_*$, let $D = \{\mathbf{x}_1, \ldots, \mathbf{x}_n\}$ be any finite subset of $n$ points in $I_1 \times \cdots \times I_d$. Let $\mathbf{p} = (p_1, \ldots, p_n) \in (\mathbb{R}_+)^n$ such that $\sum_{i=1}^n p_i = 1$. In the following, integrals in Eq. (4.2) are approximated by quadrature formulas associated to the set of nodes $D$ and to the weights $\mathbf{p}$. $c_{\mathbf{k}}(G) = \langle G, \Phi_{\mathbf{k}}\rangle$ can be approximated by

$$\widehat{c}_{\mathbf{k}}(G, D, \mathbf{p}) = \sum_{i=1}^{n} G(\mathbf{x}_i)\overline{\Phi_{\mathbf{k}}(\mathbf{x}_i)} p_i$$

and $\operatorname{Var} G(\mathbf{X})$ by

$$\widehat{V}(D, \mathbf{p}) = \sum_{i=1}^{n} \left( G(\mathbf{x}_i) - \sum_{j=1}^{n} G(\mathbf{x}_j) p_j \right)^2 p_i.$$

Let now $K_A$ be any finite subset of $\mathbb{Z}_{A,\neq 0}$, and define the following intuitive estimator for $S_A$:

$$\widehat{S}_A(K_A, D, \mathbf{p}) = \frac{\sum_{\mathbf{k} \in K_A} \widehat{c}_{\mathbf{k}}(G, D, \mathbf{p})}{\widehat{V}(D, \mathbf{p})}.$$

**Remark 20.** *As the infinite set $\mathbb{Z}_{A,\neq 0}$ is truncated, an important requirement under this way of estimating Sobol' indices is the fast decrease of the coefficients $c_{\mathbf{k}}(G)$. This is why spectral approaches are particularly well suited for regular models $G$, outperforming Monte Carlo–type approaches in this case. However, the choice of the truncation set $K_A$ is very important. Applying a too drastic truncation may lead to nonconvergent estimates.*

To end this section, we refer the reader also to Fort et al. [138], where a spectral method is studied in the framework of a general functional linear model.

### 4.1.2 ▪ About FAST and RBD

In this section, two classical approaches for SA, FAST and RBD, are first presented. The connection is then made with the previous section by proving that both these approaches are particular cases of spectral approaches.

#### Review of FAST

FAST is essentially an application of the following result due to Weyl [501] (see also Weyl's ergodic theorem in Weyl [500] (in German) or in Sinai [437]).

**Theorem 4.2 (Weyl).** *Let $g$ be a bounded Riemann integrable function on $[0,1]^d$ and, for all $i = 1, \ldots, d$, $t \in \mathbb{R}$, $x_i(t) = \{\omega_i t\}$, where the $\omega_i$'s are real numbers linearly independent over $\mathbb{Q}$ and $\{\cdot\}$ denotes the fractional part. Then*

$$\int_{[0,1]^d} g(\mathbf{x})d\mathbf{x} = \lim_{T\to\infty} \frac{1}{2T} \int_{-T}^{T} g\big(x_1(t), \ldots, x_d(t)\big)dt. \tag{4.3}$$

In particular, for any $\mathbf{k} \in \mathbb{Z}^d$ and $g : \mathbf{x} \mapsto G(\mathbf{x})\exp\big(-2i\pi\mathbf{k}\cdot\mathbf{x}\big)$, with $\mathbf{k}\cdot\mathbf{x}$ denoting the canonical scalar product on $\mathbb{R}^d$ (in order to alleviate the notation), (4.3) reads

$$\mathrm{c}_{\mathbf{k}}(G) = \lim_{T\to\infty} \frac{1}{2T} \int_{-T}^{T} G \circ \mathbf{x}(t)\exp\big(-2i\pi(\mathbf{k}\cdot\boldsymbol{\omega})t\big)dt. \tag{4.4}$$

Then FAST replaces $x_i(t) = \{\omega_i t\}$ with semiparametric functions $x_i(t) = H_i\big(\sin(\omega_i t)\big)$ [95], where the $\omega_i$'s are positive integers and the transformations $H_i$ are chosen to preserve the marginal distributions of the $X_i$'s. If the latter are uniformly distributed, it can be shown [96, 418] that $H_i(\cdot) = \frac{1}{\pi}\arcsin(\cdot) + \frac{1}{2}$. Saltelli et al. [418] also propose adding a random phase-shift $\varphi_i \in [0, 2\pi)$, which gives the semiparametric functions $x_i^*(t) = \frac{1}{\pi}\arcsin\big(\sin(2\pi\omega_i t + \varphi_i)\big) + \frac{1}{2}$. Hence, replacing $\mathbf{x}$ with $\mathbf{x}^*$ in (4.4) gives

$$\mathrm{c}_{\mathbf{k}}(G) \approx \lim_{T\to\infty} \frac{1}{2T} \int_{-T}^{T} G \circ \mathbf{x}^*(t)\exp\big(-2i\pi(\mathbf{k}\cdot\boldsymbol{\omega})t\big)dt.$$

Thus, since the functions $x_i^*$ are 1-periodic, we get

$$\mathrm{c}_{\mathbf{k}}(G) \approx \int_{0}^{1} G \circ \mathbf{x}^*(t)\exp\big(-2i\pi(\mathbf{k}\cdot\boldsymbol{\omega})t\big)dt,$$

and applying the rectangle rule to the right-hand side integral gives

$$\mathrm{c}_{\mathbf{k}}(G) \approx \widehat{\mathrm{c}}_{\mathbf{k}\cdot\boldsymbol{\omega}}(G \circ \mathbf{x}^*), \tag{4.5}$$

where

$$\widehat{\mathrm{c}}_{\mathbf{k}\cdot\boldsymbol{\omega}}(G \circ \mathbf{x}^*) = \frac{1}{n}\sum_{j=0}^{n-1} G \circ \mathbf{x}^*\Big(\frac{j}{n}\Big)\exp\Big(-2i\pi j\frac{\mathbf{k}\cdot\boldsymbol{\omega}}{n}\Big)$$

is the complex discrete Fourier coefficient of the 1D function $G \circ \mathbf{x}^*$. In the following, the dependence on $n$, $\boldsymbol{\omega}$, and $\varphi$ is generally omitted for convenience.

### Estimation

The estimators of $\mathrm{V}_A(G)$, $\mathrm{V}(G)$, and consequently of $S_A(G)$ were introduced by using the approximation in (4.5); see [95] and Appendix C in Cukier et al. [96]. On the one hand, for any nonempty subset $A \in \mathcal{P}_d$ and any finite subset $K_A \subset \mathbb{Z}_{A,\neq 0}$, Eq. (4.5) leads to the definition of the estimator of $\mathrm{V}_A(G)$:

$$\widehat{\mathrm{V}}_A^{\mathrm{FAST}}(G, K_A, \mathbf{x}^*) = \sum_{\mathbf{k} \in K_A} \left|\widehat{\mathrm{c}}_{\mathbf{k}\cdot\boldsymbol{\omega}}(G \circ \mathbf{x}^*)\right|^2. \tag{4.6}$$

On the other hand, (4.5) gives

$$\begin{aligned} \mathrm{V}(G) &= \mathrm{c}_0(G^2) - \mathrm{c}_0(G)^2 \\ &\approx \widehat{\mathrm{c}}_0(G^2 \circ \mathbf{x}^*) - \widehat{\mathrm{c}}_0(G \circ \mathbf{x}^*)^2, \end{aligned}$$

and Parseval's identity leads to the definition of the estimator of $\mathrm{V}(G)$:

$$\widehat{\mathrm{V}}^{\mathrm{FAST}}(G, \mathbf{x}^*) = \sum_{k=1}^{n-1} \left|\widehat{\mathrm{c}}_k(G \circ \mathbf{x}^*)\right|^2.$$

This naturally leads to the estimator of the variance-based sensitivity indices $S_A(G)$:

$$\widehat{S}_A^{\mathrm{FAST}}(G, K_A, \mathbf{x}^*) = \frac{\sum_{\mathbf{k} \in K_A} \left|\widehat{\mathrm{c}}_{\mathbf{k}\cdot\boldsymbol{\omega}}(G \circ \mathbf{x}^*)\right|^2}{\sum_{k=1}^{n-1} \left|\widehat{\mathrm{c}}_k(G \circ \mathbf{x}^*)\right|^2}.$$

Note that by Parseval's identity, the estimator $\widehat{\mathrm{V}}^{\mathrm{FAST}}(G, \mathbf{x}^*)$ is equal to the empirical variance $\widehat{\mathrm{V}}(G, \{\mathbf{x}^*(\frac{j}{n})\}_{j=0,\ldots,n-1})$.

**Remark 21.** *The "extended-FAST" method [418] for computing first-order and total Sobol' indices is integrated into the* sensitivity *package of* R *(*fast99() *function).*

### Choice of parameters $\omega$ and $n$

As discussed by Schaibly and Shuler [425] and Cukier et al. [97], $\omega$ and $n$ should be correctly chosen so as to minimize the cubature error in the approximation in (4.5). Let us now consider that $K_A = \mathbb{Z}_{A,\neq 0}(M) = \mathbb{Z}_{A,\neq 0} \cap (-M, M]^d$ for some positive integer $M$. In order to avoid interferences, i.e., $\mathbf{k}\cdot\boldsymbol{\omega} - \mathbf{k}'\cdot\boldsymbol{\omega} = 0$ for $\mathbf{k}, \mathbf{k}' \in \mathbb{Z}^d$, $\mathbf{k} \neq \mathbf{k}'$, and aliasing, i.e., $\mathbf{k}\cdot\boldsymbol{\omega} - \mathbf{k}'\cdot\boldsymbol{\omega} = jn$ for $\mathbf{k}, \mathbf{k}' \in \mathbb{Z}^d$, $\mathbf{k} \neq \mathbf{k}'$, and $j \in \mathbb{Z}_{\neq 0}$ (which both lead to $\widehat{\mathrm{c}}_{\mathbf{k}\cdot\boldsymbol{\omega}}(G \circ \mathbf{x}^*) = \widehat{\mathrm{c}}_{\mathbf{k}'\cdot\boldsymbol{\omega}}(G \circ \mathbf{x}^*)$), Schaibly and Shuler [425] propose choosing $\omega_1, \ldots, \omega_d$ free of interferences up to order $M$,

$$\begin{gathered} (\mathbf{k} - \mathbf{k}') \cdot \boldsymbol{\omega} \neq 0 \\ \forall \mathbf{k}, \mathbf{k}' \in \mathbb{Z}^d, \ \mathbf{k} \neq \mathbf{k}', \text{ s.t. } \sum_{i=1}^{d} |k_i - k_i'| \leq M + 1, \end{gathered} \tag{4.7}$$

and $n$ sufficiently large,

$$n \approx M \max(\omega_1, \ldots, \omega_d). \tag{4.8}$$

More recently, referring to the classical information theory, Saltelli et al. [418] (see also [301]) suggest replacing (4.8) with the Nyquist–Shannon sampling theorem:

$$n > 2M \max(\omega_1, \ldots, \omega_d). \tag{4.9}$$

The criterion stated in Eq. (4.7) should probably be replaced by

$$\begin{gathered}(\mathbf{k}-\mathbf{k}')\cdot\boldsymbol{\omega}\neq 0\\ \forall\mathbf{k},\ \mathbf{k}'\in\mathbb{Z}^d,\ \mathbf{k}\neq\mathbf{k}',\ \text{s.t.}\ \sum_{i=1}^{d}|k_i|\leq M\ \text{and}\ \sum_{i=1}^{d}|k'_i|\leq M\end{gathered}\tag{4.10}$$

since the main objective is to avoid interferences within a finite subset of $\mathbb{Z}^d$ out of which the Fourier coefficients of $G$ are likely to be negligible. In Eq. (4.10), this subset is the closed $l^1$-norm ball of radius $M$. Thus we may reformulate the two criteria stated in (4.7) and (4.9) with respect to the set $K = \sqcup_A K_A$, where the $K_A$'s are the truncation sets in the FAST estimator of $\mathrm{V}_A(G)$ given in (4.6). We propose choosing $\omega_1, \ldots, \omega_d$ free of interferences within $K$, i.e.,

$$\begin{gathered}(\mathbf{k}-\mathbf{k}')\cdot\boldsymbol{\omega}\neq 0\\ \forall\mathbf{k},\ \mathbf{k}'\in K,\ \mathbf{k}\neq\mathbf{k}'\quad\text{and}\quad n>\max_{\mathbf{k},\mathbf{k}'\in K}\left((\mathbf{k}-\mathbf{k}')\cdot\boldsymbol{\omega}\right).\end{gathered}\tag{4.11}$$

### Review of RBD

RBD makes use of the previous framework setting $\boldsymbol{\varphi} = \mathbf{0}$, $\omega_1 = \cdots = \omega_d = \omega \in \mathbb{N}^*$ (usually set to 1) and applies random permutations on the coordinates of the resulting points $\mathbf{x}^*(\frac{j}{n})$. More precisely, let $\sigma_1, \ldots, \sigma_d$ be random permutations on $[0 : n-1]$ and let $\mathfrak{S}$ denote the set of all possible $\boldsymbol{\sigma} = (\sigma_1, \ldots, \sigma_d)$. Given $\boldsymbol{\sigma} \in \mathfrak{S}$, consider the function $\mathbf{x}^\times = (x_1^\times, \ldots, x_d^\times)$ defined on $\{0, \frac{1}{n}, \ldots, \frac{n-1}{n}\}$ such that, for all $i \in [1 : d]$ and $j \in [0 : n-1]$,

$$x_i^\times\left(\frac{j}{n}\right) = \frac{1}{\pi}\arcsin\left(\sin\left(2\pi\omega\frac{\sigma_i(j)}{n}\right)\right) + \frac{1}{2}.$$

Thus, denoting by $\sigma_i^{-1}$ the inverse permutation of $\sigma_i$, define

$$\mathbf{x}^{\times,i}\left(\frac{j}{n}\right) = \mathbf{x}^\times\left(\frac{\sigma_i^{-1}(j)}{n}\right).$$

Finally, through a heuristic argument, Tarantola et al. [476] introduce the RBD estimators of $\mathrm{V}_A(G)$, $\mathrm{V}(G)$, and $S_A(G)$ for first-order terms (i.e., $A = \{i\}$, $i \in [1 : d]$). For any finite subset $K_{\{i\}} \subset \mathbb{Z}_{\{i\},\neq 0}$, we have

$$\widehat{\mathrm{V}}_i^{\mathrm{RBD}}(G, K_{\{i\}}, \mathbf{x}^\times) = \sum_{\mathbf{k}\in K_{\{i\}}} \left|\widehat{\mathrm{c}}_{k_i\omega}(G\circ\mathbf{x}^{\times,i})\right|^2,$$

$$\widehat{\mathrm{V}}^{\mathrm{RBD}}(G, \mathbf{x}^\times) = \sum_{k=1}^{n-1} \left|\widehat{\mathrm{c}}_k(G\circ\mathbf{x}^\times)\right|^2,$$

and

$$\widehat{S}_i^{\mathrm{RBD}}(G, K_{\{i\}}, \mathbf{x}^\times) = \frac{\sum_{\mathbf{k}\in K_{\{i\}}} \left|\widehat{\mathrm{c}}_{k_i\omega}(G\circ\mathbf{x}^{\times,i})\right|^2}{\sum_{k=1}^{n-1} \left|\widehat{\mathrm{c}}_k(G\circ\mathbf{x}^\times)\right|^2}.$$

As in FAST, note that by Parseval's identity the estimator $\widehat{\mathrm{V}}^{\mathrm{RBD}}(G, \mathbf{x}^\times)$ is equal to the empirical variance $\widehat{\mathrm{V}}(G, \{\mathbf{x}^\times(\frac{j}{n})\}_{j=0,\ldots,n-1})$. The parameter $\omega$ is usually set equal to 1. Criterion Eq. (4.11) can be applied to choose $n$.

**FAST and RBD revisited: The connection with spectral approaches**

Consider the particular case where, for all $j \in [1:d]$, $X_j$ is uniformly distributed on $[0,1]$. A possible choice for the basis is then that of Fourier: $\mathbf{x} \mapsto e^{i\mathbf{k}.\mathbf{x}}$, $\mathbf{k} \in \mathbb{Z}^d$, with $\mathbf{k}.\mathbf{x} = \sum_{j=1}^d k_j x_j$. For any $p \in \mathbb{Z}_{>0}$, let

$$\left\{ \begin{array}{llll} r_p: & [0,1] & \longrightarrow & [0,1], \\ & x & \longmapsto & \left\{ \begin{array}{ll} 2\{px\} & \text{if } 0 \le \{px\} < \frac{1}{2}, \\ 2-2\{px\} & \text{if } \frac{1}{2} \le \{px\} \le 1, \end{array} \right. \end{array} \right.$$

and, for any $\varphi \in [0, 2\pi)$, let

$$\left\{ \begin{array}{llll} t_\varphi: & [0,1] & \longrightarrow & [0,1], \\ & x & \longmapsto & \{x+\tilde{\varphi}\} \ \text{with } \tilde{\varphi} = \frac{1}{4} + \frac{\varphi}{2\pi}\,, \end{array} \right.$$

with, for any real number $z$, $\{z\} = z - \lfloor z \rfloor$ being the fractional part of $z \in \mathbb{R}$ and $\lfloor \cdot \rfloor$ the floor function.

Then define the linear operators $\mathcal{R}_p$ and $\mathcal{T}_\varphi$ on $L^2_{\mathbb{C}}([0,1]^d)$ (see Fig. 4.1 below) such that, for all $\mathbf{x} \in [0,1]^d$,

$$\mathcal{R}_p G(\mathbf{x}) = G\big(r_p(x_1), \dots, r_p(x_d)\big) \ \text{ and } \ \mathcal{T}_\varphi G(\mathbf{x}) = G\big(t_{\varphi_1}(x_1), \dots, t_{\varphi_d}(x_d)\big),$$

with $\mathcal{R}_p = \underbrace{\mathcal{R}_1 \circ \cdots \circ \mathcal{R}_1}_{p \text{ times}}$.

**Lemma 4.3 (Lemma 1 in Tissot and Prieur [483]).** *For any $p \in \mathbb{N}^*$ and any $\varphi \in [0,2\pi)^d$, the variance decomposition stated in Corollary 3.5 is $\mathcal{R}_p$- and $\mathcal{T}_\varphi$-invariant on $L^2_{\mathbb{C}}([0,1]^d)$.*

***Proof.*** Let $X_1, \dots, X_d$ be $d$ independent random variables uniformly distributed on $[0,1]$, and denote by $G_A(X_i, i \in A)$, $A \in \mathcal{P}_d$, the Hoeffding decomposition of $G(\mathbf{X})$. We first prove the result for the linear operator $\mathcal{R}_1$.

Let $s$ be a positive integer and $\mathcal{Q}^s$ the set of the subset Q of $[0,1)^s$ of the form $\mathrm{Q} = [q_1, q_1 + \frac{1}{2}) \times \cdots \times [q_s, q_s + \frac{1}{2})$, where $q_i \in \{0, \frac{1}{2}\}$. Note that since the Lebesgue measure is isometry-invariant, we have, for any $\mathrm{Q} \in \mathcal{Q}^s$ and any function $g \in L^2([0,1]^s)$,

$$\int_{\mathrm{Q}} \mathcal{R}_1 g(\mathbf{x}) d\mathbf{x} = \int_{[0,\frac{1}{2})^s} \mathcal{R}_1 g(\mathbf{x}) d\mathbf{x}\,.$$

Thus it becomes

$$\begin{aligned} \int_{[0,1)^s} \mathcal{R}_1 g(\mathbf{x}) d\mathbf{x} &= \sum_{\mathrm{Q} \in \mathcal{Q}^s} \int_{\mathrm{Q}} \mathcal{R}_1 g(\mathbf{x}) d\mathbf{x} \\ &= 2^s \int_{[0,\frac{1}{2})^s} \mathcal{R}_1 g(\mathbf{x}) d\mathbf{x}, \end{aligned}$$

and the definition of $\mathcal{R}_1$ gives us

$$\int_{[0,1)^s} \mathcal{R}_1 g(\mathbf{x}) d\mathbf{x} = \int_{[0,1)^s} g(\mathbf{x}) d\mathbf{x}\,. \tag{4.12}$$

Then, noting that for all $\mathbf{x} \in [0,1)^d$, $(\mathcal{R}_1 g(\mathbf{x}))^2 = \mathcal{R}_1 g^2(\mathbf{x})$, we deduce that for every set $A \in \mathcal{P}_d$,

$$\mathrm{Var}\big[\mathcal{R}_1 G_A(X_i, i \in A)\big] = \mathrm{Var}\big[G_A(X_i, i \in A)\big]. \tag{4.13}$$

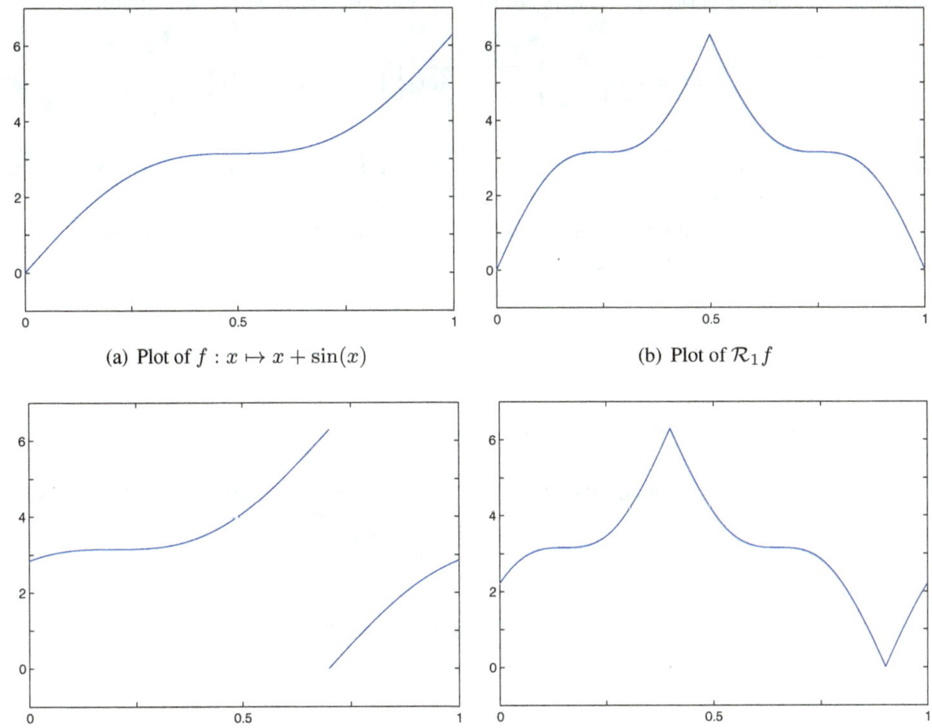


(a) Plot of $f : x \mapsto x + \sin(x)$ (b) Plot of $\mathcal{R}_1 f$

(c) Plot of $\mathcal{T}_{\pi/10} f$ (d) Plot of $(\mathcal{T}_{\pi/30} \circ \mathcal{R}_1) f$

**Figure 4.1.** *Examples of operators $\mathcal{R}_p$ and $\mathcal{T}_\varphi$ in dimension* 1.

We also deduce from (4.12) that, for every set $A \in \mathcal{P}_d$,

$$\forall \beta \subsetneq A, \;\; \mathbb{E}\big[\mathcal{R}_1 G_A(X_i, i \in A) | X_i, i \in \beta\big] = \mathbb{E}\big[G_A(X_i, i \in A) | X_i, i \in \beta\big] \,,$$

and then, by the uniqueness and the orthogonality properties of the Hoeffding decomposition (see Theorem 3.3),

$$\forall A \in \mathcal{P}_d, \;\; (\mathcal{R}_1 f)_A = \mathcal{R}_1 f_A \,. \tag{4.14}$$

Finally (4.13) and (4.14) lead to the conclusion of Lemma 4.3 for the linear operator $\mathcal{R}_1$. The proof of Lemma 4.3 for any $\mathcal{R}_p$ with $p \in \mathbb{N}^*$ and for the $\mathcal{T}_\varphi$'s is exactly the same as the previous one. It only suffices to prove that the property in (4.12) holds for any $\mathcal{R}_p$ and $\mathcal{T}_\varphi$. This property for the $\mathcal{T}_\varphi$'s is a consequence of the translation-invariance of the Lebesgue measure and is omitted here. For the $\mathcal{R}_p$'s, note that, for all $x \in [0,1]$, $r_p(x) = r_1(\{px\})$ and deduce that for all $\mathbf{x} \in [0,1]^s$, $\mathcal{R}_p g(\mathbf{x}) = \mathcal{R}_1 g(\{px_1\}, \ldots, \{px_s\})$. Hence, noting that $\mathcal{R}_p g$ is $\frac{1}{p}$-periodic in each direction, we obtain

$$\begin{aligned}
\int_{[0,1)^s} \mathcal{R}_p g(\mathbf{x}) d\mathbf{x} &= p^s \int_{[0,\frac{1}{p})^s} \mathcal{R}_p g(\mathbf{x}) d\mathbf{x} \\
&= p^s \int_{[0,\frac{1}{p})^s} \mathcal{R}_1 g(px_1, \ldots, px_s) d\mathbf{x} \\
&= \int_{[0,1)^s} g(\mathbf{x}) d\mathbf{x}. \quad \square
\end{aligned}$$

Two classical designs of experiments are introduced. For any $\boldsymbol{\omega} \in (\mathbb{Z}_{>0})^d$, denote by

$$\mathcal{G}(\boldsymbol{\omega}) = \left\{ \left( \left\{ \frac{i}{n}\omega_1 \right\}, \dots, \left\{ \frac{i}{n}\omega_d \right\} \right),\ i \in [0, n-1] \right\}$$

the cyclic subgroup (of order $n/\gcd(\omega_1, \dots, \omega_d, n)$ with $\gcd(\cdot)$ giving the greatest common factor of its elements) of the torus $\mathbb{T}^d = (\mathbb{R}/\mathbb{Z})^d \simeq [0,1)^d$ generated by $(\{\frac{\omega_1}{n}\}, \dots, \{\frac{\omega_d}{n}\})$.

Let $\sigma_1, \dots, \sigma_d$ be random permutations on $[0, n-1]$, and let $\mathfrak{S}$ denote the set of all possible $\boldsymbol{\sigma} = (\sigma_1, \dots, \sigma_d)$. For any $\boldsymbol{\sigma} \in \mathfrak{S}$ denote that

$$A(\boldsymbol{\sigma}) = \left\{ \left( \frac{\sigma_1(i)}{n}, \dots, \frac{\sigma_d(i)}{n} \right),\ i \in [0, n-1] \right\}$$

is the orthogonal array of strength 1 and index unity with elements taken from $\{0, \frac{1}{n}, \dots, \frac{n-1}{n}\}$ and based on the permutation $\sigma$.

Let $\mathbf{p}_0$ be the set of weights defined by $\mathbf{p}_0 = (\underbrace{n^{-1}, \dots, n^{-1}}_{n \text{ times}})$. FAST and RBD methods for estimating Sobol' indices are then defined by Definition 4.4 below.

**Definition 4.4.** *For any nonempty subset $A \in \mathcal{P}_d$ and any finite subset $K_A \subset \mathbb{Z}_{A,\neq 0}$, for any $\varphi \in [0, 2\pi)^d$ and any $\boldsymbol{\omega} \in (\mathbb{Z}_{>0})^d$, one has*

$$\widehat{\mathrm{S}}_A^{\mathrm{FAST}}(G, K_A, \mathbf{x}^*) = \widehat{\mathrm{S}}_A\big((\mathcal{T}_\varphi \circ \mathcal{R}_1)G, K_A, \mathcal{G}(\boldsymbol{\omega}), \mathbf{p}_0\big).$$

*For any nonempty subset $A \in \mathcal{P}_d$ and any finite subset $K_A \subset \mathbb{Z}_{A,\neq 0}$, for any $\boldsymbol{\sigma} \in \mathfrak{S}$ and any $\omega \in \mathbb{Z}_{>0}$, one has*

$$\widehat{\mathrm{S}}_A^{\mathrm{RBD}}(G, K_A, \mathbf{x}^\times) = \widehat{\mathrm{S}}_A\big((\mathcal{T}_{\tilde{\boldsymbol{\omega}}} \circ \mathcal{R}_\omega)G, \omega K_A, A(\boldsymbol{\sigma}), \mathbf{p}_0\big),$$

*where $\tilde{\boldsymbol{\omega}} = \left(\frac{(1-\omega)\pi}{2\omega}, \dots, \frac{(1-\omega)\pi}{2\omega}\right)$ and $\omega K_A = \{(\omega k_1, \dots, \omega k_d),\ \mathbf{k} \in K_A\}$.*

**Remark 22.**

- *Note that the linear operator $\mathcal{R}_p$ regularizes the function G in the sense that the coefficients $c_{\mathbf{k}}(\mathcal{R}_p G)$ decrease faster to zero than the coefficients $c_{\mathbf{k}}(G)$ as $\sum_{j=1}^d |k_j|$ tends to infinity. The other operator $\mathcal{T}_\varphi$ essentially allows to define randomized estimators in FAST.*

- *FAST was introduced in Cukier et al. [95, 97, 96] and RBD for first-order Sobol' indices in Tarantola et al. [476] and for higher-order Sobol' indices in Xu and Gertner [509]. Both methods were recently revisited in Tissot and Prieur [483], leading to Definition 4.4. Under regularity assumptions, the estimators in RBD are asymptotically unbiased.*

- *A variant of RBD called quasi-random balance design (QRBD) has recently been proposed to increase the precision and accuracy of first-order indices. The method differs from RBD in the way the sample is created. QRBD uses multidimensional quasi-random sequences to construct the permutations that are used to create the realizations of the inputs. The interested reader is referred to Tarantola et al. [477] for more details.*

#### A spectral given data method: EASI

The *effective algorithm for computing sensitivity indices* (EASI) was proposed in Plischke [370], where first-order sensitivity indices are estimated from a set of precomputed model evaluations $(\mathbf{X}^{(i)}, G(\mathbf{X})^{(i)}), i = 1, \ldots, N$, using a fast Fourier transform. The article also discusses ideas for the estimation of higher-order sensitivity indices. The sampled values of the $j$th input $X_j^{(i)}, i = 1, \ldots, N$, are increasingly ordered. Then the sampled values that correspond to all odd indices $i$ are increasingly ordered followed by all even indices in decreasing order. With this sorting and shuffling procedure, one obtains a reordered input that is triangular-shaped and can be seen as a periodic curve with frequency 1. By permuting the output values according to this rule, one looks out for resonances of period 1 and its higher harmonics in the power spectrum of the permuted output using the standard procedure implemented in both FAST and RBD. By repeating the sorting and shuffling of $X_j$ for all other $j$ on the same set of $N$ sampled values, one obtains all the other first-order sensitivity indices. Estimates produced by EASI are biased with respect to the maximal number of harmonics $M$. A bias correction formula, derived in Kelley [232], is proposed in Plischke [370].

### 4.1.3 ▪ Bias in RBD

As previously mentioned, the RBD approach is known to be biased (even if asymptotically unbiased). We therefore provide below some results on the well-known issue related to the bias of the estimates in RBD.

#### Preliminaries

We begin with the definitions of an orthogonal array and the "coincidence defect."

**Definition 4.5.** *An orthogonal array in dimension d, with q levels, strength $t \leq d$, and index $\lambda$, is a matrix with $n = \lambda q^t$ rows and d columns such that in every $n \times t$ submatrix, each of the $q^t$ possible rows (i.e., the distinct t-uples $(l_1, \ldots, l_t)$ where the $l_i$'s take their values in the set of the q levels) occurs exactly the same number $\lambda$ of times.*

**Definition 4.6.** *Let A be an orthogonal array in dimension d, with q levels, strength t, and index $\lambda$. We say that A has the coincidence defect when there exist two rows of A that do agree in $t+1$ columns; otherwise we say that A is defect-free.*

Let $\Pi(q)$ be the set of permutations on $\{0, \frac{1}{q}, \ldots, \frac{q-1}{q}\}$, $\Pi = \Pi(q, d)$ the Cartesian product $(\Pi(q))^d$, and $\mu = \mu(q, d)$ the normalized counting measure on $\Pi(q, d)$. Let $A$ be an orthogonal array in dimension $d$, with $q$ levels $\{0, \frac{1}{q} \ldots, \frac{q-1}{q}\}$, strength $t$, and index $\lambda$, and denote $n = \lambda q^t$ its number of rows. For any permutation $\boldsymbol{\pi} = (\pi_1, \ldots \pi_d) \in \Pi$, denote by $A(\boldsymbol{\pi})$ the orthogonal array obtained from $A$ after applying each permutation $\pi_j$ on the levels of the corresponding $j$th factor, i.e.,

$$\forall 1 \leq i \leq n \text{ and } 1 \leq j \leq d, \ \left(A(\boldsymbol{\pi})\right)_{ij} = \pi_j(A_{ij}) \, .$$

Note that the $A(\boldsymbol{\pi})$'s and $A$ are orthogonal arrays with the same parameters; see [186]. Conversely, it is also easy to show that if $A$ has strength and index equal to 1 (i.e., as in the classic RBD with an odd integer $n$), then any other orthogonal array $A'$ with the same parameters as $A$ is of the form $A(\boldsymbol{\pi})$ for a permutation $\boldsymbol{\pi} \in \Pi$.

**Remark 23.** *If $n$ is even, the design of experiments in RBD consists of an orthogonal array with $n/2$ levels, strength 1, and index 2 and may be faced with the coincidence defect.*

We are now interested in the quantities

$$\mathbb{E}_\mu\left[\widehat{\mathrm{V}}(G, A(\boldsymbol{\pi}))\right] \text{ and } \mathbb{E}_\mu\left[\widehat{\mathrm{V}}_A(G, K_A, A(\boldsymbol{\pi}))\right],$$

where $K_A$ is a finite subset of $\mathbb{Z}_{A,\neq 0}$.

### Bias of the estimator in RBD

Let $\widehat{\mathrm{c}}_{\mathbf{k}}(G) = \widehat{\mathrm{c}}_{\mathbf{k}}(G, D(q))$ denote the **k**th complex discrete Fourier coefficient, that is,

$$\widehat{\mathrm{c}}_{\mathbf{k}}(G, D(q)) = \frac{1}{n} \sum_{\mathbf{x}\in D(q)} G(\mathbf{x}) \exp(-2i\pi\mathbf{k}\cdot\mathbf{x}),$$

with $n = q^d$ and $D(q)$ the regular grid $\{0, \frac{1}{q}, \ldots, \frac{q-1}{q}\}^d$. Let $\mathbb{Z}_{A,\neq 0}(q) = \mathbb{Z}_A \cap \left(-\frac{q}{2}, \frac{q}{2}\right]^d$. We begin with the following important theorem.

**Theorem 4.7 (Owen [344]).** *Following the previous notation, we have*

$$\mathrm{Var}_\mu\left[\widehat{\mathrm{c}}_{\mathbf{0}}(G, A(\boldsymbol{\pi}))\right] = \frac{1}{n^2} \sum_{|A|>t} \left(\sum_{r=0}^{|A|} B(A,r)(1-q)^{r-|A|}\right)\left(\sum_{\mathbf{k}\in\mathbb{Z}_{A,\neq 0}(q)} |\widehat{\mathrm{c}}_{\mathbf{k}}(G)|^2\right),$$

*where*

$$B(A,r) = \sum_{i=1}^{n}\sum_{j=1}^{n} \mathbf{1}_{|\{l\in A,\ A_{il}=A_{jl}\}|=r}$$

*consists of the number of pairs of rows $(A_i, A_j)$ that match on exactly $r$ of the axes (columns) in $A$.*

***Proof.*** This is exactly Theorem 1 given by Owen [344]. Just note that the embedded ANOVA terms on a $q^d$ regular grid (denoted by $\beta_A$ in Owen [344]) are

$$\beta_A(\mathbf{x}) = \sum_{\mathbf{k}\in\mathbb{Z}_A^*(q)} \widehat{\mathrm{c}}_{\mathbf{k}}(G)\exp(2i\pi\mathbf{k}\cdot\mathbf{x}).$$

Indeed, for all $\mathbf{x}$ in the regular grid $\{0, \frac{1}{q}, \ldots, \frac{q-1}{q}\}^d$,

$$G(\mathbf{x}) = \sum_{A\in\mathcal{P}_d} \beta_A(\mathbf{x})$$

by a trigonometric interpolation argument, and it is also easy to show that the random variables $\beta_A(X_i,\ i\in A)$ satisfy properties 1 and 2 of Theorem 3.3 (page 49) for independent random variables $X_i$ uniformly distributed on $\{0, \frac{1}{q}\ldots, \frac{q-1}{q}\}$. $\square$

Results on the bias's behavior are established under additional regularity assumptions. To that end, the weighted Korobov spaces are introduced. For $\alpha > 1$ and $\gamma = (\gamma_A)_{A\in\mathcal{P}_d}$ with non-negative $\gamma_A$'s, define the weighted Korobov space $\mathcal{H}_{\alpha,\gamma}$ to be the Hilbert space with reproducing kernel

$$RK_{\alpha,\gamma}(\mathbf{x},\mathbf{y}) = 1 + \sum_{\mathbf{k}\in\mathbb{Z}^d_{\neq 0}} r(\mathbf{k},\alpha,\gamma)^{-1}\exp(2i\pi\mathbf{k}\cdot(\mathbf{x}-\mathbf{y})),$$

where, for any $\mathbf{k} \neq \mathbf{0}$, $r(\mathbf{k}, \alpha, \gamma) = \gamma_{A_\mathbf{k}}^{-1} \prod_{i \in A_\mathbf{k}} |k_i|^\alpha$, where $A_\mathbf{k}$ is such that $\mathbf{k} \in \mathbb{Z}_{A_\mathbf{k}, \neq 0}$. For $\mathbf{k}$ such that $\gamma_{A_\mathbf{k}} = 0$, we set by convention $r(\mathbf{k}, \alpha, \gamma) = \infty$. Thus the kernel can be rewritten as

$$RK_{\alpha,\gamma}(\mathbf{x}, \mathbf{y}) = 1 + \sum_{\substack{\mathbf{k} \in \mathbb{Z}^d_{\neq 0} \\ \gamma_{A_\mathbf{k}} \neq 0}} r(\mathbf{k}, \alpha, \gamma)^{-1} \exp(2i\pi \mathbf{k} \cdot (\mathbf{x} - \mathbf{y})),$$

and we deduce that the norm of $f \in \mathcal{H}_{\alpha,\gamma}$ satisfies

$$||f||^2_{\mathcal{H}_{\alpha,\gamma}} = \mathrm{c}_\mathbf{0}(f)^2 + \sum_{\substack{\mathbf{k} \in \mathbb{Z}^d_{\neq 0} \\ \gamma_{A_\mathbf{k}} \neq 0}} r(\mathbf{k}, \alpha, \gamma) |\mathrm{c}_\mathbf{k}(f)|^2 < +\infty$$

and consequently

$$\forall \mathbf{k} \in \mathbb{Z}^d_{\neq 0} \text{ s.t. } \gamma_{A_\mathbf{k}} \neq 0, \ |\mathrm{c}_\mathbf{k}(f)|^2 \leq \frac{\gamma_{A_\mathbf{k}} ||f||^2_{\mathcal{H}_{\alpha,\gamma}}}{\prod_{i \in A_\mathbf{k}} |k_i|^\alpha}.$$

Note that, for any $\mathbf{k} \in \mathbb{Z}^d_{\neq 0}$ such that $\gamma_{A_\mathbf{k}} = 0$, $f \in \mathcal{H}_{\alpha,\gamma}$ implies $\mathrm{c}_\mathbf{k}(f) = 0$.

In Proposition 4.8 below, the bias of the variance estimate is investigated in unweighted Korobov spaces $\mathcal{H}_\alpha = \mathcal{H}_{\alpha,\mathbf{1}}$.

**Proposition 4.8 (Tissot and Prieur [483]).** *Let $A$ be a defect-free orthogonal array in dimension $d$ with parameters $q$, $t$, and $\lambda$ in $\mathbb{Z}_{>0}$ with $t < d$. If there exists $\alpha > 2t + 1$ such that $G$ and $G^2$ are in $\mathcal{H}_\alpha$, we have*

$$\mathbb{E}_\mu \left[ \widehat{\mathrm{V}}(G, A(\boldsymbol{\pi})) \right] = \mathrm{V}(G) - \frac{1}{n} \sum_{1 \leq |A| > t} \mathrm{V}_A(G) + O\left(n^{-(1+\frac{1}{t})}\right).$$

***Proof.*** The proof can be found in Tissot and Prieur [483, Appendix A.6] (see also [481, Chapter 5]). □

As a consequence, consider the classic definition of effective dimension in the superposition sense (see [62]).

**Definition 4.9.** *The effective dimension of $G$, in the superposition sense, is the smallest $d_S(G)$ such that*

$$\sum_{1 \leq |A| \leq d_S(G)} \mathrm{V}_A(G) \geq l_S(G) \mathrm{V}(G),$$

*where $l_S(G)$ is an arbitrary constant generally set at* 0.99.

We then have the following corollary.

**Corollary 4.10 (Tissot and Prieur [483]).** *Under the assumptions of Proposition* 4.8, *let $d_S(G)$ and $l_S(G)$ be defined as in Definition* 4.9. *If $t \geq d_S$, we have*

$$\mathbb{E}_\mu \left[ \widehat{\mathrm{V}}(G, A(\boldsymbol{\pi})) \right] = \left(1 - \frac{\varepsilon}{n}\right) \mathrm{V}(G) + O\left(n^{-(1+\frac{1}{t})}\right),$$

*where $0 \leq \varepsilon \leq 1 - l_S(G)$.*

***Proof.*** The proof is straightforward from Proposition 4.8. □

Since

$$\mathbb{E}_\mu\left[\widehat{\mathrm{V}}_A(G, K_A, A(\boldsymbol{\pi}))\right] = \sum_{\mathbf{k}\in K_A} \mathbb{E}_\mu\left[\left|\widehat{\mathrm{c}}_{\mathbf{k}}(G, A(\boldsymbol{\pi}))\right|^2\right],$$

the analysis of the bias of the parts of variance estimates rests on the following result.

**Proposition 4.11 (Tissot and Prieur [483]).** *Let $A$ be a defect-free orthogonal array in dimension $d$ with parameters $q$, $t$, and $\lambda$ in $\mathbb{Z}_{>0}$ with $t < d$. Let $A$ be a nonempty subset of $[1:d]$ and $\mathbf{k} \in \mathbb{Z}_{A,\neq 0}$. If there exists $\alpha > 2t+1$ such that $G$ and $G^2$ are in $\mathcal{H}_\alpha$, we have*

$$\mathbb{E}_\mu\left[\left|\widehat{\mathrm{c}}_{\mathbf{k}}(G, A(\boldsymbol{\pi}))\right|^2\right] = \frac{n-1}{n}|\mathrm{c}_{\mathbf{k}}(G)|^2 + \frac{1}{n}\left(\mathrm{V}(G) + \mathrm{c}_{\mathbf{0}}(G)^2 - R_1 - R_2\right) + O\!\left(n^{-(1+\frac{1}{t})}\right),$$

*where*

$$R_1(q,t,\lambda,\mathbf{k}) = \sum_{\substack{1\leq |B|\leq t\\ A\cap B=\emptyset}} \sum_{\mathbf{h}\in\mathbb{Z}_{B,\neq 0}(q)} \left|\mathrm{c}_{\mathbf{k}+\mathbf{h}}(G)\right|^2$$

*consists of terms of order strictly higher than $|A|$, and*

$$R_2(q,t,\lambda,\mathbf{k}) = \sum_{\substack{1\leq |B|\leq t\\ A\cap B\neq\emptyset}} \sum_{B'\subset B} (-1)^{|B|-|B'|} \sum_{B''\subset B'} \sum_{\mathbf{h}\in\mathbb{Z}_{B'',\neq 0}(q)} \left|\mathrm{c}_{\mathbf{k}_{B'}+\mathbf{h}}(G)\right|^2,$$

*where $(\mathbf{k}_{B'})_i = 0$ if $i \in B'$, and $(\mathbf{k}_{B'})_i = k_i$ otherwise.*

***Proof.*** The proof is rather technical and is detailed in [483, Appendix A.7] (see also [481, Chapter 5]). □

From the results above, we conclude that estimators in RBD are asymptotically unbiased in unweighted Korobov spaces since

$$\mathbb{E}_\mu\left[\widehat{\mathrm{V}}(G, A(\boldsymbol{\pi}))\right] = V(G) + \frac{B_1}{n} + o(n^{-1}),$$
$$\mathbb{E}_\mu\left[\left|\widehat{\mathrm{c}}_{\mathbf{k}}(G, A(\boldsymbol{\pi}))\right|^2\right] = |\mathrm{c}_{\mathbf{k}}|^2 + \frac{B_2}{n} + o(n^{-1}),$$

where $B_1 \leq \mathrm{V}(G)$ and $B_2 \leq \mathrm{V}(G) + \mathrm{c}_{\mathbf{0}}(G)^2$, and more generally

$$\mathbb{E}_\mu\left[\widehat{\mathrm{V}}_A(G, K_A, A(\boldsymbol{\pi}))\right] = V_A(G) + \frac{B_3}{n} + \varepsilon_{trunc}(K_A) + o(n^{-1}),$$

where $B_3 \leq |K_A|(\mathrm{V}(G) + \mathrm{c}_{\mathbf{0}}(G)^2)$ and

$$\varepsilon_{trunc}(K_A) = \sum_{\mathbf{k}\in\mathbb{Z}_{A,\neq 0}\setminus K_A} |\mathrm{c}_{\mathbf{k}}(G)|^2$$

is, for instance, of order $O(M^{|A|-\alpha})$ if $K_A = \mathbb{Z}_{A,\neq 0}(M)$. Nevertheless, it is possible to reduce a part of nonasymptotic biases.

### Application to bias correction

The approach presented in this section is based on results in Tissot [481] and Tissot and Prieur [483]. We do not propose any bias correction for the variance estimates since in practice the bias of the latter is generally negligible. So, we are only interested in the bias of the parts of variance estimates

$$\widehat{\mathbb{V}}_A(K_A) = \widehat{\mathbb{V}}_A\big(G, K_A, A(\pi)\big),\ K_A \subset \mathbb{Z}_{A,\neq 0}(q)$$

under the assumptions of Proposition 4.11. In practice, one chooses $K_A = \mathbb{Z}_{A,\neq 0}(M)$, and the truncation parameter $M$, as well as the term $|K_A|^{1/|A|}$, is of order 5 or higher and is generally less than 15. Below, formula (4.15) proposes a bias correction in the case $|t| = 1$, $|A| = 1$, that is, for the estimation of first-order indices based on LHS; see [481, 483] for more general formulas. From Proposition 4.8, Corollary 4.10, and Proposition 4.11, the following corrected estimator is proposed for $1 \le i \le d$:

$$\widehat{\mathbb{V}}_i^c(M) = \frac{n}{n-(M-1)}\widehat{\mathbb{V}}_i(M) - \frac{M-1}{n-(M-1)}\widehat{\mathbb{V}}\big(G, A(\pi)\big). \tag{4.15}$$

Note that the bias correction is the same as the one suggested in Plischke [370] and proposed in Tissot and Prieur [482].

## 4.2 ▪ Metamodel-based approaches for Sobol' indices

In this section, we first introduce the main concepts of metamodeling (see Section 4.2.1), then the different ways of using metamodeling for the estimation of Sobol' indices (see Sections 4.2.2, 4.2.3, and 4.2.4). We also discuss different strategies for quantifying the impact of metamodeling in the estimation error of Sobol' indices (see Section 4.2.5).

**Remark 24.** *We only focus on Sobol' indices, while the metamodels can be useful for estimating other sensitivity measures. For example, Sudret and Mai [465] and De Lozzo and Marrel [105] have studied how to compute DGSM (see Sections* 2.3.3 *and* 3.1.4*) by using polynomial chaos and Gaussian process metamodels.*

### 4.2.1 ▪ Metamodel main principles

The metamodel approach consists in generating an approximation of the initial model (synonyms are surrogate, emulator, or proxy) that fits the initial data (using, for example, a least-squares procedure), which has good prediction capabilities and can be run with a negligible computational cost. The construction of a metamodel requires answering three practical issues:

- The choice of the metamodel, which can be of different types: linear/nonlinear, parametric/nonparametric, and so on [185]. The most used metamodels include polynomials, splines, generalized linear models, generalized additive models, kriging, neural networks, support vector machines, and boosting regression trees [436, 127]. Showing how to use different types of metamodels for SA is the main purpose of this section.
- The design of (numerical) experiments, which is denoted by

$$\chi = \boldsymbol{X}^n = \left(X_1^{(i)}, \dots, X_d^{(i)}\right)_{i=1,\dots,n}.$$

The model simulations on $\chi$ are denoted by $\boldsymbol{Y}^N = \left(Y^{(1)}, \dots, Y^{(n)}\right)^T$ with $Y^{(i)} = G(\mathbf{X}^{(i)})$, $i = 1, \dots, n$. The main qualities required for an experimental design are the robustness (ability to analyze different models), the effectiveness (optimization of a criterion), the goodness of points repartition (space-filling property), and the low cost for its construction [421, 127]. Several studies have shown the quality of different types of experimental designs with respect to the predictivity of metamodels; see [436]. Appendix A gives an overview of the types of designs for computer experiments.

- The validation of the metamodel. The usual practice is to estimate global criteria (root mean square error, absolute error, etc.) on a test basis via cross-validation or bootstrap [243, 127]. Building "good" and small-size test samples can also be considered [205].

In what follows, three metamodel-based strategies are discussed:

- If only first-order or second-order sensitivity indices are of interest, plug-in strategies based on the emulation of first-order or second-order conditional moments $\mathbb{E}(Y|X_j)$ and $\mathbb{E}(Y|X_j, X_k)$ for $1 \leq j < k \leq d$ can be employed (see Section 4.2.2).

- For some specific types of metamodels, analytical formulas can be derived for any-order Sobol' indices (see Section 4.2.3).

- For any kind of metamodel, a direct sampling–based procedure can be applied to estimate Sobol' indices of any order (see Section 4.2.4).

### 4.2.2 ▪ A plug-in approach based on the emulation of conditional moments

To estimate first- and second-order indices, the main building block consists in using input-output samples $({X_j}^{(i)}, G(\mathbf{X})^{(i)}), i = 1, \ldots, n$, or $(X_j, {X_k}^{(i)}, G(\mathbf{X})^{(i)}), i = 1, \ldots, n$, to emulate first- and second-order conditional moments of $Y$ via smoothing techniques. Such smoothing estimates are useful to produce interpretable graphs, such as

- main effect graphs between $X_j$ and $\mathbb{E}(Y|X_j) - \mathbb{E}Y$ across the whole variation domain of $X_j$ for $j = 1, \ldots, d$;

- interaction effect graphs between $(X_j, X_k)$ and $\mathbb{E}(Y|X_j, X_k) - \mathbb{E}(Y|X_j) - \mathbb{E}(Y|X_k) - \mathbb{E}Y$ across all the variation domains of $(X_j, X_k)$ for $j = 1, \ldots, d$ and $k = j+1, \ldots, d$.

Nonparametric methods (Hastie and Tibshirani [184]) are often used (moving averages, kernel methods, smoothing splines, etc.); see Storlie and Helton [461] for a comprehensive review of predictor smoothing methods for SA. Once first- and second-order conditional moments have been emulated, a Monte Carlo–based estimation procedure with a plug-in of first- and second-order conditional moments can be applied to estimate first- and second-order Sobol' indices. In the specific case of local polynomial smoothers, theoretical properties of such estimates are investigated in Da Veiga et al. [102].

### 4.2.3 ▪ Analytical formulas of Sobol' indices based on polynomial chaos and Gaussian process emulators

Two particular metamodels, namely, polynomial chaos expansion and Gaussian process emulator, allow one to derive analytical formulas of Sobol' indices.

### Polynomial chaos expansion

In the early 1990s, a new approach called the stochastic finite element method emerged, which allows one to solve partial differential equations (PDEs) with uncertain coefficients (see, e.g., [158, 159]). The key ingredient in this approach is the so-called polynomial chaos expansion, originally developed by Wiener [502], which allows one to represent a random output as a polynomial series in the input variables, where the uncertainty is lumped in the polynomials and the unknown coefficients of the decomposition become deterministic. In practice, a truncation of the polynomial series is considered.

The ingredients are very similar to the ones presented in Section 4.1. For all $j = 1, \dots, d$, we consider an orthonormal Hilbert basis of $\mathbb{L}^2_{\mathbb{R}}(\mathrm{P}_{X_j})$, denoted by $(\Phi_{j,k}\,,\ k \in \mathbb{Z})$ such that $\Phi_{j,0} = 1$. For specific distributions, it is possible to choose the basis $(\Phi_{j,k}\,,\ k \in \mathbb{Z})$ as a polynomial basis. For instance, considering uniform (resp., normal) distributions leads to Legendre (resp., Hermite) polynomials. Then, as in Section 4.1, we consider the tensorized basis of $\mathbb{L}^2_{\mathbb{R}}(\mathrm{P}_{\mathbf{X}})$, whose elements are indexed by $\mathbf{k} = (k_1, \dots, k_d) \in \mathbb{Z}^d$ and defined as the products $\Phi_{1,k_1}(\cdot) \times \dots \times \Phi_{d,k_d}(\cdot)$. We then recover, as a particular case of the results in Section 4.1, that $G(\mathbf{X})$ can be decomposed as

$$Y = G(\mathbf{X}) = \sum_{\mathbf{k} \in \mathbb{Z}^d} c_{\mathbf{k}}(G)\Phi_{1,k_1}(X_1)\dots\Phi_{d,k_d}(X_d).$$

The truncated series are then considered,

$$G^{\mathrm{PC}}(\mathbf{X}) = \sum_{\mathbf{k} \in \mathbf{K}} c_{\mathbf{k}}(G)\Phi_{1,k_1}(X_1)\dots\Phi_{d,k_d}(X_d),$$

with $K$ being a finite subset of $\mathbb{Z}^d$.

For applications with a smooth random output, one may expect good approximation results, even for $\mathbf{K}$ of small cardinality. In the framework of PDEs with random coefficients, Galerkin projection techniques are used to compute the $c_{\mathbf{k}}(G)$, $\mathbf{k} \in K$. In more recent works, the polynomial chaos expansion technique is nonintrusively applied to numerical models, and the coefficients are estimated on the basis of an *optimal* experimental design on the input by means of (sparse) linear regression techniques; see, e.g., [463, 90, 266]. The theory of polynomial chaos expansion is not further developed in this book, and the interested reader should refer to, e.g., Ghanem and Spanos [158, 159] and Le Gratiet et al. [266]. Once the metamodel is learned, that is, once the truncation set $\mathbf{K}$ is chosen and the coefficients $c_{\mathbf{k}}(G)$ are approximated/estimated by $\hat{c}_{\mathbf{k}}(G)$, Parseval's identity is applied to derive an analytical expression of Sobol' indices, namely,

$$S_A = \frac{\sum_{\mathbf{k} \subset \mathbf{K} \cap \mathbb{Z}_{A,\neq 0}} |\hat{c}_{\mathbf{k}}(G)|^2}{\sum_{\mathbf{k} \in \mathbf{K}, \mathbf{k} \neq \mathbf{0}} |\hat{c}_{\mathbf{k}}(G)|^2} \tag{4.16}$$

with $\mathbb{Z}_{A,\neq 0}$ defined as in Section 4.1 by

$$\mathbb{Z}_{A,\neq 0} = \{\mathbf{k} \in \mathbb{Z}^d \,|\, \forall j \in A,\ k_j \in \mathbb{Z}_{\neq 0} \text{ and } \forall j \notin A,\ k_j = 0\}.$$

The main difference with the spectral approach for estimating Sobol' indices presented in Section 4.1 is that the metamodel point of view is adopted here; thus the truncation set $\mathbf{K}$ is first fixed, then the metamodel is fitted and the analytical formula Eq. (4.16) is finally used to compute the Sobol' index of the fitted metamodel.

### Gaussian process emulator

The Gaussian process (GP) model [406], also called the kriging model [315, 455], supposes that the model output is a realization of a Gaussian random process defined as

$$Y(\mathbf{x}) = m(\mathbf{x}) + Z(\mathbf{x}),$$

where $m(\mathbf{x})$ is a deterministic function and $Z(\mathbf{x})$ is a centered GP. That is, $\mathbb{E}Z(\mathbf{x}) = 0$ for all $\mathbf{x}$. Moreover, for all positive integer $n$, input locations $\mathbf{x}^{(1)}, \dots, \mathbf{x}^{(n)}$, and $\lambda_1, \dots, \lambda_n \in \mathbb{R}^n$, the linear combination $\sum_{i=1}^{n} \lambda_i \mathbf{x}^{(i)}$ has a Gaussian distribution. The deterministic function $m(\mathbf{x})$ provides the mean approximation of the computer code, for example, a one-degree polynomial model $m(\mathbf{x}) = \beta_0 + \sum_{j=1}^{d} \beta_j x_j$. $Z$ is assumed to be with variance $\sigma^2$ and a covariance function $K_{\boldsymbol{\theta}}(\cdot, \cdot)$ parametrized by a vector $\boldsymbol{\theta} \in \mathbb{R}^d$ of hyperparameters. This means that the covariance of the Gaussian random process $Z$ is given, for any $\mathbf{x}, \mathbf{x}'$, by $K_{\boldsymbol{\theta}}(\mathbf{x}, \mathbf{x}') = \mathbb{E}(Z(\mathbf{x})Z(\mathbf{x}'))$. This structure allows us to provide interpolation and spatial correlation properties by modeling interactions and nonlinearities that $m(\mathbf{x})$ does not capture. Several parametric families of covariance functions can be chosen [455, 266], for example, a Matérn 5/2 parametrized by the correlation lengths $\boldsymbol{\theta}$. The regression coefficients $(\beta_j, j = 1, \dots, d)$ and the hyperparameters ($\sigma^2$ and $\boldsymbol{\theta}$) are generally estimated using a maximum likelihood optimization procedure [421, 127, 392].

The predictive distribution for $G(\mathbf{x}^*)$ ($\mathbf{x}^* \in E$) is obtained by conditioning $Y(\mathbf{x}^*)$ by $\boldsymbol{Y}^n$. $\widetilde{Y}(\mathbf{x}^*) = [Y(\mathbf{x}^*)|\boldsymbol{Y}^n]$ is also a GP [421] with the parametrization

$$\widetilde{Y}(\cdot) \sim \mathcal{N}\left(\tilde{\mu}(\cdot), \widetilde{C}(\cdot, \cdot)\right),$$

where

- $\tilde{\mu}$ is called the kriging mean (which is the metamodel predictor), which writes

$$\tilde{\mu}(\mathbf{x}) = \mathbb{E}\widetilde{Y}(\mathbf{x}) = m(\mathbf{x}) + \mathbf{k}(\mathbf{x})^T \Sigma_n^{-1} (\boldsymbol{Y}^n - m(\boldsymbol{X}^n)) \tag{4.17}$$

  with $\mathbf{k}(\mathbf{x}) = [K_{\boldsymbol{\theta}}(\mathbf{x}^{(1)}, \mathbf{x}), \dots, K_{\boldsymbol{\theta}}(\mathbf{x}^{(n)}, \mathbf{x})]^T$ and $\Sigma_n$ the $n \times n$ matrix with elements $K_{\boldsymbol{\theta}}(\mathbf{x}^{(i)}, \mathbf{x}^{(j)})$ for $i, j = 1, \dots, n$;

- $\widetilde{C}$ is called the kriging covariance, which is written as

$$\widetilde{C}(\mathbf{x}, \mathbf{x}') = \mathrm{Cov}[\widetilde{Y}(\mathbf{x}), \widetilde{Y}(\mathbf{x}')] = K_{\boldsymbol{\theta}}(\mathbf{x}, \mathbf{x}') - \mathbf{k}(\mathbf{x})^T \Sigma_n^{-1} \mathbf{k}(\mathbf{x}'). \tag{4.18}$$

The so-called kriging error at the point $\mathbf{x}$, which is the mean square error of this model, is given by the variance $\widetilde{C}(\mathbf{x}, \mathbf{x})$.

Note that Eq. (4.17) may also be interpreted as a projection in the reproducing kernel Hilbert space (RKHS) associated with the covariance kernel $K_{\boldsymbol{\theta}}$ (see Appendix B and references therein for a brief overview on RKHS). More precisely, by the representer theorem (see Theorem B.6 in Appendix B), $\mathbf{k}(\mathbf{x})^T \Sigma_n^{-1} \boldsymbol{Y}^n$ is the element of this space having minimal norm and interpolating the values $\boldsymbol{Y}^n$ on the points $\boldsymbol{X}^n$.

Focusing now on Sobol' indices estimation, the first solution is to apply the Sobol' index formula to the predictor, i.e., the mean of the conditional GP (Eq. (4.17)), which is a deterministic function of the inputs. This gives

$$S_j = \frac{\mathrm{Var}\, \mathbb{E}[\tilde{\mu}(X)|X_j]}{\mathrm{Var}\, \tilde{\mu}(X)} \text{ for } j = 1, \dots, d.$$

By using for $K_{\boldsymbol{\theta}}$ a tensorized product of monodimensional stationary correlation functions, analytical calculations of this expression can be developed. These expressions do not easily reduce and are not given here; see [79, 305]. A second solution consists in using the whole global conditional GP by taking into account not only the mean of conditional GP model but also its covariance structure [343]:

$$\tilde{S}_j = \frac{\operatorname{Var} \mathbb{E}[\widetilde{Y}(X)|X_j]}{\operatorname{Var} \widetilde{Y}(X)} \text{ for } j = 1, \dots, d. \tag{4.19}$$

In this case, the Sobol' index $\tilde{S}_j$ is a random variable. Its expectation can then be considered as the sensitivity index. Its variance and, more generally, its distribution can also be used as an indicator of the sensitivity index accuracy; see [310]. Closed-form expressions of the expectation and the variance of the numerator in Eq. (4.19) in terms of integrals can be obtained (once again using a tensorized structure for the correlation function); see [343, 305, 309].

Another line of work focuses on computing Sobol' indices based on $\tilde{\mu}$ but with special kernels, called therein ANOVA kernels, which allows for more amenable closed-form formulas as proposed by Durrande et al. [120]. For the sake of simplicity, we assume here that, for all $x$, $m(x) = 0$. Then assume that $K_{\boldsymbol{\theta}}$ is the product

$$K_{\boldsymbol{\theta}}(\mathbf{x}, \mathbf{x}') = K_{\boldsymbol{\theta},1}(x_1, x_1') \dots K_{\boldsymbol{\theta},d}(x_d, x_d'),$$

where, for $j = 1, \dots, d$, $K_{\boldsymbol{\theta},j}$ is a 1D kernel. Set further

$$\overline{K_{\boldsymbol{\theta},j}}(x_j, x_j') = \frac{\mathbb{E}K_{\boldsymbol{\theta},j}(X_j, x_j)\mathbb{E}K_{\boldsymbol{\theta},j}(X_j, x_j')}{\mathbb{E}K_{\boldsymbol{\theta},j}(X_j, X_j')},$$

where we recall that $X_j'$ is an independent copy of $X_j$. The ANOVA kernel associated to $K_{\boldsymbol{\theta}}$ is then

$$K_{\boldsymbol{\theta},\text{ANOVA}}(\mathbf{x}, \mathbf{x}') = \prod_{j=1}^{d} (1 + K_{\boldsymbol{\theta},j}(x_j, x_j') - \overline{K_{\boldsymbol{\theta},j}}(x_j, x_j')).$$

For this very special kernel, the Hoeffding decomposition and all the Sobol' indices of the function $\tilde{\mu}$ have explicit expressions. We refer the reader to Durrande et al. [120] for these closed forms. To end this subsection, let us note that the previous decomposition is related to Hoeffding decomposition with regularization in RKHS. We refer the reader to Lin and Zhang [283] and Touzani [486] for statistical and optimization aspects of these regularization methods, including sparse model topics.

## 4.2.4 ▪ Direct sampling estimation

In all the estimation formulas of the Sobol' indices given in Sections 3.2 and 4.1, a direct replacement of the model $G$ by a fitted metamodel is always possible, as in the early works of Marseguerra et al. [313] with neural networks and Iooss et al. [214] with higher-order polynomials. As the metamodel has a negligible CPU time cost, simpler formulas, such as the pick-freeze ones (see Section 3.2.1), are preferred. Indeed, accurate confidence intervals can be obtained by randomly repeating the estimation procedure instead of using asymptotic formulas or bootstrap replicates.

For the specific case of the random forest metamodel, a permuted column sampling strategy such as the one detailed in Section 3.2.2 to estimate total sensitivity indices actually corresponds to the well-known permutation importance measure (or permutation-based variable importance measure) of random forests, as proved in Wei et al. [498]. Another equality link between a permutation-based measure and the first-order Sobol index is also developed in Benoumechiara [30].

As already mentioned, using a surrogate model biases the estimation of the Sobol' index, as what is actually estimated is sensitivity of surrogate output and not the full one. More precisely, if $\widehat{\widetilde{S}}_A$ denotes the Sobol' index associated to $\mathbf{X}_A$ for the metamodel $\widetilde{G}$, then the bias is equal to $\widehat{\widetilde{S}}_A - S_A$. It justifies the consideration of metamodels whose quality of approximation directly depends on the (learning) sample size $n$. Considering a metamodel $\widetilde{G}_n$ such that there exists $c \in \mathbb{R}$ such that

$$\widetilde{G}_n(x) - G(x) \xrightarrow[n \to +\infty]{\mathbb{L}^2(\mathrm{P}_X)} c$$

and that

$$n\mathrm{Var}\left(\widetilde{G}_n(X) - G(X)\right) \xrightarrow[n \to +\infty]{} 0,$$

the asymptotic normality still holds, even by replacing the true model with the surrogate in the Monte Carlo estimator. We refer the reader to Janon et al. [218, Corollary 1] for further details.

## 4.2.5 ▪ Error estimation

Monte Carlo–based procedures, such as the ones presented above, are robust and popular ways to estimate Sobol' indices. The aim is now to quantify the loss of accuracy due to the use of a metamodel combined with a Monte Carlo method to compute sensitivity indices. The sensitivity index produced by Monte Carlo estimation with a surrogate model is tainted by a twofold error. First, our Monte Carlo sampling procedure assimilates the whole (generally infinite) population of possible inputs with the finite, randomly chosen sample; this produces *sampling* or *Monte Carlo error*. Second, using a surrogate model biases the estimation of the Sobol' index, as what is actually estimated is sensitivity of surrogate output instead of the actual one; we call this bias the *metamodel error*.

### Bootstrap

A variation on the bootstrap, which addresses sampling error as well as metamodel error, has been proposed in Storlie et al. [462]; in this work, the authors propose using a bootstrap strategy on the metamodel residuals to estimate metamodel error and Monte Carlo error simultaneously. Let us consider a metamodel $\widetilde{G}$ built on the learning sample $\mathcal{X} = \{(\mathbf{x}^{(1)}, G(\mathbf{x}^{(1)})), \dots, (\mathbf{x}^{(n)}, G(\mathbf{x}^{(n)}))\}$. We define $\mathcal{E} = \{e^{(1)}, \dots, e^{(n)}\} = \{\widetilde{G}(\mathbf{x}^{(1)}) - G(\mathbf{x}^{(1)}), \dots, \widetilde{G}(\mathbf{x}^{(n)}) - G(\mathbf{x}^{(n)})\}$.

Let $R \in \mathbb{Z}_{>0}$. Let $j \in [1:d]$. To generate a bootstrapped sample of $\widehat{S}_j$, the following steps are repeated for $r = 1, \dots, R$:

1. We sample with replacement $\mathbf{x}_*^{(1)}, \dots, \mathbf{x}_*^{(n)}$ from $\mathbf{x}^{(1)}, \dots, \mathbf{x}^{(n)}$.
2. We sample with replacement from $\mathcal{E}$ and get $(e^{(1),(r)}, \dots, e^{(n),(r)})$, then compute a noisy version of the sample $\mathcal{X}_* = \{(\mathbf{x}_*^{(1)}, G_*(\mathbf{x}_*^{(1)})), \dots, (\mathbf{x}_*^{(n)}, G_*(\mathbf{x}_*^{(n)}))\}$ with, for $i = 1, \dots, n$,

$$G_*(\mathbf{x}_*^{(k)}) = \widetilde{G}(\mathbf{x}_*^{(k)}) + e^{(k),(r)}.$$

   We now have a sample $\mathcal{X}_*^{(r)} = \{(x_*^{(1)}, G_*(\mathbf{x}_*^{(1)}), \dots, (x_*^{(n)}, G_*(\mathbf{x}_*^{(n)})\}$, from which we construct a new metamodel $\widetilde{G}_*^{(r)}$.

3. We compute a bootstrapped replication $\widehat{S_{j*}}^{(r)}$ of $\widehat{S_j}$ by using, e.g., (3.15) and (3.16) with the metamodel $\widetilde{G}_*^{(r)}$ instead of $\widetilde{G}$.

For $r = 1, \ldots, R$, let $Z_*^{(r)} = \widehat{S_{j*}}^{(r)} - \widehat{S_j}$. Then find the $\alpha/2$ and $(1-\alpha/2)$ quantiles from the collection $\mathcal{Z} = \{Z_*^{(1)}, \ldots, Z_*^{(R)}\}$ ($\xi_*^{\alpha/2}$, $\xi_*^{1-\alpha/2}$, resp.). Then the $100(1-\alpha)\%$ confidence interval for $S_j$ is provided by $\left[\widehat{S_j} - \xi_*^{1-\alpha/2}, \widehat{S_j} - \xi_*^{\alpha/2}\right]$. A technical detail is that the lower (or upper) limit of the confidence interval can be less than zero (or greater than one). When this happens, the endpoint of the confidence interval could simply be truncated so that the confidence interval is a subset of $[0, 1]$.

### Particular case of Gaussian process emulators

The probabilistic nature of the GP metamodel gives a direct access to the metamodel approximation error (Eq. (4.18)) (see Section 4.2.3). The different quantities of interests (such as the variance, quantile, maximal value, etc., of $Y$) can then be estimated with some associated confidence intervals, which are useful in many engineering studies [136]. In particular, simulation techniques of conditional GP, well known in geostatistics [81], are useful when the quantity of interest is a nonlinear operator.

For estimating the distributions of the Sobol' indices $S_A^{\text{clos}}$ and $S_A^{\text{T}}$, this approach has been proposed by Le Gratiet et al. [264]; see also [266]. Set $R$ the number of samples for the GP conditional simulations of $\widetilde{Y}$ (for evaluating the uncertainty due to the GP approximation) and $B$ the number of bootstrap samples (for evaluating the uncertainty due to Monte Carlo integration). The algorithm is as follows:

1. Generate the $N$-size pick-freeze samples $\mathbf{X}$ and $\mathbf{X}^{\sim A}$ of the input vector.

2. For $r = 1, \ldots, R$,

   (a) sample a realization $\widetilde{y}(\mathbf{x})$ of $\widetilde{Y}(\mathbf{x})$ with $\mathbf{x} = (\mathbf{X}, \mathbf{X}^{\sim A})$;

   (b) for $b = 1, \ldots, B$,

   - sample with replacements two samples $\mathbf{U}$ and $\mathbf{U}^{\sim A}$ from $\mathbf{X}$ and $\mathbf{X}^{\sim A}$;
   - compute $\widehat{S_{A,r,b}^{\text{clos}}}$ and $\widehat{S_{A,r,b}^{\text{T}}}$ from $\widetilde{y}(\mathbf{x}^B)$ with $\mathbf{x}^B = (\mathbf{U}, \mathbf{U}^{\sim A})$.

3. Return the samples (of size $R \times B$) of Sobol' indices: $\left(\widehat{S_{A,r,b}^{\text{clos}}}\right)_{\substack{r=1,\ldots,R\\ b=1,\ldots,B}}$ and $\left(\widehat{S_{A,r,b}^{\text{T}}}\right)_{\substack{r=1,\ldots,R\\ b=1,\ldots,B}}$.

The mean and variance of each of these two latter samples provide the estimate and the associated error of $S_A^{\text{clos}}$ and $S_A^{\text{T}}$. In Le Gratiet et al. [264], the variances of the estimators are decomposed in the variances coming from the GP approximation and the ones coming from the Monte Carlo integration, omitting the covariance term (added in Menz et al. [322] in another estimation context).

In practice, this simulation-based method is much more usable than the semianalytical ones cited in Section 4.2.3 (which involve numerical discretization of random integrals; see [310]).

### Certifying sensitivity analysis from a posteriori error bounds

We now assume that we have a punctual error bound for our metamodel: We know $\varepsilon$ such that

$$\forall \mathbf{x} \in \mathcal{X}, \ \left|G(\mathbf{x}) - \widetilde{G}(\mathbf{x})\right| \leq e(\mathbf{x}),$$

with $\mathcal{X}$ being the input space. For efficiency reasons, we require that $e$ can be evaluated at location $\mathbf{x}$ with a cost similar to the one needed to evaluate the metamodel $\widetilde{G}$ at the same location $\mathbf{x}$.

As a first example, let us consider an RKHS interpolator as metamodel $\widetilde{G}$; see [421, 424, 428]. The interpolator depends on a learning sample $\{(\mathbf{u}^{(1)}, G(\mathbf{u}^{(1)})), \ldots, (\mathbf{u}^{(n)}, G(\mathbf{u}^{(n)}))\}$, where the design points $\mathcal{D} = \{\mathbf{u}^{(i)}\}_{i=1,\ldots,n} \subset \mathcal{X}$ are generally chosen as a space-filling design, for instance, the so-called maximin LHS (see Appendix A). The error analysis of the RKHS method [424, 298] shows that there exist positive constants $\mathcal{C}$ and $\mathcal{K}$, depending on $G$, such that

$$\forall \mathbf{x} \in \mathcal{X}, \quad \left|G(\mathbf{x}) - \widetilde{G}(\mathbf{x})\right| \le \mathcal{C} \exp(-\mathcal{K}/h_{\mathcal{D},\mathcal{X}}),$$

where

$$h_{\mathcal{D},\mathcal{X}} = \sup_{\mathbf{x}\in\mathcal{X}} \min_{\mathbf{u}\in\mathcal{D}} \|\mathbf{u} - \mathbf{x}\|$$

for a given norm $\|\cdot\|$ on $\mathcal{X}$.

The quantity $h_{\mathcal{D},\mathcal{X}}$ can be linked to the number of points $n^*(\varepsilon)$ in an optimal covering of $\mathcal{D}$ defined as the minimum over the $p \in \mathbb{Z}_{>0}$ such that

$$\exists\, (\mathbf{u}^{(1)}, \ldots, \mathbf{u}^{(p)}) \in \mathcal{X} \text{ s.t. } \forall \mathbf{x} \in \mathcal{X}, \exists i \in [1:p] \text{ s.t. } \|\mathbf{x} - \mathbf{u}^{(i)}\| \le \varepsilon.$$

In other words, $n^*(\varepsilon)$, known as the *covering number of* $\mathcal{X}$, is the smallest size of a design $\mathcal{D}$ satisfying $h_{\mathcal{D},\mathcal{X}} \le \varepsilon$.

It is known that, when $\mathcal{X}$ is a compact subset of $\mathbb{R}^d$, there exist constants $A$ and $B$ such that

$$A\varepsilon^{-d} \le n^*(\epsilon) \le B\varepsilon^{-d}.$$

Hence, assuming that an optimal design of size $n$ is chosen, we have, for a constant $B'$,

$$h_{\mathcal{D},\mathcal{X}} \le B' n^{-1/d},$$

and we have the following pointwise metamodel error bound for constants $C$ and $\mathcal{K}'$:

$$\forall \mathbf{x} \in \mathcal{X}, \quad \left|G(\mathbf{x}) - \widetilde{G}(\mathbf{x})\right| \le \mathcal{C} \exp(-\mathcal{K}' n^{1/d}).$$

This obviously leads to an integrated error bound on the variance of the metamodel error

$$\operatorname{Var} \delta \le C \exp(-k n^{1/d})$$

for suitable constants $C$ and $k$. For that example, the metamodel error bound, denoted $e(\cdot)$, is chosen independent of $\mathbf{x} \in \mathcal{X}$. It is not possible to get an explicit formulation for the constants $C$ and $k$. An indicator of the error bound (denoted $\widehat{e}$) is thus proposed:

$$\forall \mathbf{x} \in \mathcal{X}, \; \widehat{e}(\mathbf{x}) = \widehat{C} \exp(-\widehat{k} n^{1/d}).$$

The estimates $\widehat{C}$ and $\widehat{k}$ of the constants are derived from the following procedure: We plot an estimation (based on a sample of, e.g., 1000 metamodel errors) of the variance of the metamodel error against $n^{1/d}$, with $n$ the number of points in $\mathcal{D}$. Then, using an exponential regression, we obtain approximated values for $C$ and $k$.

Another example is the one of reduced basis (RB); see [295] and references therein. We now consider the case where the initial model is defined as the solution of a PDE. The RB method is such a procedure; we split the overall computation into two successive parts: the *offline* phase, which makes use of the standard, computationally intensive numerical procedure used to solve the PDE to gather "knowledge" about solutions of the latter, and the *online* phase, where we rely on data collected during the offline phase to compute, for each desired instance of the parameter, a good approximation of the solution for a per-instance cost that is orders of magnitude smaller than the cost of one run of the standard numerical code. The advantage is that, for a sufficiently large number of online evaluations, the fixed cost of the offline phase is strongly dominated by the reduction in the marginal cost provided by the online procedure. This cost reduction is made possible by the fact that, in most cases, the desired solutions of the PDE, for all the considered values of the parameter, lie in some manifolds of functions that are "close" to a low-dimensional linear subspace. One goal of the offline phase is to find such a suitable subspace so that the online procedure can look for the solution of the PDE as an element of the subspace. Then it reduces the number of degrees of freedom and thus the computational cost. One interesting feature of the RB approach is that it comes with an online error bound, that is, a (provably) certified, natural norm, fast-computed (i.e., almost of the same complexity of the online phase) upper bound of the distance between the solution provided by the online phase (called the *reduced* or *online* solution) and the one given by the standard, expensive numerical procedure (called the *full* or *reference* solution).

We assume now that we have such an error bound $e(\cdot)$, and we describe the methodology proposed in Janon et al. [219] to assess SA performed from the metamodel.

### Methodology

We first draw two $n$-samples $\mathbf{x}^{(1)}, \dots, \mathbf{x}^{(n)}$ and $\mathbf{x}'^{(1)}, \dots, \mathbf{x}'^{(n)}$ from the distribution of inputs. We then compute from these samples

$$\widehat{y}^{(k)} = \widetilde{G}(\mathbf{x}^{(k)}), \quad \widehat{y'}^{(k)} = \widetilde{G}(x'^{(k)}_1, \dots, x'^{(k)}_{j-1}, x^{(k)}_j, x'^{(k)}_{j+1}, \dots, x^{(k)}_d),$$

$$e^{(k)} = e(\mathbf{x}^{(k)}), \quad e'^{(k)} = e(x'^{(k)}_1, \dots, x'^{(k)}_{j-1}, x^{(k)}_j, x'^{(k)}_{j+1}, \dots, x^{(k)}_d),$$

$k = 1, \dots, n$.

Recall that we have assumed that all these quantities can be evaluated in a reasonable time (which does not depend on the time needed for one evaluation of the true model $G$), at least once the metamodel $\widetilde{G}$ has been fitted.

The central idea is then to lower and upper bound $\widehat{S}_j$ with two estimators $\widehat{S}_j^{\text{lo}}$ and $\widehat{S}_j^{\text{up}}$, which only depend on $\left(\widehat{y}^{(k)}, \widehat{y'}^{(k)}, e^{(k)}, e'^{(k)}, k = 1, \dots, n\right)$.

We then draw $R$ bootstrap replications for both the lower and the upper bounds. From these bootstrap replications, we deduce a $100(1-\alpha)\%$ confidence interval for $\widehat{S}_j^{\text{lo}}$, $[I_1^{\text{lo}}, I_2^{\text{lo}}]$ and a $100(1-\alpha)\%$ confidence interval for $\widehat{S}_j^{\text{up}}$, $[I_1^{\text{up}}, I_2^{\text{up}}]$.

## 4.2.6 ▪ Examples in R

We first describe the metamodel-based approaches to compute Sobol' indices which are integrated into the sensitivity package of R:

- estimation of the Sobol' first-order indices with B-spline smoothing [393] (sobolSmthSpl() function);

- estimation of the Sobol' first-order and total indices with a Gaussian process (kriging metamodel)–based global sensitivity analysis [264] (see Section 4.2.5 and sobolGP() function).

Using the sobolGP() function requires a previously built Gaussian process metamodel (using the km() function of the DiceKriging package [404]). It then applies the algorithm described in Section 4.2.5. The Ishigami function (see Section C.4 with independent $X_j \sim \mathcal{U}(-\pi, \pi)$)

is studied with a learning sample (space-filling design coming from the DiceDesign package [119]) of $N = 50$ points:

```
> set.seed(12345) # To reproduce the same results
> d <- 3
> n <- 50
> #
> # Building a space-filling design of experiment
> library(DiceDesign)
> Xinit <- lhsDesign(n, d)$design
> X <- maximinSA_LHS(Xinit, T0 = 10)$design
> X <- X * 2 * pi - pi
> Y <- ishigami.fun(X) # Evaluation of the function
> #
> # Fitting the Gaussian process metamodel
> library(DiceKriging)
> gp <- km(design = X, response = Y)
> #
> # Sensitivity analysis
> library(sensitivity)
> N <- 1000
> X1 <- data.frame(matrix(runif(d*N, min = -pi, max = pi), nrow=N))
> X2 <- data.frame(matrix(runif(d*N, min = -pi, max = pi), nrow=N))
> sa <- sobolGP(gp, MCmethod = "soboljansen", X1 = X1, X2 = X2,
+                nsim = 100, nboot = 100)
> plot(sa)
```

Figure 4.2 has been obtained by the last command of the previous block of instructions. The left plot gives for each input (numbered from 1 to $d$) the first-order Sobol' index ("main effect") and the total Sobol' index ("total effect"). Then the middle and right plots allow us to have the

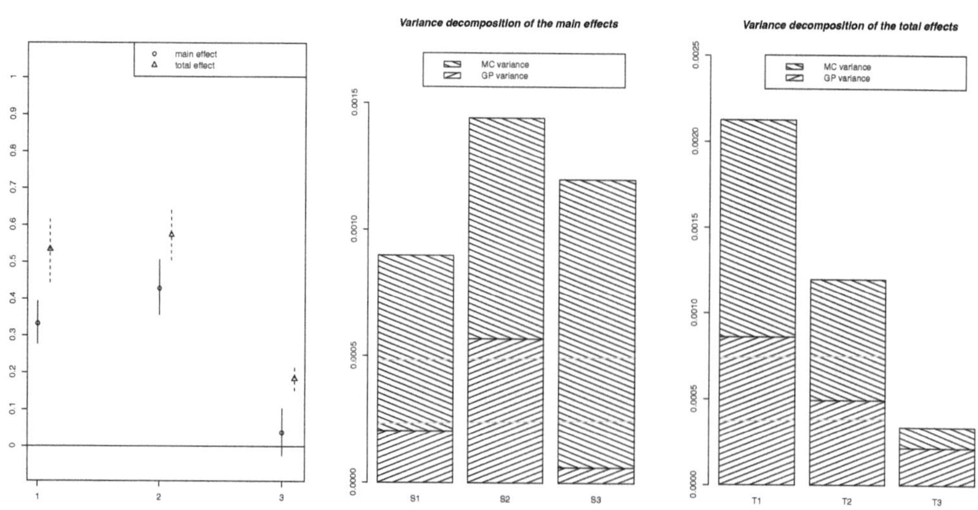


**Figure 4.2.** *Plotted results of the sobolGP() function on the Ishigami model.*

full estimation variance of each Sobol' index and to decompose it into that due to the Gaussian process model approximation and that due to the pick-freeze estimation scheme.

## 4.3 ▪ Further references

Other recent developments deal with spectral expansions for computing Sobol' indices. In Pronzato [376], the author considers spectral bases built on an RKHS. Using the connection between such kernels and the Gaussian processes, Pronzato [376] proposes an enlightening new point of view for the Bayesian estimation of Sobol' indices. Following Tissot and Prieur [483] (see also Section 4.1 above), Roustant et al. [403] consider generalized chaos expansions built on a general tensor Hilbert basis and derive general lower bounds for total Sobol' indices (see the PoincareChaosSqCoef() function of the sensitivity R package). Choosing for the basis the eigenfunction system associated with a Poincaré differential operator leads to lower bounds involving the derivatives of the model.

## 4.4 ▪ Summary

Figure 4.3 offers a classification grid of the Sobol' index estimation methods (presented in this chapter and in Chapter 3) that we should recommend. It positions each method in terms of its computational cost (linearly dependent on $d$ or $d$-independent) and the Sobol' indices it allows to be estimated (first-order, second-order, and/or total indices). This grid brings an engineering viewpoint based on several rules of thumb (coming from the authors' practice on many real applications) concerning the computational cost indicated for each method. For example, the number of simulations required to fit a metamodel with a sufficient accuracy cannot be known a priori, but a standard practice in a first approach is to consider a design of (possibly space-filling) experiments of $n = 10d$ runs.

DGSM bounds are interesting, as they inform on total Sobol' indices. If an adjoint model is available, it could be the most powerful method for high-dimensional input space applications, as all model derivatives can then be evaluated at an approximate cost of 5 times the cost of evaluating the direct model. Moreover, a sample of model derivatives comes most often with a sample of model outputs allowing one to compute first-order Sobol' indices from given-data methods in addition to DGSM bounds.

If one is interested in estimating first-order and second-order Sobol' indices, up to $d \approx 100$, our recommendation is to use replicated LHS and replicated designs based on a randomized orthogonal array coded in sobolroalhs() of the sensitivity R package. A recursive version of the procedure is implemented in sobolrec(). For higher input space dimensions, we recommend combining Janon–Monod's scheme [324, 218] with Saltelli's trick [409]. This approach is coded in sobolSalt(). If we suspect interactions of order higher than 3, then we recommend estimating first-order, second-order, and total Sobol' indices. For small to moderate input space dimensions (up to 100), we recommend the combination of replicated designs with Saltelli's trick (see Gilquin et al. [160]) coded in sobolrep(). If the input space dimension is higher than 100, we recommend the function sobolSalt(). All these recommendations are robust to the regularity of the model. However, under regularity assumptions on the model, other approaches may be much more efficient. For instance, if the model presents a rapid decay of its Fourier spectrum, then the FAST approach can achieve rates of order $\mathcal{O}\left(n^{-1}\right)$.

All the aforementioned methods are based on a specific design of experiments. It happens that we just have at our disposal an input-output sample that we would like to reuse for sensitivity analysis. One sometimes deals with the terminology *given-data*. The spectral given-data method EASI [370] has already been mentioned in this chapter. Other given-data approaches

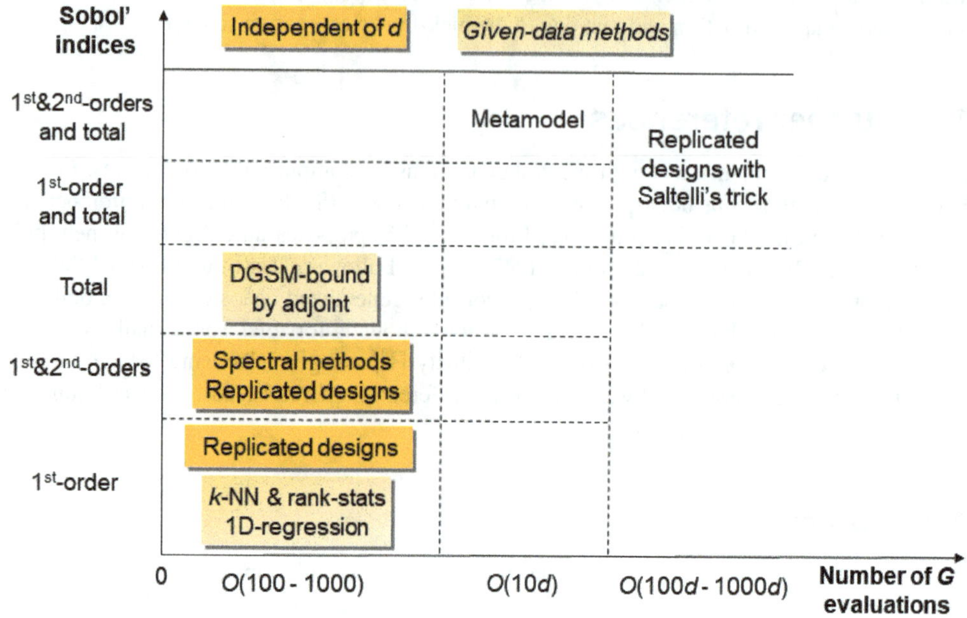


**Figure 4.3.** *Classification grid (valid for* d *up to* 100*) of Sobol' indices estimation methods in terms of the required number of model evaluations and provided information.*

based on classical statistical tools, such as $k$-nearest neighbors, rank statistics, or parametric/nonparametric unidimensional regression, have also been proposed in the literature (see, e.g., [146, 56, 191] and references therein). At the present time, most of these methods are limited to first-order indices, and the corresponding theoretical results should be strengthened. Most of the time, these methods require regularity assumptions on the model. Therefore, even if we believe that these approaches are promising, only a few of them are implemented in the sensitivity package of R.

Chapter 5

# Variance-based sensitivity measures with dependent inputs

In the previous chapters, the uncertainty on the input parameters is modeled by a product probability measure. More precisely, we assumed $\mathbf{X} = (X_1, \dots, X_d)$ with the vector $\mathbf{X}$ distributed as $P_{\mathbf{X}} = \prod_{j=1}^{d} P_{X_j}$. However, in applications to computer experiments, it is common that the input space is constrained to a nonrectangular region, or that the input variables have some other known form of dependence, such as a general Gaussian distribution. Addressing dependence in SA is not an easy task. The main issue relies on the difficulty of discriminating between dependencies (e.g., correlations) and interactions. Correlation between two variables means the values of one variable relate in some way to the values of the other. Interaction is different, as it describes a situation in which the simultaneous influence of two variables on a model output is not additive.

As a very simple example in dimension $d = 2$, let us consider the model

$$y = G(\mathbf{x}) = x_1 + x_2,$$

with $\mathbf{x} \in [-1, 1]^2$. This model is additive; it does not present any interaction term. Let us assume now that the input vector $\mathbf{X}$ is described by two components $X_1$ and $X_2$, both uniformly distributed on $[-1, 1]$, and related by the relation $X_2 = -X_1$ almost surely. Under this assumption, the random variables $X_1$ and $X_2$ are perfectly dependent. We have $\mathrm{Cov}(X_1, X_2) = -1/3$, and $Y$ is almost surely equal to zero. The fact that two variables are correlated or not does not say anything on the additive property of the model. Likewise, if a model is nonadditive, the input variables may or may not be dependent.

In this chapter, we first review briefly the different approaches to be found in the literature (Section 5.1). We provide more details on SA based on groups of dependent inputs (Section 5.2) and covariance-based SA (Section 5.3). Then Section 5.4 presents the Sobol' indices formulation via deconditioning techniques, while Section 5.5 describes the recent solution based on Shapley values. Examples are handled with the package sensitivity of R in Section 5.6.

## 5.1 ▪ State of the art: Difficulties and drawbacks

In many applications, it is common that the input variables have a statistical dependence structure imposed by a probabilistic dependence function [255] (e.g., a copula function) or by physical constraints on the input or the output space [366, 161, 250, 292]. As shown in different studies, estimating and interpreting Sobol' indices with correlated inputs is not trivial [416, 102].

Many propositions exist in the literature, but associated sensitivity indices are not always easy to interpret.

One strategy proposed by Jacques et al. [216] is to evaluate the Sobol' indices of subsets of inputs which are correlated within the subsets but assuming that the subsets of inputs are independent of each other. This strategy is detailed in Section 5.2. Its main drawback is that it does not allow one to compute Sobol' indices at a finer scale than each subset of variables.

An alternative strategy, given in Section 5.3, aims to decompose the partial variance of an input into a correlated part and an uncorrelated part. Such an approach allows one to exhibit inputs that have an impact on the output only through their strong correlation with other inputs. Xu and Gertner [508] investigated linear models with linear dependencies, while Li et al. [278] extended this approach to more general models, using the concept of high-dimensional model representation (HDMR) [277]. HDMR is based on a hierarchy of component functions of increasing dimensions (truncation of Hoeffding decomposition in the case of independent variables). The component functions are then approximated by expansions in terms of some suitable basis functions (e.g., polynomials [464, 388, 390], splines, etc.). However these works do not give an exact and unambiguous definition of the functional ANOVA for correlated inputs such as the one provided by the Hoeffding decomposition when inputs are independent (see Section 5.3 for more details). Chastaing et al. [75] have applied the works of Stone [459] and Hooker [198], who define an ANOVA for dependent inputs and then define variable importance through that generalization of ANOVA. However, the dependent-variable ANOVA leads to importance measures with two conceptual problems underlined in Owen and Prieur [353]: The inputs can get negative influences, and the approach places strong restrictions on the joint probability distribution of the inputs.

Another method, proposed by Mara and Tarantola [303] in the setting of linear dependencies, is to consider a first step of decorrelation of the inputs (with the Gram–Schmidt procedure) and a second step which uses these new inputs to perform the ANOVA-HDMR. At the same time, Kucherenko et al. [253] proposed a nonparametric procedure to estimate first-order and total Sobol' indices in the presence of dependencies, not necessarily of linear type. Their methodology requires the knowledge of the conditional probability densities and the ability to draw random samples from those. Later, in Mara et al. [304], the authors established a link between the approaches in Kucherenko et al. [253] and Mara and Tarantola [303], allowing the distinction between the independent contributions of inputs to the response variance and their mutual dependent contributions via the estimation of four sensitivity indices for each input, namely, full and independent first-order indices and full and independent total indices. This approach is detailed in Section 5.4.

The most recent solution, proposed by Owen [350], is to adapt the Shapley values [431], well known in game theory and economics, to the variance-based SA issues. Instead of passing by the model output variance decomposition (FANOVA), this method consists in a direct allocation of a part of the variance of the output to each input. Section 5.5 is devoted to the properties and estimation procedures of the Shapley effects (i.e., variance-based Shapley values in the context of SA of model outputs). To carry out an SA in the case where the model inputs exhibit dependencies among them, recent works have demonstrated great interest in the Shapley effects [447, 353, 211, 31, 56, 57, 386].

**Remark 25.** *As a different approach, Borgonovo [42] and Borgonovo et al. [44] initiated the construction of novel generalized moment-free sensitivity indices. Based on some geometrical considerations, these indices measure the shift area between the outcome density and this same density conditioned to a parameter. Thanks to the properties of these new indices, a methodology is available to obtain them analytically through test cases. This approach is not specific to dependent inputs and is handled as a particular case in Chapter* 6 *(Section* 6.1*).*

## 5.2 ▪ Grouping

This section assumes that $\mathbf{X}$ is a vector of $d$ inputs among which one or multiple groups of variables are correlated. Variables belonging to the same group are dependent, while variables of different groups are independent. This implies that one variable cannot appear in two groups. Each group is denoted by a multidimensional variable. $\mathbf{X}$ can be rewritten as

$$\begin{aligned}\mathbf{X} = (&\underbrace{X_1, \dots, X_{\ell_1}}_{\vec{X}_1}, \dots, \underbrace{X_{L_{k-1}+1}, \dots, X_{L_{k-1}+\ell_k}}_{\vec{X}_k}, \\ &\dots, \underbrace{X_{L_{m-1}+1}, \dots, X_d}_{\vec{X}_m}),\end{aligned} \tag{5.1}$$

where each $\vec{X}_k$ contains $\ell_k$ variables, and with $L_k = \sum_{l=1}^{k} \ell_l$. Naturally, $\sum_{l=1}^{m} \ell_l = d$. As in Chapter 3, we assume for the sake of clarity in the presentation that, for any $j = 1, \dots, d$, $X_j$ is valued in $(\mathbb{R}, \mathcal{B}(\mathbb{R}))$. More generally, we could assume that $X_j$ is valued in $(E_j, \mathcal{B}(E_j))$ with $E_j$ an abstract Polish measurable space endowed with its Borel $\sigma$-field.

When inputs are dependent, the FANOVA decomposition stated in Theorem 3.3 (Section 3.1.1, page 49) does not hold anymore. Indeed, the elements in the decomposition (3.6) (page 49) are no longer mutually orthogonal. The correlation of variables within $\vec{X}_k$ implies that a part of the sensitivity of one of these variables is explained by the other correlated variables. For any $A \in \mathcal{P}_d$, it is possible to compute the Sobol' index $S_A$, but a clear interpretation of this index does not exist. An alternative proposed in Jacques et al. [216] consists in defining grouped Sobol' indices. More precisely, let us consider $\mathbf{X}$ defined as in (5.1). For any $k \in [1 : m]$, we define $A_k = \{L_{k-1} + 1, \dots, L_{k-1} + \ell_k\}$ and $\mathbf{X}_{A_k} = (X_{L_{k-1}+1}, \dots, X_{L_{k-1}+\ell_k}) = \vec{X}_k$ with the convention $L_0 = 0$. Let us recall that $V = \operatorname{Var} G(\mathbf{X})$. First-order grouped Sobol' indices are defined as

$$\vec{S}_k = S^{\text{clos}}_{A_k} = \frac{\operatorname{Var} \mathbb{E}[G(\mathbf{X}) | \vec{X}_k]]}{V} \; \forall k \in [1 : m] \,.$$

If $\vec{X}_k$ is 1D, then $\vec{X}_k = X_{L_k}$ and $\vec{S}_k$ coincides with $S_{L_k}$.

Higher-order grouped Sobol' indices can also be defined. In particular, second-order grouped Sobol' indices are given by

$$\begin{aligned}\vec{S}_{k,k'} &= S^{\text{clos}}_{A_k \cup A'_k} - S^{\text{clos}}_{A_k} - S^{\text{clos}}_{A'_k} \\ &= \frac{\operatorname{Var} \mathbb{E}[G(\mathbf{X}) | \vec{X}_k, \vec{X}'_k] - \operatorname{Var} \mathbb{E}[G(\mathbf{X}) | \vec{X}_k] - \operatorname{Var} \mathbb{E}[G(\mathbf{X}) | \vec{X}'_k]}{V} \,.\end{aligned}$$

Similarly, closed second-order grouped Sobol' indices are defined by

$$\begin{aligned}\vec{S}^{\text{clo}}_{k,k'} &= S^{\text{clos}}_{A_k \cup A'_k} \\ &= \frac{\operatorname{Var} \mathbb{E}[G(\mathbf{X}) | \vec{X}_k, \vec{X}'_k]}{V} \,.\end{aligned}$$

The index $\vec{S}_{k,k'}$ measures the impact of the interaction between $\vec{X}_k$ and $\vec{X}'_k$ on the output $G(\mathbf{X})$, while the closed index $\vec{S}^{\text{clo}}_{k,k'}$ measures the impact of $\vec{X}_k$ and $\vec{X}'_k$ on the output $G(\mathbf{X})$, by themselves or in interaction with each other. The information provided by grouped Sobol' indices is interesting, but it does not allow one to understand what happens at a finer scale than the groups, contrarily to the covariance-based procedure presented in the next section.

The estimation of grouped Sobol' indices does not raise any new issues. We just consider as inputs the independent vectors $\vec{X}_1, \ldots, \vec{X}_m$ and apply any of the procedures presented in Section 3.2 of Chapter 3. Let us illustrate below this strategy on a toy model, the Bratley function (see Section C.2).

**Example 5.1.** We assume that the input space dimension is $d = 4$ and that $X_1$ and $X_2$ are independent inputs, uniformly distributed on $[0, 1]$. Moreover, we assume that $(X_1, X_2)$ is independent from $(X_3, X_4)$ and that $(X_3, X_4)$ is uniformly distributed on the triangle $\Delta_2 = \{(x_3, x_4) \in [0, 1]^2,\ x_3 \leq x_4\}$. We thus identify $m = 3$ groups: $\vec{X}_1 = X_1$, $\vec{X}_2 = X_2$, and $\vec{X}_3 = (X_3, X_4)$. The notation $\mathbf{X} \sim P_{\mathbf{X}}^{\text{dep}}$ is used. We consider the Bratley function, whose definition appears in Appendix C.2. Recall that $G$ is defined on $[0, 1]^4$ by $G(x_1, x_2, x_3, x_4) = -x_1 + x_1x_2 - x_1x_2x_3 + x_1x_2x_3x_4$. Since $G$ has an analytical expression, Sobol' indices can be computed by symbolic integrals to evaluate their theoretical value.

Three of the estimation procedures presented in Chapter 3 were tested in Gilquin et al. [161]: the standard pick-freeze procedure (see Section 3.2.1), the replication procedure (see Section 3.2.2), and Saltelli's approach (see Section 3.2.3). The results for the estimation of the three first-order grouped Sobol' indices and of the three closed second-order grouped Sobol' indices are presented in Figs. 5.1 and 5.2. For each estimation procedure and each index, boxplots were drawn from $r = 100$ repetitions and for different sample sizes $N$ of the design. In each subfigure,

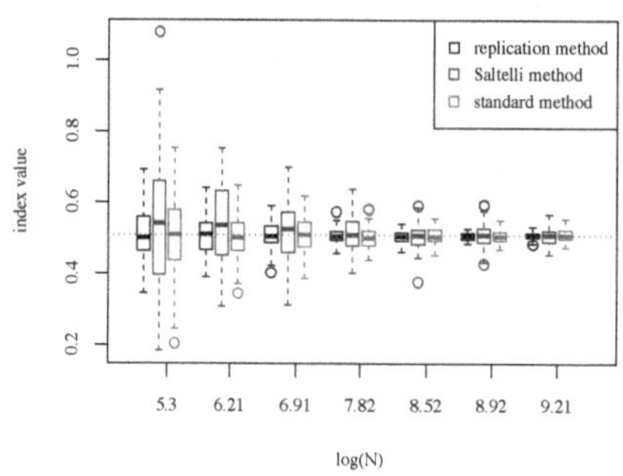


(a) Sobol' index $\vec{S}_1$

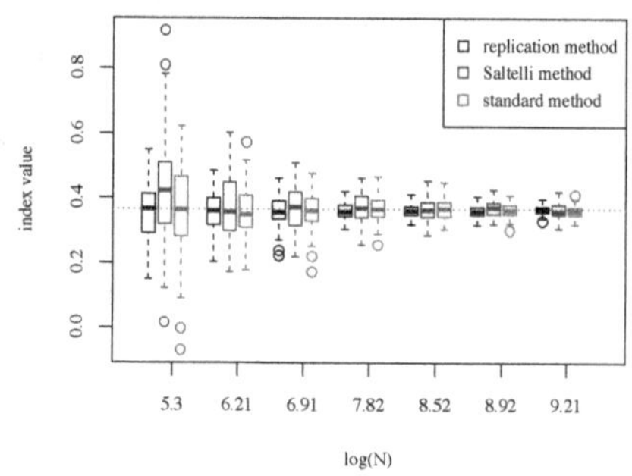


(b) Sobol' index $\vec{S}_2$

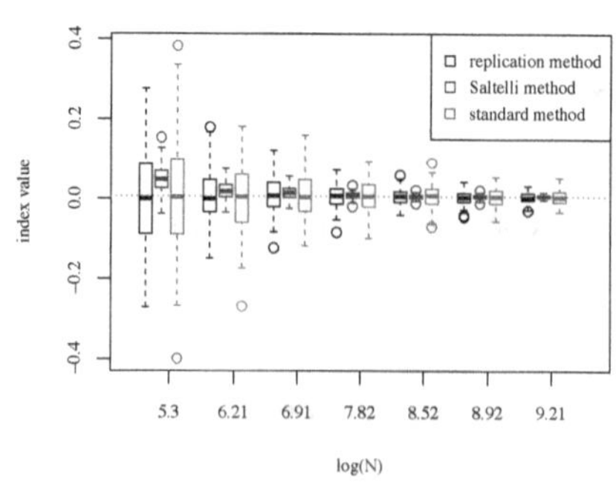


(c) Sobol' index $\vec{S}_3 = S_{3,4}$

**Figure 5.1.** *Estimation of first-order grouped Sobol' indices given by the three methods for different sizes $N$ of the design, using a boxplot representation based on $r = 100$ repetitions. In each subfigure, the red dashed horizontal line corresponds to the theoretical value of the corresponding first-order grouped Sobol' index.*

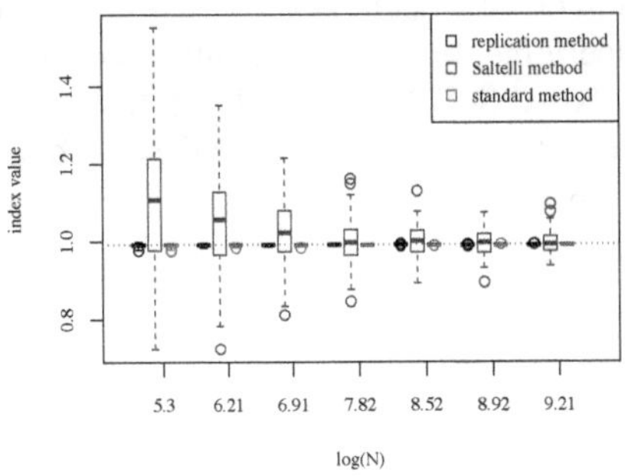


(a) Sobol' index $S^{\text{clos}}_{1,2}$

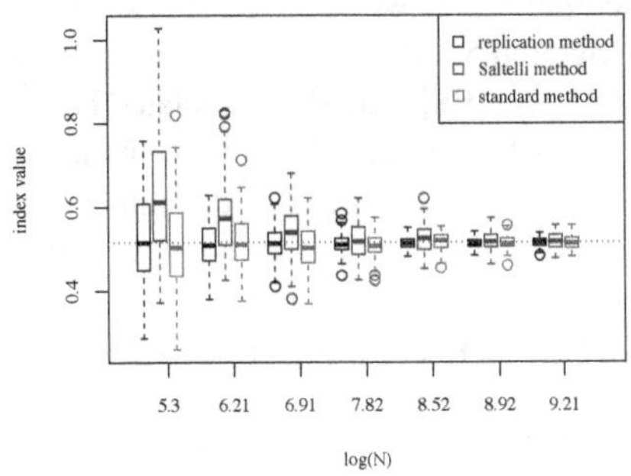


(b) Sobol' index $\vec{S}^{\text{clo}}_{1,3}$

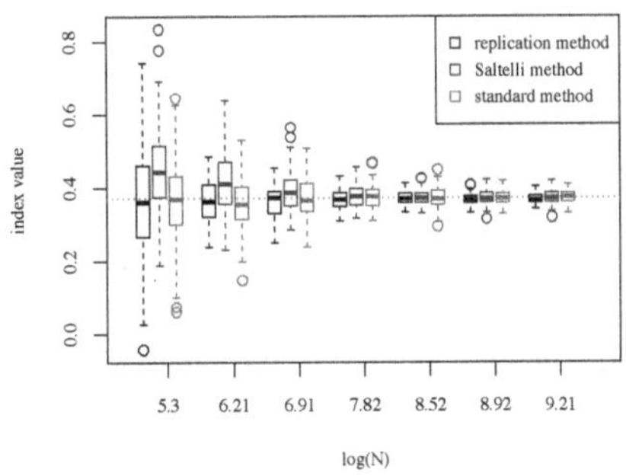


(c) Sobol' index $\vec{S}^{\text{clo}}_{2,3}$

**Figure 5.2.** *Estimation of closed second-order grouped Sobol' indices given by the three methods for different sizes $N$ of the design, using a boxplot representation based on $r = 100$ repetitions. In each subfigure, the red dashed horizontal line corresponds to the theoretical value of the corresponding closed second-order grouped Sobol' index.*

the red dashed line corresponds to the theoretical value of the corresponding Sobol' index. We can measure the impact of the dependence between $X_3$ and $X_4$ on the theoretical values of first-order grouped Sobol' indices by comparing the values given by the red dashed lines in Fig. 5.1 with the theoretical values computed by assuming independence of the four components, that is, by assuming $\mathbf{X} \sim \mathcal{U}\left([0,1]^4\right)$: $S_1 = 675/1111 \approx 0.6075608$, $S_2 = 243/1111 \approx 0.21872187$, $S^{\text{clos}}_{34} = 171/1111 \approx 0.1539154$.

The results in Fig. 5.1 (resp., Fig. 5.2) show that the three procedures provide coherent estimates of first-order (resp., closed second order) grouped Sobol' indices. The replication procedure is particularly interesting for moderate sample sizes. ∎

## 5.3 ▪ Covariance-based sensitivity analysis

This section introduces covariance-based sensitivity indices. The starting ingredient to define such covariance-based indices is a *meaningful* decomposition of $G$ on which the computation of $V = \operatorname{Cov}(G(\mathbf{X}), G(\mathbf{X}))$ will be based. Due to the dependencies among inputs, not only variance terms but also covariance terms will appear, leading to the definition of two kinds of indices, summarizing on the one hand the structural contribution of a set of inputs $\mathbf{X}_A$ on the output $Y$ and on the other its correlative contribution.

In Section 5.3.1, we describe this strategy for the choice of the Hoeffding decomposition. Then, in Section 5.3.2, we present the generalized Hoeffding decomposition introduced in Stone [459] (see also [198]) and the associated covariance-based Sobol' indices. In Section 5.3.4, we compare the set of indices obtained from the usual Hoeffding decomposition to the ones obtained from the generalized Hoeffding decomposition on a particular example to underscore that the choice of the decomposition has a real impact on the results.

### 5.3.1 ▪ Using the Hoeffding decomposition

Let us consider the usual Hoeffding decomposition of $G$ defined as (see Theorem 3.3, page 49)

$$\begin{aligned} G(\mathbf{X}) &= G_\emptyset + \sum_j G_{\{j\}}(X_j) + \sum_{j<j'} G_{\{j,j'\}}(X_j, X_j') + \cdots + G_{[1:d]}(X_1, \ldots, X_d) \\ &= \sum_{A \in \mathcal{P}_d} G_A(\mathbf{X}_A) \end{aligned}$$

with, for any $A \in \mathcal{P}_d$, $G_A(\mathbf{X}_A) = \sum_{B \subset A} (-1)^{|A|-[B|} \mathbb{E}[G(\mathbf{X})|\mathbf{X}_B]$. Recall that $\mathbb{E}[G(\mathbf{X})|\mathbf{X}_B]$ can still be interpreted as the best approximation of $G(\mathbf{X})$ by a measurable function of $\mathbf{X}_B$ in the $\mathbb{L}^2$ sense.

The elements in the decomposition are not mutually orthogonal anymore due to dependencies between input variables. Thus, the variance decomposition provided in Corollary 3.5 (page 52) no longer holds true. However, it is still possible to write some covariance decomposition

$$V = \operatorname{Var} G(\mathbf{X}) = \operatorname{Cov}(G(\mathbf{X}), G(\mathbf{X})) = \sum_{A \in \mathcal{P}_d} \operatorname{Cov}(G_A(\mathbf{X}_A), G(\mathbf{X})).$$

Then, for any $A \in \mathcal{P}_d$,

$$\operatorname{Cov}(G_A(\mathbf{X}_A), G(\mathbf{X})) = \operatorname{Var} G_A(\mathbf{X}_A) + \sum_{A' \neq A} \operatorname{Cov}(G_A(\mathbf{X}_A), G_{A'}(\mathbf{X}_{A'})). \tag{5.2}$$

From (5.2), it is possible to define, for any $A \in \mathcal{P}_d$, two indices:

$$S_A^{\mathrm{U}} = \frac{\operatorname{Var} G_A(\mathbf{X}_A)}{V},$$

$$S_A^{\mathrm{C}} = \frac{\operatorname{Cov}(G_A(\mathbf{X}_A), \sum_{A' \in \mathcal{P}_d, A' \neq A} G_{A'}(\mathbf{X}_{A'}))}{V}.$$

These indices correspond, respectively, to the structural and correlative contributions for $\mathbf{X}_A$ to $G(\mathbf{X})$. In Hart and Gremaud [182], the authors consider as Sobol' index associated to $\mathbf{X}_A$ the quantity

$$S_A^{\mathrm{U}} + S_A^{\mathrm{C}} = \frac{\operatorname{Cov}(G_A(\mathbf{X}_A), G(\mathbf{X}))}{V}.$$

It is obvious that $\sum_{A \in \mathcal{P}_d} (S_A^{\mathrm{U}} + S_A^{\mathrm{C}}) = 1$. However, these indices may be negative.

It is also possible to define the total Sobol' index associated to $A$ by

$$S_A^{\mathrm{T}} = \frac{\mathbb{E} \operatorname{Var}[G(\mathbf{X})|\mathbf{X}_{\bar{A}}]}{V}.$$

Then we get, for any $A \in \mathcal{P}^d$, $A \neq \emptyset$,

$$S_A^{\mathrm{T}} = \sum_{A' \in \mathcal{P}_d, A' \cap A \neq \emptyset} (S_{A'}^{\mathrm{U}} + S_{A'}^{V}).$$

This identity has to be compared with the corresponding identity in the independent framework written in Eq. (3.10) (page 54) of Remark 3.8 in Section 3.1.2. Equation (3.10) is not true anymore if the input vector has nonindependent components.

An interesting identity written in Hart and Gremaud [182] is recalled in Proposition 5.2 below, which is true whenever the dependence structure of the input vector $\mathbf{X}$.

**Proposition 5.2 (Hart and Gremaud [182]).** *We have*

$$S_A^{\mathrm{T}} = \frac{\mathbb{E}\left[\left(G(\mathbf{X}) - \mathbb{E}[G(\mathbf{X})|\mathbf{X}_{\bar{A}}]\right)^2\right]}{V},$$

*where $\mathbb{E}[G(\mathbf{X})|\mathbf{X}_{\bar{A}}]$ is the best approximation of $G(\mathbf{X})$ by a measurable function of $\mathbf{X}_{\bar{A}}$ in the $\mathbb{L}^2$ sense.*

***Proof.*** From the total variance theorem (see Lemma 3.7, page 53) we know that

$$V = \mathbb{E}\,\mathrm{Var}\,[G(\mathbf{X})|\mathbf{X}_{\bar{A}}] + \mathrm{Var}\,\mathbb{E}[G(\mathbf{X})|\mathbf{X}_{\bar{A}}].$$

Thus

$$\begin{aligned}\mathbb{E}\,\mathrm{Var}\,[G(\mathbf{X})|\mathbf{X}_{\bar{A}}] &= V - \mathrm{Var}\,\mathbb{E}[G(\mathbf{X})|\mathbf{X}_{\bar{A}}] \\ &= \mathbb{E}[G^2(\mathbf{X}) - (\mathbb{E}[G(\mathbf{X})|\mathbf{X}_{\bar{A}}])^2] \\ &= \mathbb{E}\left[\left(G(\mathbf{X}) - \mathbb{E}[G(\mathbf{X})|\mathbf{X}_{\bar{A}}]\right)^2\right],\end{aligned}$$

which concludes the proof. The last equality is a direct application of the Pythagorean theorem. □

We deduce from Proposition 5.2 that a negligible total Sobol' index $S_A^{\mathrm{T}}$ indicates that it is possible to build a good approximation in the $\mathbb{L}^2$ sense of $Y = G(\mathbf{X})$ as a measurable function of $\mathbf{X}_{\bar{A}}$. The theoretical best approximation is defined as $\mathbb{E}[G(\mathbf{X})|\mathbf{X}_{\bar{A}}]$. Note that this conclusion is different from the conclusion that inputs indexed by $A$ can be fixed. The conclusion driven by Proposition 5.2 leads to a dimension reduction for metamodeling purposes, which is valid even if the inputs are not independent.

The definition of indices $S_A^{\mathrm{U}}$ and $S_A^{\mathrm{C}}$ is based on a decomposition whose components are not orthogonal. The choice of this decomposition is thus arbitrary, and one can wonder if another choice could be meaningful. In Section 5.3.2, we present covariance-based sensitivity indices defined from the generalized Hoeffding decomposition introduced in Stone [459], which is particularly interesting because it is less arbitrary.

### 5.3.2 ▪ Using a generalized Hoeffding decomposition

Chastaing et al. [75] introduce covariance-based sensitivity indices from the generalized Hoeffding decomposition introduced in Stone [459] and further developed in Hooker [198]. Let $\nu$ be a $\sigma$-finite measure on $(\mathbb{R}^d, \mathcal{B}(\mathbb{R}^d))$. Assume that $\mathrm{P}_{\mathbf{X}}$ is absolutely continuous with respect to $\nu$, and let $p_{\mathbf{X}}$ be the density of $\mathrm{P}_{\mathbf{X}}$ with respect to $\nu$, that is, $p_{\mathbf{X}} = \frac{d\mathrm{P}_{\mathbf{X}}}{d\nu}$. Also, assume that $G \in \mathbb{L}^2(\mathrm{P}_{\mathbf{X}})$. The associated inner product of this Hilbert space is

$$\langle h_1, h_2 \rangle = \int h_1(\mathbf{x}) h_2(\mathbf{x}) p_{\mathbf{X}} d\nu(\mathbf{x}) = \mathbb{E}[h_1(\mathbf{X}) h_2(\mathbf{X})].$$

For $A \in \mathcal{P}_d$, let $H_A$ be the set of all measurable and square integrable functions depending only on $\mathbf{X}_A$, and let $H_\emptyset$ be the set of constants. Then $H_A^0$ and $H^0$ are defined as follows:

$$H_A^0 = \left\{h_A(\mathbf{X}_A) \in H_A,\ \langle h_A, h_B\rangle = 0\ \forall\, B \subset A, \forall\, h_B \in H_B^0\right\},$$

$$H^0 = \left\{h(\mathbf{X}) = \sum_{A \in \mathcal{P}_d} h_A(\mathbf{X}_A), h_A \in H_A^0\right\}.$$

Note that $H_\emptyset$ is identical to $H_\emptyset^0$. $\mathrm{P}_{\mathbf{X}}$ is not assumed to be a product measure. Proposition 5.3 introduces the generalized Hoeffding decomposition and establishes its unicity.

**Proposition 5.3 (Chastaing et al. [75]).** *Let us assume that*

(i) $\nu(dx) = \nu_1(dx_1) \otimes \cdots \otimes \nu_d(dx_d)$*;*

(ii) $\exists\, 0 < M \leq 1\ \forall\, A \in \mathcal{P}_d,\quad p_{\mathbf{X}} \geq M p_{\mathbf{X}_A} p_{\mathbf{X}_{\bar{A}}}$ *$\nu$-almost everywhere,*

*where $\bar{A}$ still denotes the complement set of $A$ in $[1:d]$, $p_{\mathbf{X}_A}$ and $p_{\mathbf{X}_{\bar{A}}}$ the marginal densities of $\mathbf{X}_A$ and $\mathbf{X}_{\bar{A}}$.*

*Then there exist functions $\widetilde{G}_\emptyset, \widetilde{G}_1, \ldots, \widetilde{G}_{[1:d]} \in H_\emptyset \times H_1^0 \times \cdots H_{[1:d]}^0$ such that the following equality holds:*

$$\begin{aligned} G(X_1, \ldots, X_d) &= \widetilde{G}_\emptyset + \sum_i \widetilde{G}_i(X_i) + \sum_{i,j} \widetilde{G}_{ij}(X_i, X_j) + \cdots + \widetilde{G}_{[1:d]}(X_1, \ldots, X_d) \\ &= \sum_{A \in \mathcal{P}_d} \widetilde{G}_A(\mathbf{X}_A). \end{aligned}$$

*Moreover, this decomposition is unique, and the elements in the decomposition satisfy the property of hierarchical orthogonality. That is, for any $A \in \mathcal{P}_d$, $\widetilde{G}_A$ is orthogonal to any $\widetilde{G}_B$, $B \subset A$ in the sense that $\mathbb{E}[\widetilde{G}_A(\mathbf{X}_A)\widetilde{G}_B(\mathbf{X}_B)] = 0$.*

Note that the proof of Proposition 5.3 relies on the assumptions (i) and (ii), which are discussed in Chastaing et al. [75].

**Remark 26.** *Proposition* 5.3 *can be compared to Theorem* 3.3 *(page* 49*), which states the uniqueness of the usual Hoeffding decomposition in the framework of independent inputs. It is important to note that the generalized Hoeffding decomposition coincides with the usual Hoeffding decomposition if the inputs are independent.*

Chastaing et al. [75] define the structural and correlative contributions for $\mathbf{X}_A$ to $G(\mathbf{X})$ using the generalized Hoeffding decomposition instead of the classical one. Then, due to the hierarchical orthogonality, they get

$$\widetilde{S}_A^{\mathrm{U}} = \frac{\operatorname{Var} \widetilde{G}_A(\mathbf{X}_A)}{V},$$

$$\widetilde{S}_A^{\mathrm{C}} = \frac{\operatorname{Cov}(\widetilde{G}_A(\mathbf{X}_A), \sum_{B \subset A, B \neq A} \widetilde{G}_B(\mathbf{X}_B))}{V}.$$

We see that the definition of $\widetilde{S}_A^{\mathrm{C}}$ is slightly different from the one of $S_A^{\mathrm{C}}$ in Section 5.3.1 due to the property of hierarchical orthogonality of the terms in the generalized Hoeffding decomposition.

## 5.3.3 ▪ Estimation procedures

The definition of indices $S_A^{\mathrm{U}}$ and $S_A^{\mathrm{C}}$ in Section 5.3.1 (resp., $\widetilde{S}_A^{\mathrm{U}}$ and $\widetilde{S}_A^{\mathrm{C}}$ in Section 5.3.2) is based on the Hoeffding decomposition (resp., on the generalized Hoeffding decomposition).

Li et al. [278] propose to estimate $S_A^{\mathrm{U}}$ and $S_A^{\mathrm{C}}$ from a truncation of the Hoeffding decomposition. The component functions are then determined sequentially or simultaneously by least-squares regression on a suitable basis. As the number of decomposition components exponentially grows with the model dimension, it is often assumed that only the low-order interaction terms contain the major part of the model behavior. Forward or backward stepwise selection may be used to search for the significant component functions. Li et al. [278] recommend backfitting for high-dimensional systems. Closely related results can be read in Caniou and Sudret [66] and Sudret and Caniou [464] based on polynomial chaos expansions. We can also cite the early papers by Xu and Gertner [507, 508], which consider a simple linear approximation for $G$.

Later, Li and Rabitz [276] propose an estimation procedure of $\widetilde{S}_A^{\mathrm{U}}$ and $\widetilde{S}_A^{\mathrm{C}}$ based on extended bases and D-MORPH regression (see [276] and references therein).

Chastaing et al. [75] propose a new approach for the estimation of $\widetilde{S}_A^{\mathrm{U}}$ and $\widetilde{S}_A^{\mathrm{C}}$ based on local linear regression to approximate conditional expectations. Examples are provided in small dimensions only. In higher dimensions, assuming a sparsity of the significant components, Chastaing et al. [76, 73] propose estimation procedures based on penalized regression, such as lasso or $\mathbb{L}^2$-boosting.

The next section provides a simple example to illustrate the approaches presented in Sections 5.3.1 and 5.3.2.

## 5.3.4 ▪ Comparison of an additive model

Let us consider the additive model $G(\mathbf{x}) = G(x_1, x_2) = x_1 + x_2$. We assume that the random vector $(X_1, X_2)$ has normal distribution with zero mean and variance-covariance matrix $\left(\begin{smallmatrix} 1 & \rho \\ \rho & 1 \end{smallmatrix}\right)$ with $\rho \in [0, 1]$. We consider the usual Hoeffding decomposition of $G$ recalled in Section 5.3.1. Straightforward calculations lead to $G_\emptyset = 0$, $G_1(x_1) = x_1(1+\rho)$, $G_2(x_2) = x_2(1+\rho)$, $G_{1,2}(x_1, x_2) = -\rho(x_1 + x_2)$.

Then we compute the indices corresponding, respectively, to the structural and correlative contributions for $\mathbf{X}_A$ to $G(\mathbf{X})$, with $A \subset \{1, 2\}$:

$$\begin{aligned}
S_1^{\mathrm{U}} &= \frac{\operatorname{Var} G_1(X_1)}{V} = \frac{1+\rho}{2}, \\
S_1^{\mathrm{C}} &= \frac{\operatorname{Cov}[G_1(X_1), G_2(X_2) + G_{1,2}(X_1, X_2)]}{V} = 0, \\
S_2^{\mathrm{U}} &= \frac{\operatorname{Var} G_2(X_2)}{V} = \frac{1+\rho}{2}, \\
S_2^{\mathrm{C}} &= \frac{\operatorname{Cov}[G_2(X_2), G_1(X_1) + G_{1,2}(X_1, X_2)]}{V} = 0, \\
S_{1,2}^{\mathrm{U}} &= \frac{\operatorname{Var} G_{1,2}(X_1, X_2)}{V} = \rho^2, \\
S_{1,2}^{\mathrm{C}} &= \frac{\operatorname{Cov}[G_{1,2}(X_1, X_2), G_1(X_1) + G_2(X_2)]}{V} = -\rho(1+\rho).
\end{aligned}$$

We recover for $\rho = 0$ (independent case) the usual Sobol' indices. Moreover, even in the presence of dependencies among input variables, we have

$$S_1^{\mathrm{U}} + S_1^{\mathrm{C}} + S_2^{\mathrm{U}} + S_2^{\mathrm{C}} + S_{1,2}^{\mathrm{U}} + S_{1,2}^{\mathrm{C}} = 1.$$

Now let us turn to the approach based on the generalized Hoeffding decomposition, presented in Section 5.3.2. For our simple example, it is obvious to see that $\widetilde{G}_\emptyset = 0 \in H_\emptyset$, $\widetilde{G}_1(x_1) = x_1 \in H_1^0$, $\widetilde{G}_2(x_2) = x_2 \in H_2^0$, and $\widetilde{G}_{1,2}(x_1, x_2) = 0 \in H_{1,2}^0$. This decomposition may not be unique, as assumption (ii) in Proposition 5.3 is violated for the normal distribution. Following the strategy in Chastaing et al. [75], we get the uniqueness as the system

$$\begin{pmatrix} \mathrm{Id} & \mathrm{Proj}_{H_1^0} \\ \mathrm{Proj}_{H_2^0} & \mathrm{Id} \end{pmatrix} \begin{pmatrix} g_1 \\ g_2 \end{pmatrix} = \begin{pmatrix} \mathbb{E}[G(\mathbf{X})|X_1] - \mathbb{E}G(\mathbf{X}) \\ \mathbb{E}[G(\mathbf{X})|X_2] - \mathbb{E}G(\mathbf{X}) \end{pmatrix},$$

with $\mathrm{Proj}_{H_i^0}$ the orthogonal projector on $H_i^0$, has a unique solution.

With the generalized decomposition, we then compute the indices corresponding, respectively, to the structural and correlative contributions for $\mathbf{X}_A$ to $G(\mathbf{X})$, with $A \subset \{1, 2\}$:

$$\begin{aligned}
\widetilde{S}_1^{\mathrm{U}} &= \frac{\operatorname{Var}\widetilde{G}_1(X_1)}{V} = \frac{1}{2(1+\rho)}, & \widetilde{S}_1^{\mathrm{C}} &= \frac{\operatorname{Cov}[\widetilde{G}_1(X_1), \widetilde{G}_2(X_2)]}{V} = \frac{\rho}{2(1+\rho)},\\
\widetilde{S}_2^{\mathrm{U}} &= \frac{\operatorname{Var}\widetilde{G}_2(X_2)}{V} = \frac{1}{2(1+\rho)}, & \widetilde{S}_2^{\mathrm{C}} &= \frac{\operatorname{Cov}[\widetilde{G}_2(X_2), \widetilde{G}_1(X_1)]}{V} = \frac{\rho}{2(1+\rho)},\\
\widetilde{S}_{1,2}^{\mathrm{U}} &= \frac{\operatorname{Var}\widetilde{G}_{1,2}(X_1, X_2)}{V} = 0, & \widetilde{S}_{1,2}^{\mathrm{C}} &= 0.
\end{aligned}$$

Once more, we recover for the independent case ($\rho = 0$) the usual Sobol' indices. Moreover, even in the presence of dependencies among input variables, we have

$$\widetilde{S}_1^{\mathrm{U}} + \widetilde{S}_1^{\mathrm{C}} + \widetilde{S}_2^{\mathrm{U}} + \widetilde{S}_2^{\mathrm{C}} + \widetilde{S}_{1,2}^{\mathrm{U}} + \widetilde{S}_{1,2}^{\mathrm{C}} = 1.$$

Finally, we have $\widetilde{S}_{1,2}^{\mathrm{C}} = 0$, which is inherent to the choice of the generalized Hoeffding decomposition and not to the model itself.

### 5.3.5 ▪ Conclusion and further references

To estimate $\widetilde{S}_A^{\mathrm{U}}$ and $\widetilde{S}_A^{\mathrm{C}}$, Chastaing and Le Gratiet [77] propose substituting the model $G$ with a Gaussian process regression model. The main advantage of their approach is that it does not rely on the assumption that only a few interactions are significant. Their approach relies on a specific choice of the covariance kernel that should belong to a special class of ANOVA kernels studied in Berlinet and Thomas-Agnan [34] and in Durrande et al. [120] (see Section 4.2.3) and on a functional decomposition of a Gaussian process distributed with respect to the posterior distribution. Although interesting in terms of the curse of dimensionality, their approach is strongly dependent on the choice of the kernel in the ANOVA class.

The main issue of the covariance-based SA strategy is the interpretation of the indices. Indeed, the values of the indices depend on the underlying decomposition of $G$ as illustrated in our simple example. Moreover, even if the equality $\sum_{A \in \mathcal{P}_d} S_A = 1$ is still true, indices may be negative (see, e.g., [75]). For that reason, another strategy, based on the estimation of four sensitivity indices, was proposed in the literature [253, 304]. It still consists in defining variance-based sensitivity indices which allow distinguishing the independent contributions of the inputs to the response variance from their mutual dependent contributions. Further details are provided in Section 5.4 below.

## 5.4 ▪ Independent and full Sobol' indices

Once more in this section, the variables $X_1, \dots, X_d$ are not necessarily independent. The joint probability measure $\mathrm{P}_{\mathbf{X}}$ is assumed to be absolutely continuous with respect to the Lebesgue measure on $\mathbb{R}^d$. Thus it admits a joint probability density function, denoted by $p$ in the following.

### 5.4.1 ▪ Main principles

Defined from conditional probability distributions, the Rosenblatt transform is the main ingredient of this section. We refer the reader to Lévy [274] (pages 121–123) and Rosenblatt [398] for a complete description of the construction of this transformation. The Rosenblatt transform $T^1$ associated to $\mathbf{X} = (X_1, \dots, X_d)$ is defined by

$$T^1(x_1, \dots, x_d) = \begin{pmatrix} T_1^1(x_1) \\ \cdot \\ \cdot \\ \cdot \\ T_k^1(x_1, \dots, x_k) \\ \cdot \\ \cdot \\ \cdot \\ T_d^1(x_1, \dots, x_d) \end{pmatrix} = \begin{pmatrix} \mathbb{P}(X_1 \leq x_1) \\ \cdot \\ \cdot \\ \cdot \\ \mathbb{P}(X_k \leq x_k | X_1 = x_1, \dots, X_{k-1} = x_{k-1}) \\ \cdot \\ \cdot \\ \cdot \\ \mathbb{P}(X_d \leq x_d | X_1 = x_1, \dots, X_{d-1} = x_{d-1})) \end{pmatrix}.$$

It is proved (see, e.g., [398, 115]) that $\mathbf{U} = T^1(\mathbf{X})$ is uniformly distributed on $[0,1]^d$. For any $j = 1, \dots, d$, let $P_j$ be the circular permutation which maps $\mathbf{X}$ to $(X_j, X_{j+1}, \dots, X_d, X_1, \dots, X_{j-1})$. In a similar manner, we define the Rosenblatt transform $T^j$ associated to $P_j(\mathbf{X})$. We then have $\mathbf{U}^j = T^j \circ P_j(\mathbf{X}) \sim \mathcal{U}\left([0,1]^d\right)$. Let us use the convention that $\mathbf{U}^{d+1} = \mathbf{U}^1$. Let $(T^j)^{\rightarrow}$ denote the generalized inverse of $T^j$ and $G^j = G \circ P_j^{-1} \circ (T^j)^{\rightarrow}$. Then we have $G(\mathbf{X}) = G(X_1, \dots, X_d) = G^j(\mathbf{U}^j)$.

Now we can define the four following indices associated to $X_j$:

$$\begin{cases} S_{(j)} = \dfrac{\operatorname{Var} \mathbb{E}[G^j(\mathbf{U}^j) | U_1^j]}{\operatorname{Var} G^j(\mathbf{U}^j)}, \\[2ex] S_{(j)}^{\mathrm{T}} = \dfrac{\mathbb{E} \operatorname{Var}[G^j(\mathbf{U}^j) | \mathbf{U}_{-1}^j]}{\operatorname{Var} G^j(\mathbf{U}^j)}, \\[2ex] S_{(j)}^{\mathrm{ind}} = \dfrac{\operatorname{Var} \mathbb{E}[G^{j+1}(\mathbf{U}^{j+1}) | U_d^{j+1}]}{\operatorname{Var} G^{j+1}(\mathbf{U}^{j+1})}, \\[2ex] S_{(j)}^{\mathrm{T,ind}} = \dfrac{\mathbb{E} \operatorname{Var}[G^{j+1}(\mathbf{U}^{j+1}) | \mathbf{U}_{-d}^{j+1}]}{\operatorname{Var} G^{j+1}(\mathbf{U}^{j+1})}. \end{cases} \tag{5.3}$$

**Remark 27.** *Note that the previous indices can be extended to indices for groups of inputs [304].*

**Remark 28.** *We clearly get from the above definitions and from the independence between the components of the vector $\mathbf{U}^j$, respectively, $\mathbf{U}^{j+1}$, the relation*

$$0 \leq S_{(j)} \leq S_{(j)}^{\mathrm{T}} \leq 1, \quad \textit{resp.}, \; 0 \leq S_{(j)}^{\mathrm{ind}} \leq S_{(j)}^{\mathrm{T,ind}} \leq 1.$$

*However, there are no obvious relationships between $(S_{(j)}, S_{(j)}^{\mathrm{T,ind}})$, $(S_{(j)}, S_{(j)}^{\mathrm{ind}})$, $(S_{(j)}^{\mathrm{ind}}, S_{(j)}^{\mathrm{T}})$, and $(S_{(j)}^{\mathrm{T,ind}}, S_{(j)}^{\mathrm{T}})$.*

It is important to note that

$$S_{(j)} = \frac{\operatorname{Var}\mathbb{E}[G^j(\mathbf{U}^j)|U_1^j]}{\operatorname{Var} G^j(\mathbf{U}^j)} = \frac{\operatorname{Var}\mathbb{E}[G(\mathbf{X})|X_j]}{V} = S_j,$$

$$S_{(j)}^{\mathrm{T,ind}} = \frac{\mathbb{E}\operatorname{Var}[G^{j+1}(\mathbf{U}^{j+1})|\mathbf{U}_{-d}^{j+1}]}{\operatorname{Var} G^{j+1}(\mathbf{U}^{j+1})} = \frac{\mathbb{E}\operatorname{Var}[G(\mathbf{X})|\mathbf{X}_{-j}]}{V} = S_j^{\mathrm{T}}.$$

Thus we recover the usual first-order and total Sobol' indices associated to $X_j$.

By definition, $S_{(j)}$ and $S_{(j)}^{\mathrm{T}}$ correspond to the first-order and total sensitivity indices associated to $U_1^j = F_j(X_j)$ with $F_j$ the unconditional cumulative distribution function of $X_j$. These indices thus include the effects of the dependence of $X_j$ with other inputs. For this reason, Mara and Tarantola [303] call them the full sensitivities of $X_j$.

We have assumed that the random vector $\mathbf{X}$ is defined by a joint probability density function $p(\mathbf{x})$. Thus, for any $j = 1, \ldots, d$, $p(\mathbf{x})$ can be decomposed as

$$\underbrace{p(x_{j+1})p(x_{j+2}|x_{j+1})\ldots p(x_d|\mathbf{x}_{[j+1:d-1]})p(x_1|\mathbf{x}_{[j+1:d]})\ldots p(x_{j-1}|\mathbf{x}_{[j+1:d]\cup[1:j-2]})}_{\mathbf{U}_{-d}^{j+1}}\underbrace{p(x_j|\mathbf{x}_{-(j)})}_{\mathbf{U}_d^{j+1}},$$

where, for $k, i_1, \ldots, i_\ell \in [1 : d]$, $p(x_k|\mathbf{x}_{[i_1,\ldots,i_\ell]})$ denotes the probability density function of $X_k|\mathbf{X}_{[i_1,\ldots,i_\ell]} = \mathbf{x}_{[i_1,\ldots,i_\ell]}$. From that last equality, we conclude that $\mathbf{U}_d^{j+1}$ is independent of $\mathbf{U}_{-d}^{j+1}$. Thus, the indices $S_{(j)}^{\mathrm{ind}}$ and $S_{(j)}^{\mathrm{T,ind}}$ measure, respectively, the first-order and total independent contributions of $X_j$ as stated in Mara et al. [304] (see also Mara and Tarantola [303]).

### 5.4.2 ▪ Estimation procedures

The estimation of the four indices defined by (5.3) is based on integral formulations (see [304]) recalled hereafter:

$$S_{(j)} = \frac{1}{V}\int_{\mathbb{R}^d} G(x_j, \mathbf{x}_{-j})p(x_j, \mathbf{x}_{-j})d\mathbf{x}_{-j}$$
$$\left(\int_{\mathbb{R}^{d-1}} G(x_j, \overline{\mathbf{x}}'_{-j})p(\overline{\mathbf{x}}'_{-j}|x_j)d\overline{\mathbf{x}}'_{-j} - \int_{\mathbb{R}^d} G(x'_j, \mathbf{x}'_{-j})p(x'_j, \mathbf{x}'_{-j})d\mathbf{x}'_{-j}dx'_j\right) dx_j,$$

$$S_{(j)}^{\mathrm{T,ind}} = \frac{1}{2\,V}\int_{\mathbb{R}^{d+1}} \left(G(x'_j, \mathbf{x}'_{-j}) - G(\overline{x}_j, \mathbf{x}'_{-j})\right)^2 p(x'_j, \mathbf{x}'_{-j})p(\overline{x}_j|\mathbf{x}'_{-j})dx'_j d\overline{x}_j d\mathbf{x}'_{-j},$$

$$S_{(j)}^{\mathrm{ind}} = \frac{1}{V}\int_{\mathbb{R}^d} G(\overline{x}_j, \mathbf{x}_{-j})p(\overline{x}_j|\mathbf{x}_{-j})p(\mathbf{x}_{-j})d\mathbf{x}_{-j}$$
$$\left(\int_{\mathbb{R}^{d-1}} p(\overline{x}_j, \mathbf{x}'_{-j})p(\mathbf{x}'_{-j})d\mathbf{x}'_{-j} - \int_{\mathbb{R}^d} G(x'_j, \mathbf{x}'_{-j})p(x'_j, \mathbf{x}'_{-j})dx'_j d\mathbf{x}'_{-j}\right) d\overline{x}_j,$$

$$S_{(j)}^{\mathrm{T}} = \frac{1}{2\,V}\int_{\mathbb{R}^{d+1}} \left(G(x'_j, \overline{\mathbf{x}}'_{-j}) - G(x_j, \overline{\mathbf{x}}'_{-j})\right)^2$$
$$p(\overline{\mathbf{x}}'_{-j}|x'_j)p(\overline{\mathbf{x}}'_{-j}|x_j)p(x'_j)p(x_j)d\overline{\mathbf{x}}'_{-j}dx'_j dx_j.$$

Following Mara et al. [304], we now explain why, for $j \in [1 : d]$, four samples of size $n$ are necessary to compute sensitivity indices $(S_{(j)}, S_{(j)}^{\mathrm{T,ind}}, S_{(j)}^{\mathrm{ind}}, S_{(j)}^{\mathrm{T}})$. Two independent uniformly distributed samples $\mathbf{u}$ and $\mathbf{u}'$ of size $n$ are created. Then we generate four samples each of length

$N$ from $\mathbf{u}$ and $\mathbf{u}'$ by the generalized inverse Rosenblatt transform $(T^j)^{\to}$ as follows:

$$
\begin{aligned}
(u_1,\ldots,u_d) &\xrightarrow{(T^j)^{\to}} (x_j,\ldots,x_d,x_1,\ldots,x_{j-1}),\\
(u'_1,\ldots,u'_d) &\xrightarrow{(T^j)^{\to}} (x'_j,\ldots,x'_d,x'_1,\ldots,x'_{j-1}),\\
(u_1,u'_2,\ldots,u'_d) &\xrightarrow{(T^j)^{\to}} (x_j,\overline{x}'_{j+1},\ldots,\overline{x}'_d,\overline{x}'_1,\ldots,\overline{x}'_{j-1}) \sim p(x_j)p(\overline{\mathbf{x}}'_{-j}|x_j),\\
(u'_1,\ldots,u'_{d-1},u_d) &\xrightarrow{(T^j)^{\to}} (x'_j,x'_{j+1},\ldots,x'_{j-2},\overline{x}_{j-1}) \sim p(\mathbf{x}'_{-(j-1)})p(\overline{x}_{j-1}|\mathbf{x}'_{-(j-1)}).
\end{aligned}
$$

Finally, we estimate the indices by approaching the integral formulations by Monte Carlo formulations. This procedure is thus based on $6 \times n$ evaluations of our model: $G(x_j,\mathbf{x}_{-j})$, $G(x_j,\overline{\mathbf{x}}'_{-j})$, $G(x'_j,\mathbf{x}'_{-j})$, $G(\overline{x}_j,\mathbf{x}'_{-j})$, $G(\overline{x}_j,\mathbf{x}_{-j})$, and $G(x'_j,\overline{\mathbf{x}}'_{-j})$ for $j = 1,\ldots,N$.

Note that the procedure described above is based on the Rosenblatt transformation and requires knowledge of conditional probability densities. Such information is not always available. Iman and Conover [203] propose a distribution-free transformation to induce rank correlation among input variables. Mara et al. [304] suggest using the Iman and Conover transformation in place of the Rosenblatt one. However, such a procedure is only well suited if the dependence structure of vector $\mathbf{X}$ is characterized (or at least well summarized) by its rank correlation matrix.

### 5.4.3 ▪ Illustration on a few examples

First, let us provide the example of an interactive model with two inputs, which has been handled in Benoumechiara and Elie-Dit-Cosaque [31].

**Example 5.4.** We consider a pure interactive model

$$G(\mathbf{X}) = Y = \beta_1\beta_2 X_1 X_2 \tag{5.4}$$

with $\mathbf{X} \sim \mathcal{N}((0,0)^T,\Sigma)$, with variance-covariance matrix

$$\Sigma = \begin{pmatrix} \sigma_1^2 & \rho\sigma_1\sigma_2 \\ \rho\sigma_1\sigma_2 & \sigma_2^2 \end{pmatrix}, \qquad -1 \le \rho \le 1, \sigma_1 > 0, \sigma_2 > 0.$$

Then it is possible to derive analytical formulas for the indices defined by (5.3) (see Table 5.1, which is reproduced from Benoumechiara and Elie-Dit-Cosaque [31]). These formulas are also plotted, as a function of $\rho$, in Fig. 5.3 (no distinction is made between indices related to $X_1$ or $X_2$ because they are equal). In Fig. 5.3 are also plotted the Shapley effects, which will be introduced in Section 5.5. Note that the indices do not depend on $\beta_1$, $\beta_2$, $\sigma_1$, and $\sigma_2$: They only depend on $\rho$.

In the independent model ($\rho = 0$), the first-order Sobol' indices are null because there is no dependence and the inputs have no marginal contribution. Moreover, in that independence setting, the total Sobol' indices of $X_1$ and $X_2$ are equal and represent the variability in the model which is due to the interaction. If correlated inputs ($\rho \neq 0$) are considered, we still get $S^{\text{ind}}_{(j)} = 0$ (for $j = 1, 2$), as the model is a pure interaction model. However, the full first-order indices $S_{(j)}$ (for $j = 1, 2$) are nonnull and represent marginal contribution due to the dependence between the two inputs. The extreme case is reached when $|\rho| = 1$ (total dependence between $X_1$ and $X_2$), where the independent first-order and total Sobol' indices are zero and the full first-order and total Sobol' indices are one. ■

**Table 5.1.** *Formulas of independent and full first-order and total Sobol' indices for the pure interactive model given by Eq.* (5.4).

| **Model variance** $\sigma^2 = \mathrm{Var}\, Y = V = (1+\rho^2)\beta_1^2\beta_2^2\sigma_1^2\sigma_2^2$ |
|---|
| **Independent first-order Sobol' indices** $S_{(1)}^{\text{ind}} = S_{(2)}^{\text{ind}} = 0$ |
| **Independent total Sobol' indices** $\sigma^2 S_{(1)}^{\text{T,ind}} = \sigma^2 S_{(2)}^{\text{T,ind}} = (1-\rho^2)\beta_1^2\beta_2^2\sigma_1^2\sigma_2^2$ |
| **Full first-order Sobol' indices** $\sigma^2 S_{(1)} = \sigma^2 S_{(2)} = 2\rho^2\beta_1^2\beta_2^2\sigma_1^2\sigma_2^2$ |
| **Full total Sobol' indices** $\sigma^2 S_{(1)}^{\text{T}} = \sigma^2 S_{(2)}^{\text{T}} = (1+\rho^2)\beta_1^2\beta_2^2\sigma_1^2\sigma_2^2$ |

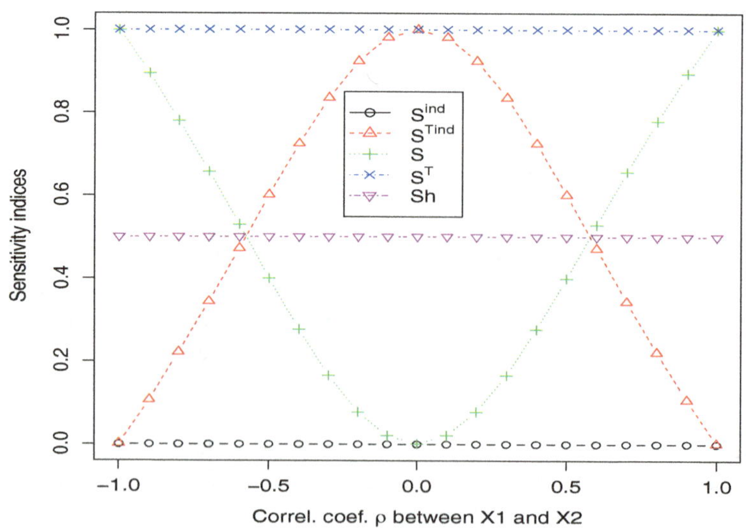


**Figure 5.3.** *Sobol' indices and Shapley effects for the pure interactive model given by Eq.* (5.4).

We also consider a 3D example below.

**Example 5.5.** Let us consider the model in dimension $d = 3$ defined by

$$G(\mathbf{X}) = Y = X_1 + X_2X_3 \tag{5.5}$$

with

$$\mathbf{X} = \begin{pmatrix} X_1 \\ X_2 \\ X_3 \end{pmatrix} \sim \mathcal{N}\left( \begin{pmatrix} 0 \\ 0 \\ 0 \end{pmatrix}, \Sigma \right), \Sigma = \begin{pmatrix} \sigma_1^2 & 0 & \rho\sigma_1\sigma_3 \\ 0 & \sigma_2^2 & 0 \\ \rho\sigma_1\sigma_3 & 0 & \sigma_3^2 \end{pmatrix}, \quad -1 \le \rho \le 1.$$

Let $\sigma^2 = V = \operatorname{Var} Y$. We have $\sigma^2 = \sigma_1^2 + \sigma_2^2\sigma_3^2$. It is also possible to compute analytically the full and independent Sobol' indices for this example (see, e.g., [211] for the computation of full first-order and independent total Sobol' indices). All the formulas are given in Table 5.2 and plotted, as a function of $\rho$, in Fig. 5.4 (for the case $\sigma_1 = \sigma_2 = \sigma_3 = 1$).

**Table 5.2.** *Formulas of independent and full first-order and total Sobol' indices given by* (5.3) *for the* $3D$ *model given by Eq.* (5.5).

| |
|---|
| **Model variance**<br>$\sigma^2 = \operatorname{Var} Y = V = \sigma_1^2 + \sigma_2^2\sigma_3^2$ |
| **Independent first-order Sobol' indices**<br>$\sigma^2 S_{(1)}^{\text{ind}} = \sigma_1^2(1-\rho^2), \sigma^2 S_{(2)}^{\text{ind}} = 0, \sigma^2 S_{(3)}^{\text{ind}} = 0$ |
| **Independent total Sobol' indices**<br>$\sigma^2 S_{(1)}^{\text{T,ind}} = \sigma_1^2(1-\rho^2), \sigma^2 S_{(2)}^{\text{T,ind}} = \sigma_2^2\sigma_3^2, \sigma^2 S_{(3)}^{\text{T,ind}} = \sigma_2^2\sigma_3^2(1-\rho^2)$ |
| **Full first-order Sobol' indices**<br>$\sigma^2 S_{(1)} = \sigma_1^2, \sigma^2 S_{(2)} = 0, \sigma^2 S_{(3)} = \sigma_1^2\rho^2$ |
| **Full total Sobol' indices**<br>$\sigma^2 S_{(1)}^{\text{T}} = \sigma_1^2 + \sigma_2^2\sigma_3^2\rho^2, \sigma^2 S_{(2)}^{\text{T}} = \sigma_2^2\sigma_3^2, \sigma^2 S_{(3)}^{\text{T}} = \sigma_2^2\sigma_3^2 + \sigma_1^2\rho^2$ |

From $S_{(2)}^{\text{ind}} = 0$ and $S_{(2)} = 0$, we deduce that $X_2$ only acts on $Y$ through interaction. Then, from $S_{(3)}^{\text{ind}} = 0$ and $\sigma^2 S_{(3)} = \sigma_1^2\rho^2$, we deduce that the main effect of $X_3$ on $Y$ is only due to the correlation between $X_1$ and $X_3$. No full total Sobol' index is null so that all the inputs influence $Y$ (maybe not that much, depending on the values of $\sigma_1^2$, $\sigma_2^2$, and $\sigma_3^2$). We also note that $S_{(2)}^{\text{T}} = S_{(2)}^{\text{T,ind}}$, which is due to the independence of $X_2$ from $(X_1, X_3)$. ∎

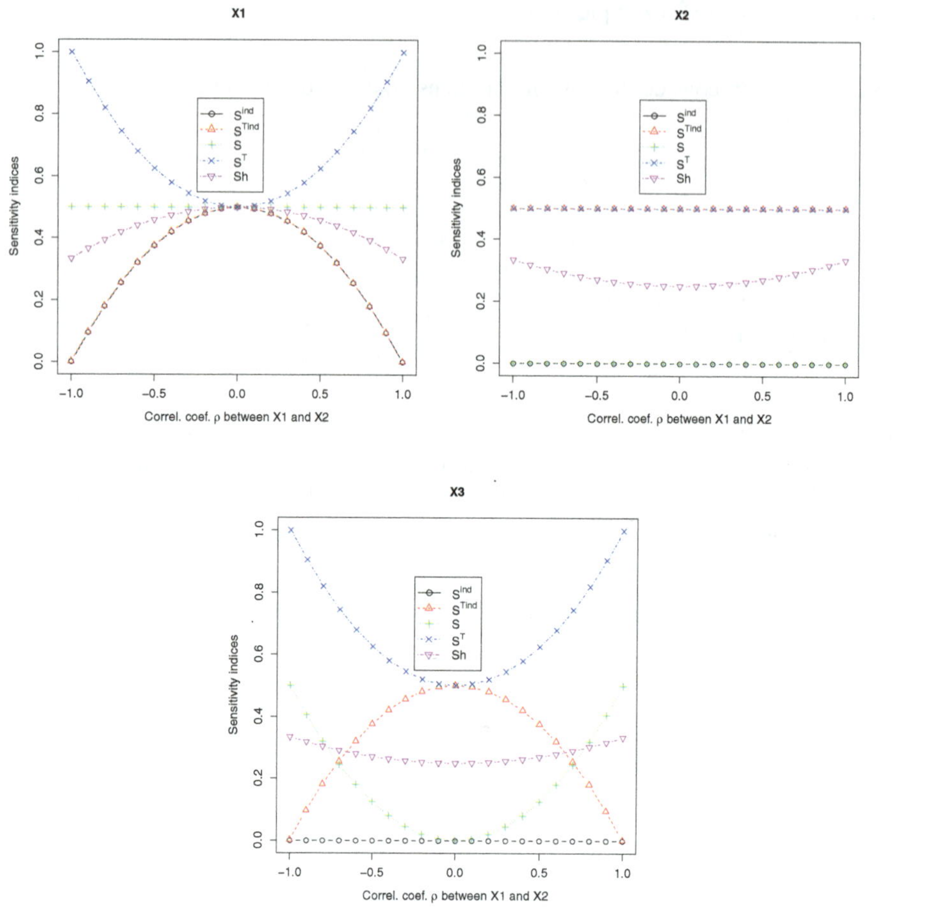


**Figure 5.4.** *Sobol' indices and Shapley effects for each input $X_1$, $X_2$, and $X_3$ for the 3D model given by Eq.* (5.5).

## 5.4.4 ▪ Conclusion

The approach based on the estimation of the four indices defined by (5.3) for each input variable, although informative and interesting, does not allow one to allocate part of the global variance of the output to each input variable. An interesting alternative was introduced in Owen [350] in the framework of independent input variables. It is based on the theory of Shapley values and can be extended to the framework of dependent input variables (see [447, 353, 211, 31]). This strategy is presented in Section 5.5 below.

# 5.5 ▪ Shapley effects

An alternative way to define variable importance consists in dealing with Shapley values that provide a way to attribute the value created by a team to its individual members. For SA of model output, the members are individual input variables $X_1, \ldots, X_d$, and the value created by a team is related to its explanatory power. In that context, we deal with Shapley effects.

### 5.5.1 ▪ Main principles

Shapley values [431], well known in game theory and economics, have been proposed by Owen [350] in the framework of SA of model outputs. Their use and interpretation do not require the probability distribution of the input vector $\mathbf{X}$ to be absolutely continuous with respect to the Lebesgue measure. The only assumption is that the model $G$ belongs to $\mathbb{L}^2(\mathrm{P}_\mathbf{X})$. Contrarily to covariance-based sensitivity analysis (see Section 5.3), Shapley effects result from a direct allocation of a part of the variance of the output at each input and provide a nonnegative importance measure for each input variable. Moreover, the sum of the $d$ Shapley effects associated to the $d$ input variables $X_1, \dots, X_d$ is worth the variance of the output $Y$. The allocation rule is based on an equitable principle, ensuring in particular that an interaction effect is equally distributed between each input involved in the interaction.

Although Shapley effects solve the conceptual problems, computational problems remain a serious challenge (see [68]). An algorithm based on random permutations was introduced in Song et al. [447]. The use of metamodel inside this algorithm was discussed and studied in Iooss and Prieur [211] and Benoumechiara and Elie-Dit-Cosaque [31].

Let us now get into more details. Consider a function (known as a characteristic function) val : $\mathcal{P}_d \to \mathbb{R}_+$ such that val$(\emptyset) = 0$. Based on the characteristic function, a value $\phi_j$ is attributed to each input variable $X_j$ (to each player $j$ in a game theory context). There are four very compelling properties that an attribution method should have (see, e.g., [504]):

(1) (Efficiency) $\sum_{j=1}^d \phi_j = \mathrm{val}([1:d])$. The resources available to the grand coalition of all the input variables are distributed among themselves.

(2) (Symmetry) If $\mathrm{val}(A \cup \{i\}) = \mathrm{val}(A \cup \{j\})$ for all $A \subset -\{i,j\}$, then $\phi_i = \phi_j$. If two input variables have the same marginal contribution to any coalition, they have the same attributed value.

(3) (Dummy) If $\mathrm{val}(A \cup \{i\}) = \mathrm{val}(A)$ for all $A \in \mathcal{P}_d$, then $\phi_i = 0$. The input variables whose marginal contribution is null to every coalition have null values.

(4) (Additivity) If val and val$'$ have Shapley values $\phi$ and $\phi'$, respectively, then the game with value val + val$'$ has Shapley value $\phi_j + \phi'_j$ for $j \in [1:d]$. The value is an additive operator on the space of all games.

Axioms (1)–(3) are fully intuitive and essential for SA, which aims to attribute a value (that will be translated in terms of sensitivity index) for each input variable. Axiom (1) means that the value of all the inputs on the model output has to be shared between each input. Axiom (2) means that if two inputs have the same effects on the model output, their values are equal. Axiom (3) means that the values of inputs which have no effect on the model output are zero.

For $j = 1, \dots, d$, the Shapley values are then given by the following theorem.

**Theorem 5.6 (Shapley [431]).** *The unique valuation $\phi$ that satisfies axioms* (1)–(4) *attributes value*

$$\phi_j = \frac{1}{d} \sum_{A \subset -j} \binom{d-1}{|A|}^{-1} \left(\mathrm{val}(A \cup \{j\}) - \mathrm{val}(A)\right) \tag{5.6}$$

*to variable $X_j$.*

***Proof.*** We refer the interested reader to Shapley [431]. □

The Shapley values can then be adapted in the context of SA.

**Definition 5.7 (Shapley effect).** *For any $j = 1, \dots, d$, we define the Shapley effect for $X_j$ as*

$$Sh_j = \frac{1}{d} \sum_{A \subset -j} \binom{d-1}{|A|}^{-1} (S^{\text{clos}}_{A\cup\{j\}} - S^{\text{clos}}_A). \tag{5.7}$$

*This definition corresponds to the index $\phi_j$ obtained by defining the value function* val $: \mathcal{P}_d \to \mathbb{R}_+$ *as*

$$\text{val}(A) = S^{\text{clos}}_A = \frac{\text{Var}\,\mathbb{E}[G(\mathbf{X})|\mathbf{X}_A]}{V}. \tag{5.8}$$

From (5.7), we see that the Shapley effect is defined for any function for which $\text{Var}\,\mathbb{E}[G(\mathbf{X})|\mathbf{X}_A]$ exists. The components $X_j$ do not have to be real valued, though $G(\mathbf{X})$ must be.

The property of fundamental importance for SA is given by the following proposition.

**Proposition 5.8 (Shapley effects decomposition).** *If the characteristic function* val *is defined by Eq.* (5.8), *then we get the following decomposition:*

$$1 = \sum_{j=1}^{d} Sh_j.$$

***Proof.*** The proof is straightforward from the efficiency axiom (4) and from the definition of the characteristic function associated to Shapley effects. □

**Remark 29.** *Note that the Shapley effect simplifies for independent inputs [350]*

$$Sh_j = \sum_{A\in\mathcal{P}_d,\ j\in A} \frac{S_A}{|A|}.$$

*This equation clearly shows that the interaction effects are fairly distributed to their corresponding inputs: A second-order interaction is partitioned into two parts, a third-order one is partitioned into three parts, etc. With Sobol' indices, an interaction effect is fully allocated to each input which is involved inside.*

The following proposition is useful for the estimation of the Shapley effects (instead of using Eq. (5.8)).

**Proposition 5.9 (Song et al. [447]).** *Let* $\text{val}' : \mathcal{P}_d \to \mathbb{R}_+$ *be defined as*

$$\text{val}'(A) = \frac{\mathbb{E}\,\text{Var}\,[G(\mathbf{X})|\mathbf{X}_{\bar{A}}]}{V}. \tag{5.9}$$

*Choosing as characteristic function the function defined by Eq.* (5.9) *also leads to $\phi_j = Sh_j$, $j = 1, \dots, d$.*

***Proof.*** First, we get from the total variance theorem (see Lemma 3.7, page 53)

$$V = \mathbb{E}\,\text{Var}\,[G(\mathbf{X})|\mathbf{X}_A] + \text{Var}\,\mathbb{E}[G(\mathbf{X})|\mathbf{X}_A].$$

Thus, for any $j = 1, \dots, d$, for any $A \subset -j$,

$$\begin{aligned}\text{val}(A\cup\{j\}) - \text{val}(A) &= \text{Var}\,\mathbb{E}[G(\mathbf{X})|\mathbf{X}_{A\cup\{j\}}] - \text{Var}\,\mathbb{E}[G(\mathbf{X})|\mathbf{X}_A] \\ &= \mathbb{E}\,\text{Var}\,[G(\mathbf{X})|\mathbf{X}_A] - \mathbb{E}\,\text{Var}\,[G(\mathbf{X})|\mathbf{X}_{A\cup\{j\}}].\end{aligned} \tag{5.10}$$

Then note that

$$\{A \in \mathcal{P}_d\,,\ A \subset -j\} = \{-(C \cup \{j\})\ ,\ C \in \mathcal{P}_d\,,\ C \subset -j\}.$$

Now, replacing $A \subset -j$ with $-(C \cup \{j\})$, $C \subset -j$ in Eq. (5.10), we get

$$\begin{aligned}\text{val}(A \cup \{j\}) - \text{val}(A \cup \{j\}) &= \mathbb{E}\,\text{Var}\left[G(\mathbf{X})|\mathbf{X}_{-(C\cup\{j\})}\right] - \mathbb{E}\,\text{Var}\left[G(\mathbf{X})|\mathbf{X}_{-C}\right] \\ &= \text{val}'(C \cup \{j\}) - \text{val}'(C),\end{aligned}$$

which concludes the proof of Proposition 5.9. □

That provides an alternative way to compute Shapley effects.

### 5.5.2 ▪ Estimation procedures

As noticed in Song et al. [447], one barrier to using the Shapley value (Shapley effect in our framework) is its computational complexity, as all possible subsets of the players (input variables in our framework) need to be considered. One way to enumerate these subsets is to first consider all possible permutations of the players (see [68, 447]).

Given a permutation $\pi$ of all $d$ players, define the set $P_j(\pi)$ as the players that precede player $j$ in $\pi$. For instance, if $d = 4$ and $\pi = \{1, 3, 4, 2\}$, then $P_4(\pi) = \{1, 3\}$. Hence, the incremental cost of including player $j$ in $P_j(\pi)$ is $\text{val}(P_j(\pi) \cup \{j\}) - \text{val}(P_j(\pi))$. Taking all possible permutations $\Pi_d$ of $d$ players into consideration, the Shapley value $\phi_j$ in Eq. (5.6) can be rewritten as

$$Sh_j = \sum_{\pi \in \Pi_d} \frac{1}{d!} \left(\text{val}(P_j(\pi) \cup \{j\}) - \text{val}(P_j(\pi))\right). \tag{5.11}$$

To compute the Shapley effect from Eq. (5.11), we essentially need to evaluate $\text{val}(A)$ for all $A \in \mathcal{P}_d$ (i.e., $2^d - 1$ variance components) and $d$ permutations.

Instead of the exact calculation, Castro et al. [68] provide an approximation algorithm, which is modified in Song et al. [447] to increase its efficiency and to include the Monte Carlo simulation of the characteristic function val. In Song et al. [447], the authors consider the estimation of Shapley effects, working with the characteristic function defined by Eq. (5.9). Indeed, they point out that the double Monte Carlo estimator with the characteristic function defined by Eq. (5.8) can suffer from a nonneglectable bias if the inner loop sample is small, while in contrast the double Monte Carlo estimator with the characteristic function defined by Eq. (5.9) is unbiased for any sample size. For large dimension $d$ of the input space, they draw uniformly and independently $m$ permutations $\pi_1, \pi_2, \ldots, \pi_m$ in $\Pi_d$, and they estimate $Sh_j$ by a Monte Carlo scheme

$$\widehat{Sh}_j = \frac{1}{m} \sum_{\ell=1}^{m} \left(\text{val}'(P_j(\pi_\ell) \cup \{j\}) - \text{val}'(P_j(\pi_\ell))\right) \tag{5.12}$$

with val′ defined in Eq. (5.9). Then if the incremental cost

$$\Delta_j \text{val}'(\pi_\ell) = \text{val}'(P_j(\pi_\ell) \cup \{j\}) - \text{val}'(P_j(\pi_\ell))$$

is calculated for each $j$, redundant calculations occur. For instance, if $d = 4$ and $\pi_\ell = \{1, 3, 4, 2\}$, the initial algorithm in Castro et al. [68] involves the computation of

$$\Delta_1 \mathrm{val}'(\pi_\ell) = \mathrm{val}'(\{1\}) - \mathrm{val}'(\emptyset),$$

$$\Delta_2 \mathrm{val}'(\pi_\ell) = \mathrm{val}'(\{1,3,4,2\}) - \mathrm{val}'(\{1,3,4\}),$$

$$\Delta_3 \mathrm{val}'(\pi_\ell) = \mathrm{val}'(\{1,3\}) - \mathrm{val}'(\{1\}),$$

$$\Delta_4 \mathrm{val}'(\pi_\ell) = \mathrm{val}'(\{1,3,4\}) - \mathrm{val}'(\{1,3\}).$$

To avoid these redundancies, Song et al. [447] propose calculating the costs from the smallest subset of $\pi$ to the largest one and subtracting the previous set cost to obtain the marginal cost, i.e.,

$$\begin{aligned}\Delta_{\pi_\ell(j)} \mathrm{val}'(\pi_\ell) &= \mathrm{val}'(P_{\pi_\ell(j)}(\pi_\ell) \cup \{\pi_\ell(j)\}) - \mathrm{val}'(P_{\pi_\ell(j)}(\pi_\ell)) \\ &= \mathrm{val}'(P_{\pi_\ell(j+1)}(\pi_\ell)) - \mathrm{val}'(P_{\pi_\ell(j)}(\pi_\ell)).\end{aligned}$$

We get on our example

$$\Delta_{\pi_\ell(1)} \mathrm{val}'(\pi_\ell) = \Delta_1 \mathrm{val}'(\pi_\ell) = \mathrm{val}'(\{1\}) - \mathrm{val}'(\emptyset),$$

$$\Delta_{\pi_\ell(2)} \mathrm{val}'(\pi_\ell) = \Delta_3 \mathrm{val}'(\pi_\ell) = \mathrm{val}'(\{1,3\}) - \mathrm{val}'(\{1\}),$$

$$\Delta_{\pi_\ell(3)} \mathrm{val}'(\pi_\ell) = \Delta_4 \mathrm{val}'(\pi_\ell) = \mathrm{val}'(\{1,3,4\}) - \mathrm{val}'(\{1,3\}),$$

$$\Delta_{\pi_\ell(4)} \mathrm{val}'(\pi_\ell) = \Delta_2 \mathrm{val}'(\pi_\ell) = \mathrm{val}'(\{1,3,4,2\}) - \mathrm{val}'(\{1,3,4\}).$$

Thus, ignoring the estimation of val$'$ for the moment, the algorithm in Song et al. [447] saves the cost of the current set to *prev* to calculate the next marginal cost. For $\pi_\ell = \{1,3,4,2\}$, the algorithm performs as

$$\Delta_{\pi_\ell(1)} = \Delta_1 = \mathrm{val}'(\{1\}) - \mathrm{val}'(\emptyset) = \mathrm{val}'(\{1\}),\ \text{set}\ prev = \mathrm{val}'(\{1\}),$$

$$\Delta_{\pi_\ell(2)} = \Delta_3 = \mathrm{val}'(\{1,3\}) - prev,\ \text{set}\ prev = \mathrm{val}'(\{1,3\}),$$

$$\Delta_{\pi_\ell(3)} = \Delta_4 = \mathrm{val}'(\{1,3,4\}) - prev,\ \text{set}\ prev = \mathrm{val}'(\{1,3,4\}),$$

$$\Delta_{\pi_\ell(4)} = \Delta_2 = \mathrm{val}'(\{1,3,4,2\}) - prev,\ \text{set}\ prev = \mathrm{val}'(\mathcal{P}_4) = 1,$$

where val$'(\pi_\ell)$ has been omitted in the notation for the sake of concision.

Taking the estimation of val$'$ into account, the total computation budget to estimate $Sh_j$ grows by the factor of the sampling budget for the Monte Carlo variance estimation, which can be a substantial increase. Note that the evaluations of the characteristic function val$'$ are variances estimated from Monte Carlo simulations. For each sampled permutation $\pi_\ell$, we need to evaluate the total variance $V$ and the expectation of the conditional variances $\mathbb{E}\,\mathrm{Var}\,[G(\mathbf{X})|\mathbf{X}_{-P_{\pi_\ell(j)}}]$, $j = 2, \ldots, d$. Then, for each term to be estimated, the algorithm proceeds in two loops. Getting back to our example, let us consider the estimation of

$$\mathbb{E}\,\mathrm{Var}\,[G(\mathbf{X})|\mathbf{X}_{-P_{\pi_\ell(4)}}] = \mathbb{E}\,\mathrm{Var}\,[G(\mathbf{X})|\mathbf{X}_{-P_2}] = \mathbb{E}\,\mathrm{Var}\,[G(\mathbf{X})|X_2].$$

In an outer loop, the sample $X_2^{(1)}, \ldots, X_2^{(N_\mathrm{O})}$ is generated first. Then, in the $k$th inner loop, $k = 1, \ldots, N_\mathrm{O}$, conditional on $X_2^{(k)}$, $\mathbf{X}_{\{1,3,4\}}^{(k,1)}, \ldots, \mathbf{X}_{\{1,3,4\}}^{(k,N_\mathrm{I})}$ are sampled. Thus, the total computational budget for the algorithm is $N_\mathrm{V} + m\,N_\mathrm{I}\,N_\mathrm{O}(d-1)$, with $N_\mathrm{V}$ the size of the Monte

Carlo sample used for the estimation of $V$. The combination of Eq. (5.12) with the estimation of the expectation of the conditional variances is referred to as the random permutation aggregation procedure in Broto et al. [56]. Adding a bootstrap procedure for the computation of confidence intervals associated to the estimation of each $Sh_j$, the algorithm summarizes as follows [31]:

1. Generate a sample $\mathbf{x}^{(1)}$ of size $N_\mathrm{V}$ from the random vector $\mathbf{X}$.
2. Compute $\mathbf{y}^{(1)} = G(\mathbf{x}^{(1)})$ to estimate $V$.
3. Generate a sample $\mathbf{x}^{(2)}$ of size $m(d-1)N_\mathrm{O}N_\mathrm{I}$ from the different conditional laws necessary to estimate the $\mathbb{E}\operatorname{Var}\left[G(\mathbf{X})|\mathbf{X}_{\bar{A}}\right]$, $A \in \mathcal{P}_d$.
4. Compute $\mathbf{y}^{(2)} = G(\mathbf{x}^{(2)})$.
5. Compute $\widehat{Sh}_j$ with Eq. (5.12).
6. For $b = 1$ to $B$,
   - sample with replacement a realization $\tilde{\mathbf{y}}^{(1)}$ of $\mathbf{y}^{(1)}$ to compute $V$;
   - sample by block with replacement a realization $\tilde{\mathbf{y}}^{(2)}$ of $\mathbf{y}^{(2)}$;
   - compute $\widehat{Sh}_j^{(b)}$ with Eq. (5.12) from $\{\tilde{\mathbf{y}}^{(1)}, \tilde{\mathbf{y}}^{(2)}\}$;

   end.
7. Compute confidence intervals for $Sh_j$ (see Section 4.2.5).

### 5.5.3 ▪ Beyond Song's algorithm [447]

Broto et al. [56] have developed an alternative to the random permutation aggregation procedure: the subset aggregation procedure. For any $A \subset \mathcal{P}_d$, let $W_A$ denote either $\operatorname{Var}\mathbb{E}(Y|\mathbf{X}_A)$ or $\mathbb{E}\operatorname{Var}\left[Y|\mathbf{X}_{\bar{A}}\right]$, depending on the characteristic function we consider, based on Eq. (5.8) or Eq. (5.9). One ingredient of both the random permutation and the subset aggregation procedures is the estimation of $W_A$. Broto et al. [56] propose either estimating $\mathbb{E}\operatorname{Var}\left[Y|\mathbf{X}_{\bar{A}}\right]$ with a double Monte Carlo procedure, as in Song et al. [447], or estimating $\operatorname{Var}\mathbb{E}(Y|\mathbf{X}_A)$ with a pick-freeze procedure. More precisely, the double Monte Carlo estimate of $\mathbb{E}\operatorname{Var}\left[Y|\mathbf{X}_{\bar{A}}\right]$ is defined as

$$\frac{1}{N_A}\sum_{n=1}^{N_A}\frac{1}{N_I-1}\sum_{k=1}^{N_I}\left(G\left(\mathbf{X}_{\bar{A}}^{(n)},\mathbf{X}_A^{(n,k)}\right)-\overline{G\left(\mathbf{X}_{\bar{A}}^{(n)}\right)}\right) \tag{5.13}$$

with $\overline{G(\mathbf{X}_{\bar{A}}^{(n)})} = \frac{1}{N_I}\sum_{k=1}^{N_I} G(\mathbf{X}_{\bar{A}}^{(n)}, \mathbf{X}_A^{(n,k)})$, where $(\mathbf{X}_{\bar{A}}^{(n)})_{1\leq n\leq N_A}$ is an i.i.d. sample from the distribution of $\mathbf{X}_{\bar{A}}$ and $(\mathbf{X}_A^{(n,k)})_{1\leq k\leq N_I}$ conditionally to $\mathbf{X}_{\bar{A}}^{(n)}$ is i.i.d. from the distribution of $\mathbf{X}_A$ conditionally to $\mathbf{X}_{\bar{A}} = \mathbf{X}_{\bar{A}}^{(n)}$. The pick-freeze estimate of $\operatorname{Var}\mathbb{E}(Y|\mathbf{X}_A)$ is defined as

$$\frac{1}{N_A}\sum_{n=1}^{N_A} G\left(\mathbf{X}_A^{(n)},\mathbf{X}_{\bar{A}}^{(n,1)}\right) G\left(\mathbf{X}_A^{(n)},\mathbf{X}_{\bar{A}}^{(n,2)}\right) - (\mathbb{E}Y)^2, \tag{5.14}$$

where $(\mathbf{X}_A^{(n)})_{1\leq n\leq N_A}$ is an i.i.d. sample from the distribution of $\mathbf{X}_A$, and $\mathbf{X}_{\bar{A}}^{(n,1)}$ and $\mathbf{X}_{\bar{A}}^{(n,2)}$ conditionally to $\mathbf{X}_A^{(n)}$ are independent from the distribution of $\mathbf{X}_{\bar{A}}$ conditionally to $\mathbf{X}_A = \mathbf{X}_A^{(n)}$. Note that $\mathbb{E}Y$ can be replaced by the empirical mean. The subset aggregation procedure consists in computing once for all the estimates of $W_A$ for all $A \in \mathcal{P}_d$ and to store them. When the number of inputs $d$ is large, it may appear that it is too costly to estimate $W_A$ for all $A \in \mathcal{P}_d$. To tackle this issue, Broto et al. [56] suggest not to estimate all the $(W_A)_{A\in\mathcal{P}_d}$ with the same accuracy. When the number of variables is large, most of the $(W_A)_{A\in\mathcal{P}_d}$ are approximated by 0 (that is, $N_A$ is fixed to 0), and the subset aggregation procedure works even for large value of $d$. More precisely, if the $(W_A)_{A\in\mathcal{P}_d}$ are estimated independently from each other, at a cost $\kappa N_A$ and with a similar variance of estimation, they prove that the optimal set $(N_A)_{A\in\mathcal{P}_d}$ can be approximated by $N_A^* = \text{Round}\left(N_{\text{tot}}\kappa^{-1}\binom{d}{|A|}^{-1}(d-1)^{-1}\right)$, where Round is the nearest integer function. Here optimal is to be understood in the sense that the estimation variance for Shapley effects is minimal under the constraint $N_{tot} = \kappa\sum_{A\in\mathcal{P}_d} N_A$. The integer $\kappa$ is related to the number of model evaluations: $\kappa = 2$ for the pick-freeze estimate and $\kappa = N_I$ for the double loop estimate. The total cost $N_{\text{tot}}$ is fixed by the practitioner and corresponds to an upper bound on the computational budget.

**Remark 30.** *Note that it is possible to prove that choosing $N_I = 2$ in the definition of the double loop estimate leads to the pick-freeze estimate. Song et al. [447] prove that the variance of estimation is smaller by choosing $N_I = 3$ rather than $N_I = 2$ and that the gain from increasing $N_I$ from 3 decreases as $N_I$ increases. Also, Benoumechiara and Elie-Dit-Cosaque [31] show on numerical examples that choosing $N_I = 3$ leads to accurate results both in terms of accuracy and the probability of coverage of bootstrap confidence intervals.*

**Remark 31.** *Among the assumptions listed above, that is, the $(W_A)_{A\in\mathcal{P}_d}$ are estimated independently of each other, at a cost $\kappa N_A$ and with a similar variance of estimation, the last one is rather strong. However, Broto et al. [56] argue that the optimal allocation $N_A^*$, $A \in \mathcal{P}_d$ works well in practice, even if it is not satisfied.*

## 5.5.4 ▪ More recent estimation procedures

More recently, Plischke et al. [372] have proposed an algorithm based on the Möbius inverses to estimate Shapley effects. Implicitly, the Möbius inverses mob$(A)$ of the value functions val$(A)$ are defined by $\text{val}(A) = \sum_{B\subseteq A}\text{mob}(B)$ and therefore $\text{mob}(A) = \sum_{B\subseteq A}(-1)^{|A|+|B|}\text{val}(B)$ (see, e.g., [400]). Plischke et al. [372] start with the relation

$$\Phi_j(\text{val}) = \sum_{A\in\mathcal{P}_d,\, j\in A}\frac{\text{mob}(A)}{|A|}. \tag{5.15}$$

Let us recall the interpretation provided by Plischke et al. [372]: Player $j$ gets full credit for the games won by itself, half credit for those where the player is teamed in a pair, etc. From Eq. (5.15) with the value function defined by Eq. (5.8), it is possible to compute Shapley effects by computing the $2^d - 1$ value functions val$(A)$, $A \in \mathcal{P}_d$, and by solving a system of equations defined with a sparse $(2^d-1)\times(2^d-1)$ matrix. More precisely, let $H_\ell = \text{val}(A_\ell)$, with $A_\ell$, $\ell = 1,\ldots,2^d-1$ denoting all the $2^d-1$ nonvanishing subsets of $\mathcal{P}_d$. Let the matrix $Z$ code the partial ordering, that is, $Z_{k,\ell} = 1$ if $A_k \subseteq A_\ell$, and 0 otherwise. The Möbius inverses $M_\ell = \text{mob}(A_\ell)$ are then obtained by $M^T = H^T Z^{-1}$. Forming and inverting the $Z$ matrix altogether may be out of reach, even if $Z$ is sparse. The trick in Plischke et al. [372] is to generate the power set $\mathcal{P}_d$ via binary coding so that the $Z$ matrix coding the subset inclusion is Pascal's triangle modulo 2, for which we can construct the column by an exclusive-or operation. We refer the interested reader to Plischke et al. [372] for more details on the implementation. The main advantage of this algorithm is that it does not make any assumption on the terms that can be neglected in the representation given by Eq. (5.7) and that it does not rely on the random selection of permutations as described in Eq. (5.12).

Among recent development for the estimation of Shapley effects, let us mention a procedure based on independent groups of dependent inputs in the particular case of block-additive models introduced in Broto et al. [57]. In this specific framework, most of the Sobol' indices are equal to zero, and Shapley effects can be estimated more efficiently. More precisely, let us come back to the description of inputs in Eq. (5.1),

$$\mathbf{X} = (\underbrace{X_1, \dots, X_{\ell_1}}_{\vec{X}_1}, \dots, \underbrace{X_{L_{k-1}+1}, \dots, X_{L_{k-1}+\ell_k}}_{\vec{X}_k},$$
$$\dots, \underbrace{X_{L_{m-1}+1}, \dots, X_d}_{\vec{X}_m}),$$

where the $\vec{X}_k$ are independent groups of dependent inputs. We assume that $G$ is of the form $G(\mathbf{X}) = H(\vec{X}_1, \dots, \vec{X}_m) = \sum_{k=1}^{m} H_k(\vec{X}_k)$. For any $j \in \{1, \dots d\}$, $k(j) = k$ if and only if $X_j$ is one component of $\vec{X}_k$. Broto et al. [57] have established the relation $Sh_j = S^H_{k(j)}\, Sh_j^{H_{k(j)}}$, where $S^H_{k(j)}$ denotes the Sobol' index associated to input $\vec{X}_{k(j)}$ for the model $(\vec{X}_1, \dots, \vec{X}_m) \mapsto H(\vec{X}_1, \dots, \vec{X}_m)$ and $Sh_j^{H_{k(j)}}$ denotes the Shapley effect associated to input $X_j$ for model $(X_{L_{k(j)-1}+1}, \dots, X_{L_{k(j)-1}+\ell_{k(j)}}) \mapsto H_{k(j)}(X_{L_{k(j)-1}+1}, \dots, X_{L_{k(j)-1}+\ell_{k(j)}})$. Thus, the sum to compute the Shapley effect $Sh_j$ in block-additive models reduces from $\mathcal{P}_d \setminus \{j\}$ to all the subsets of $\{L_{k(j)-1}+1, \dots, L_{k(j)-1}+\ell_{k(j)}\} \setminus \{j\}$ of size $2^{\ell_{k(j)}} - 1$. Then each of the Sobol' indices implied in the definition of $Sh_k$ can be estimated, e.g., with a pick-freeze scheme.

**Remark 32.** *Another efficient scheme is proposed in Goda [165]. However, this scheme is only presented for independent inputs and has not been extended to dependent inputs yet. The algorithm combines a pick-freeze scheme directly applied to estimate the differences* $\mathrm{val}(A \cup j) - \mathrm{val}(A)$ *in Eq. (5.7) with random permutation arguments of Song et al. [447]. The cost is then reduced to* $(d+1)n$ *to estimate the full set of Shapley effects* $Sh_1, \dots, Sh_d$.

**Remark 33.** *The estimation procedures described in this section are based on the assumption that it is possible to sample from* $\mathbf{X}_{\bar{A}}$ *conditionally to* $\mathbf{X}_A = \mathbf{x}_A$. *In many cases, it is, however, not possible to sample from this distribution. Therefore, Broto et al. [56] introduced a given-data procedure, as described in the next section.*

### 5.5.5 ▪ A one-sample algorithm

In this section, we focus on a one-sample algorithm introduced in Broto et al. [56]. The main motivation for such an algorithm is to go beyond the restrictive assumption that we can generate an i.i.d. sample from the distribution $\mathbf{X}_{\bar{A}}$ conditionally to $\mathbf{X}_A = \mathbf{x}_A$. The algorithm is based on nearest neighbors as described hereafter. We consider an i.i.d. sample $\mathbf{X}^{(i)}$, $i = 1, \dots, n$, from the distribution of the input vector $\mathbf{X}$. We assume that for all $j \in [1:d]$, $(\mathcal{X}_j, d_j)$ is a Polish space with metric $d_j$ and $\mathbf{X}$ has a density $f_{\mathbf{X}}$ with respect to a finite measure $\mu = \otimes_{j=1}^{d} \mu_j$. This assumption allows us to consider some continuous variables but also categorical variables in ordered or unordered sets as far as variables in a separable Hilbert space. We consider $(s(\ell))_{1 \le \ell \le N_A}$ a sample of uniformly distributed integers in $\{1, \dots, n\}$, with or without replacement, independent of the other random variables. For $\emptyset \subsetneq B \subsetneq [1:d]$, let us define by $k_n^B(s(\ell), i)$ the index such that $\mathbf{X}_B^{(k_n^B(s(\ell),i))}$ is the (or one of the) $i$th closest element to $\mathbf{X}_B^{(s(\ell))}$ in $(\mathbf{X}_B^{(h)})_{h \in [1:n]}$ and such that $(k_n^B(s(\ell), i))_{i \in [1:N_I]}$ are two by two distinct. The algorithm consists either in replacing the double Monte-Carlo estimate defined in Eq. (5.13) by

$$\frac{1}{N_A}\sum_{\ell=1}^{N_A}\frac{1}{N_I-1}\sum_{i=1}^{N_I}\left(G\left(\mathbf{X}^{(k_n^{\bar{A}}(s(\ell),i))}\right)-\frac{1}{N_I}\sum_{h=1}^{N_I}G\left(\mathbf{X}^{(k_n^{\bar{A}}(s(\ell),h))}\right)\right)^2$$

or in replacing the pick-freeze estimate defined in Eq. (5.14) by

$$\frac{1}{N_A}\sum_{\ell=1}^{N_A}G\left(\mathbf{X}^{(k_n^{A}(s(\ell),1))}\right)G\left(\mathbf{X}^{(k_n^{\bar{A}}(s(\ell),2))}\right)-\left(\mathbb{E}\left[Y\right]\right)^2.$$

It is proved in Broto et al. [56] that under the more restrictive assumption that $f_{\mathbf{X}}$ is bounded and $P_{\mathbf{X}}$-almost everywhere continuous, both estimates are consistent as $n$ and $N_A$ tend to infinity. Under regularity assumptions on $f_{\mathbf{X}}$, it is also possible to prove rates of convergence. The interested reader is referred to Broto et al. [56] and the associated supplementary material for more details and proofs. This algorithm is implemented in the function shapleySubsetMc() of the sensitivity package.

**Remark 34.** *The function* sobolshap-knn() *of the* sensitivity *package implements the estimation of Shapley effects among several sensitivity indices with a one-sample algorithm, either adapting a recent procedure based on ranking introduced in Gamboa et al. [146] and inspired by Chatterjee [78] or the one in Broto et al. [56] based on nearest neighbor (see also Azadkia and Chatterjee [14]). For large sample sizes, the nearest-neighbor algorithm can be significantly accelerated by using approximate nearest-neighbor search.*

**Remark 35.** *This approach has been extended in Heredia et al. [190] to handle vectorial or functional outputs by defining and estimating aggregated Shapley effects (see also Section* 3.3.2 *in Chapter* 3*).*

### 5.5.6 ▪ Illustration of a few examples

Let us now illustrate the procedure based on Shapley effects in Example 5.4.

**Example 5.4 continued.** Let us come back to the pure interaction model handled in Example 5.4 of the previous section. Recall that we consider a centered random input vector $(X_1, X_2)$ with $\operatorname{Var} X_1 = \sigma_1^2$, $\operatorname{Var} X_2 = \sigma_2^2$, $\operatorname{Cov}(X_1, X_2) = \rho \in [-1, 1]$. Table 5.3 (see also Fig. 5.3) provides the Shapley effects for $X_1$ and $X_2$ related to this model.

**Table 5.3.** *Formulas of the Shapley effects for the pure interactive model.*

| **Model variance** |
|---|
| $\sigma^2 = \operatorname{Var} Y = V = (1+\rho^2)\beta_1^2\beta_2^2\sigma_1^2\sigma_2^2$ |
| **Shapley effects** |
| $\sigma^2 Sh_1 = \sigma^2 Sh_2 = \frac{1}{2}(1+\rho^2)\beta_1^2\beta_2^2\sigma_1^2\sigma_2^2$ |

We can see that the Shapley effects award fairly the interaction and the correlation, allowing to compare the effect of $X_1$ with the one of $X_2$. We also recover an exact decomposition: $Sh_1 + Sh_2 = 1$. Moreover, in this case, Shapley effects do not depend on $\rho$ and are both equal to 0.5. ∎

**Example 5.5 continued.** Let us come back to the model introduced in Example 5.5,

$$G(\mathbf{X}) = Y = X_1 + X_2 X_3$$

with

$$\mathbf{X} = \begin{pmatrix} X_1 \\ X_2 \\ X_3 \end{pmatrix} \sim \mathcal{N}\left( \begin{pmatrix} 0 \\ 0 \\ 0 \end{pmatrix}, \Sigma \right) \text{ and } \Sigma = \begin{pmatrix} \sigma_1^2 & 0 & \rho\sigma_1\sigma_3 \\ 0 & \sigma_2^2 & 0 \\ \rho\sigma_1\sigma_3 & 0 & \sigma_3^2 \end{pmatrix}, \quad -1 \leq \rho \leq 1.$$

Let $\sigma^2 = V = \operatorname{Var} Y$. It is possible to compute analytically the Shapley effects for this example [211]:

$$\begin{aligned} \sigma^2 Sh_1 &= \sigma_1^2 \left(1 - \frac{\rho^2}{2}\right) + \frac{\sigma_2^2\sigma_3^2}{6}\rho^2, \\ \sigma^2 Sh_2 &= \frac{\sigma_2^2\sigma_3^2}{6}(3 + \rho^2), \\ \sigma^2 Sh_3 &= \frac{\rho^2\sigma_1^2}{2} + \frac{\sigma_2^2\sigma_3^2}{6}(3 - 2\rho^2), \\ \sigma^2 &= \sigma_1^2 + \sigma_2^2\sigma_3^2. \end{aligned}$$

As we can see in the above formulas and in Fig. 5.4, Shapley effects share the correlation effect between $X_1$ and $X_3$, while at the same time it takes into account the interaction effect between $X_2$ and $X_3$, which is also shared with the fair allocation principle underlying Shapley's theory. Shapley effects seem therefore eligible for factor fixing, contrarily to all the other strategies presented in Sections 5.2, 5.3, and 5.4. ∎

To complete this section, we propose to illustrate one last example what happens in the presence of a correlated input which is not included in the model.

**Example 5.10.** Let us consider $Y = G(X_1, X_2) = X_1$ with $(X_1, X_2)$ two dependent standard Gaussian variables with a correlation coefficient $\rho$.

We compute the Shapley effects and get

$$Sh_1 = 1 - \rho^2/2\,, \; Sh_2 = \rho^2/2.$$

If $\rho$ is close to zero, $Sh_2$ is small and $X_2$ can be fixed without changing the output variance (factor fixing setting).

We can also compute Sobol' indices and get

$$S_{(1)} = 1\,, \; S_{(1)}^{\text{ind}} = 1 - \rho^2\,, \; S_{(1)}^{\text{T}} = 0\,, \; S_{(1)}^{\text{T,ind}} = 1 - \rho^2,$$

$$S_{(2)} = \rho^2\,, \; S_{(2)}^{\text{ind}} = 0\,, \; S_{(2)}^{\text{T}} = \rho^2\,, \; S_{(2)}^{\text{T,ind}} = 0.$$

We can deduce from these computations that $X_2$ is only important because of its correlation with $X_1$ (factor prioritization setting). This means that it is possible to evaluate the uncertainty of $Y$ accurately by only accounting for the uncertainty in $X_1$.

More generally, for a black-box model, if $S_{(j)}^{\text{T,ind}} = 0$, the model output is a measurable function of $(X_1, \ldots, X_{j-1}, X_{j+1}, \ldots, X_d)$ only. Then if $S_j > 0$, $X_i$ is correlated to $(X_1, \ldots, X_{j-1}, X_{j+1}, \ldots, X_d)$. ∎

### 5.5.7 ▪ Conclusion and further references

Shapley effects provide a synthetic information compared to the one contained in the full and independent first-order and total Sobol' indices. The main advantage is that the Shapley effects can be easily compared to each other due to the Shapley allocation principle that leads to a fair sharing of the variance among input variables. Their interpretation is then facilitated. However, Shapley effects do not provide any sensitivity measure for input interactions. In Rabitti and Borgonovo [386], a so-called Shapley–Owen index is introduced for quantifying the sensitivity to any input interaction. The idea is interesting and deserves to be investigated further from both interpretation and estimation perspectives.

As mentioned in Owen and Prieur [353], the Shapley value theory for linear functions has historically been used to partition the $R^2$ quantity (proportion of sample variance explained) from a regression on $d$ variables among those $d$ variables. Taking the value of a subset $u$ of variables to be $R_A^2$, the $R^2$ value when regressing a response on predictors $X_j$ for $j \in A$ yields Shapley value

$$\phi_j = \frac{1}{d} \sum_{A \subseteq -j} \binom{d-1}{|A|}^{-1} (R^2_{A\cup\{j\}} - R^2_A).$$

This is the LMG measure of variable importance, named after the authors' initials of Lindeman et al. [284] (see also [155]), included in the function lmg() of the sensitivity package. Grömping [175] cites several more references on partitioning $R^2$ in regression and discusses alternative measures and criteria for choosing (see also Remark 4 in Section 2.2.2).

Current research attempts to put other cost functions than Sobol' indices in the Shapley value formulation of Eq. (5.6) in order to perform non–variance-based SA. Sarazin et al. [422] extend Shapley-based SA with a dissimilarity measure (such as those developed in Section 6.1) and use copula concepts to define efficient estimators. Il Idrissi et al. [202] address the Shapley-based target SA in the same way as in Section 6.3.1 (by using Sobol' indices on the indicator function). Another method, explored in Elie-Dit-Cosaque [123], defines quantile-oriented Shapley effects by using the contrast function framework (see Section 6.3.3).

In a parallel line of work, the machine learning community recently investigated quite intensively Shapley values in the context of the interpretability of black-box learning algorithms. In this framework, the aim is to explain a machine learning model prediction by identifying the input features that drive most of the prediction mechanism. Some authors focused on explaining individual predictions, thus relying on *local* feature importance (see, for example, Lundberg and Lee [294] and references therein). But lately the need for a feature's *global* importance has emerged to understand the impact of one feature across a whole dataset when building a learning model. In particular, Covert et al. [89] introduced SAGE, for Shapley Additive Global Importance, in an attempt to propose a general view of global feature importance for different types of learning models and loss functions. Interestingly, Shapley effects are obtained as a notable specific case of SAGE. On the other hand, Williamson and Feng [503] proposed a computationally efficient procedure for estimating what they call the *Shapley Population Variable Importance Measure*, which is actually a Shapley effect with any value function.

## 5.6 ▪ Examples in R

In this section, by using the sensitivity package, we illustrate the estimation of the Sobol' indices for groups of inputs (see Section 5.2) and the estimation of the Shapley effects (see Section 5.5). The sobol() function allows to compute Sobol' indices by group by defining a list of vectors (each vector representing a group of inputs) in its argument "order." The shapleyPermEx() and shapleyPermRand() functions compute the Shapley effects via, respectively, the exact permutation and the random permutation algorithms, while the shapleySubsetMC() and sobolshap_knn() use the nearest-neighbor algorithm.

**Remark 36.** *The shapleyLinearGaussian() and shapleyBlockEstimation() functions implement the analytical computations of the Shapley effects in the linear Gaussian framework (linear model and Gaussian variables) and uses block-diagonal covariance structure to deal with high-dimensional cases; see e.g., [57, 55].*

### 5.6.1 ▪ Linear model example

We first study a linear model $Y = X_1 + X_2 + X_3$ with $\mathbf{X} \sim \mathcal{N}((0,0,0)^T, \Sigma)$ and

$$\Sigma = \begin{pmatrix} 1 & 0 & 0 \\ 0 & 1 & 0.9 \\ 0 & 0.9 & 1 \end{pmatrix}.$$

```
> modlin <- function(X) apply(X, 1, sum) # Linear model
> mu <- rep(0, 3) # Mean vector
> sig <- c(1, 1, 2) # Variance vector
> ro <- 0.9 # correlation coefficient (between X2 and X3)
> Cormat <- matrix(c(1,0,0,0,1,ro,0,ro,1),3,3) # Correlation matrix
> Covmat <- ( sig %*% t(sig) ) * Cormat # Covariance matrix
```

Sobol' indices by groups can be obtained by the following:

```
> set.seed(12345) # To reproduce the same results
> n <- 1e4
> library(mvtnorm) # Multivariate Gaussian variables
> Xall <- function(n) rmvnorm(n, mu, Covmat) # Sampling function
> #
> X1 <- data.frame(Xall(n)) ; X2 <- data.frame(Xall(n))
> library(sensitivity)
> sa <- sobol(modlin, X1=X1, X2=X2, order=list(1,2:3), nboot=100)
> print(sa$V[-1,]/sa$V[1,1]) # Divide by variance to get Sobol'

          original         bias std. error  min. c.i. max. c.i.
X1      0.08845611 0.0008270262 0.01010859 0.06623788 0.1084258
X2*X3 0.90469011 0.0004547900 0.01280654 0.87553648 0.9269086
```

In order to use the Shapley effects estimation functions, we first need to define the following conditional generation function:

```
> library(condMVNorm) # Conditional multivar. Gaussian variables
> Xset <- function(n, Sj, Sjc, xjc){
+   if (is.null(Sjc)){
+     if (length(Sj) == 1){ rnorm(n, mu[Sj], sqrt(Covmat[Sj,Sj]))
+     }
+     else{ rmvnorm(n, mu[Sj], Covmat[Sj,Sj])}
+   }
+   else{
+     rcmvnorm(n,mu,Covmat,dependent.ind=Sj,given.ind=Sjc,X.given=xjc)
+     }}
```

Figure 5.5 shows the results of the exact permutation algorithm for Shapley effects estimation, obtained by

```
> library(sensitivity)
> sa <- shapleyPermEx(modlin, Xall, Xset, d=3, Nv=1e4, No=1e3, Ni=3)

> library(ggplot2) ; ggplot(sa, ylim=c(0,1.5))
```

The Shapley effects give the influence hierarchy between the inputs: $X_3$ in first, $X_2$ in second, then $X_1$, which has a much smaller effect. As a by-product of the Shapley effects estimation algorithm, the first and total Sobol' indices correspond to the full first-order Sobol' index and the independent total Sobol' index (see Section 5.4). The large full first-order and negligible independent total Sobol' indices of $X_2$ and $X_3$ means that they have strong first-order effects (without any interaction) in the model, supported by a high correlation between these two inputs.

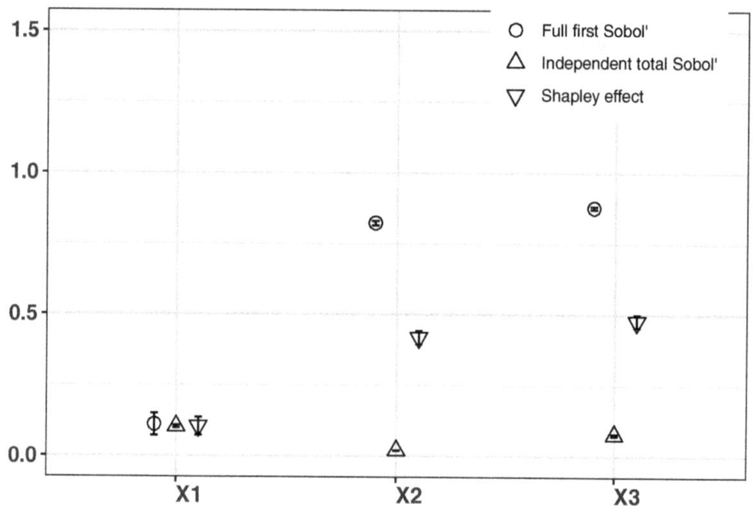


**Figure 5.5.** *Plotted results of the* shapleyPermEx() *function on the linear model.*

We compare this result with the nearest-neighbor procedure of Broto et al. [56] implemented in the function shapleySubsetMc(). An approximated and strongly accelerated version of this algorithm is available in the function sobolshap_knn(). It just requires a sample of input-output values, here of size 1000:

```
> n <- 1000
> X <- Xall(n) ; Y <- modlin(X)
> Shap <- sobolshap_knn(model = NULL, X = X, return.shap = TRUE)

> tell(Shap,Y)

> print(Shap$Shap)

             X1        X2        X3
[1,] 0.09253387 0.4234256 0.4840405
```

One can observe the efficiency of this algorithm, which provides Shapley effects estimations very close to the ones obtained with the exact permutation algorithm (with a cost of 46000 model evaluations):

```
> print(sa$Shapley$original)

[1] 0.1048334 0.4181758 0.4769908
```

### 5.6.2 ▪ Ishigami function

We study at present the Ishigami function (see Section C.4), where $X_j \sim \mathcal{U}[-\pi, \pi]$ for $j = 1, \dots, 3$ and $\mathbf{X}$ has a Gaussian copula–based dependence structure with covariance

$$\Sigma = \begin{pmatrix} 1 & 0 & 0.9 \\ 0 & 1 & 0 \\ 0.9 & 0 & 1 \end{pmatrix}.$$

```
> ishigami <- function (X){ # Ishigami function
+   X <- X * 2 * pi - pi     # transform to [-pi,pi]^3
+   sin(X[, 1]) + 7 * sin(X[, 2])^2 + 0.1 * X[, 3]^4 * sin(X[, 1])
+ }
> # Simulation function for U[0,1] variables (with Gaussian copula)
> library(copula)
> Xall <- function(n) rCopula(n, norm.cop)
> ro <- 0.9 # Correlation coefficient (between X1 and X3)
> Cormat <- matrix(c(1,0,ro,0,1,0,ro,0,1),3,3) # Correlation matrix
> norm.cop <- normalCopula(c(0,ro,0), dim = 3, dispstr = "un")
```

Sobol' indices by groups can be obtained by the following:

```
> set.seed(12345) # To reproduce the same results
> library(sensitivity)
> n <- 1e4
> X1 <- data.frame(Xall(n)) ; X2 <- data.frame(Xall(n))
> sa <- sobol(ishigami,X1=X1,X2=X2,order=list(c(1,3),2),nboot=100)
> print(sa$V[-1,]/sa$V[1,1]) # Divide by variance to get Sobol'

        original          bias std. error min. c.i. max. c.i.
X1*X3 0.4242746 -0.0001535507 0.01702974 0.3964756 0.4681101
X2    0.6025167 -0.0007775155 0.01239455 0.5770246 0.6298705
```

In order to use the Shapley effects estimation functions, we first need to define the following conditional generation function:

```
> library(gtools)
> d <- 3
> Xset <- function(n, Sj, Sjc, xjc){
+   nj <- length(Sj) ; njc <- length(Sjc)
+   if (is.null(Sjc)){
+     if (nj == 1){ runif(n)
+     } else{
+       Cormat2 <- Cormat[Sj,Sj]
+       a2 <- combinations(nj, 2, 1:nj)
```

```
+         corp2 <- Cormat2[matrix(a2[1:dim(a2)[[1]],],nr=dim(a2)[[1]])]
+         norm2.cop <- normalCopula(corp2, dim = nj, dispstr = "un")
+         rCopula(n,norm2.cop)
+       }
+     } else{
+       xx <- cbind(matrix(xjc,nc=njc,nr=n,byrow=T), matrix(runif(n*nj),nc=nj))
+       cc <- c(Sjc,Sj) ; a2 <- NULL
+       for (i in 1:(d-1)) for (j in (i+1):d) a2 <- rbind(a2,c(cc[i],cc[j]))
+       corp2 <- Cormat[matrix(a2[1:dim(a2)[[1]],], nr=dim(a2)[[1]])]
+       norm2.cop <- copula::normalCopula(corp2, dim = d, dispstr = "un")
+       x <- cCopula(xx, copula = norm2.cop, inverse = TRUE)
+       x[,((njc+1):(njc+nj))]
+     }}
```

Figure 5.6 shows the results of the random permutation algorithm for Shapley effects estimation, obtained by

```
> library(sensitivity)
> sa <- shapleyPermRand(ishigami,Xall,Xset,d=3,Nv=1e4,No=1,Ni=3,m=1e3)

> library(ggplot2) ; ggplot(sa)
```

The Shapley effects and their confidence intervals allow us to draw the influence hierarchy between the inputs: $X_2$ has around a 60% influence, while $X_1$ and $X_3$ have a similar influence of 20%. The equality between the full first-order and independent total Sobol' indices of $X_2$ means that this input has neither correlation nor interaction effects with the other inputs.

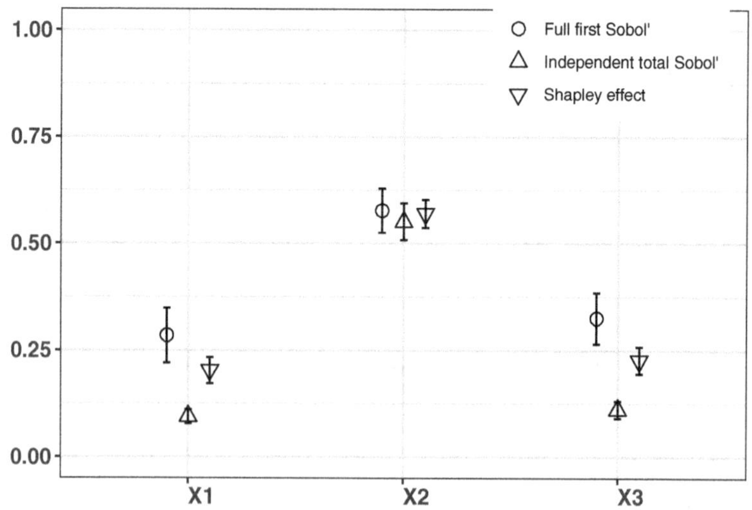


**Figure 5.6.** *Plotted results of the* shapleyPermRand() *function on the Ishigami model.*

## 5.7 ▪ Summary

In this chapter, four different approaches for deriving variance-based sensitivity indices in the case of dependent inputs were presented. The information provided by grouped Sobol' indices is interesting, but it does not allow us to understand what happens on a finer scale than the groups.

This approach is useful, as a preliminary SA, in a high-dimensional context with several blocks of dependent inputs. The covariance-based SA strategy has the significant drawback of providing difficult-to-interpret indices. In particular, these indices may be negative. The decomposition in independent and full Sobol' indices provides rich information on the influential inputs and model behavior. However, it is a complex approach and the most difficult to implement in practice. Moreover, the interpretation of independent and full Sobol' indices is not straightforward. The method based on Shapley effects is easy to understand: Each input is fairly allocated a part of the variance of the output. A practical challenge, however, is to reduce the prohibitive computational cost of Shapley effects for applications with a high-dimensional input space.

## Chapter 6

# Beyond variance-based indices

Sobol' indices are based on the second-order moment (i.e., the variance) of the output distribution. However, in some cases, variance poorly represents the variability of the distribution. In Sections 6.1 and 6.2, we present different ways to consider other features of the output, providing generalization frameworks for Sobol' indices. It also happens that the QoI is not related to the overall variability of the model output. In such cases, SA techniques must then be adapted: Goal-oriented SA tools are presented in Section 6.3. Finally, the so-called robustness analysis methods, introduced in Section 1.4.4 of Chapter 1, are developed in Section 6.4. All along the present chapter, illustrations are provided with the R software on simple analytical functions taken from Appendix C.

## 6.1 ▪ Moment-independent importance measures

So far, we have focused on sensitivity measures, which only study the impact of the input parameters on the variance of the output. Since this is a restricted summary of the output distribution, this point of view happens to be inadequate in some practical cases. Some authors have then introduced the so-called moment-independent importance measures. Instead of relying on moments only, the idea here is to investigate how the whole output distribution changes with respect to the input parameters. In this class of moment-independent measures, previous works include, for example, entropy-based indices [248, 286, 13] or distribution-based indices [42, 44, 45, 169, 368]. It has been shown that these indices can provide complementary information to Sobol' indices.

In the following, a general framework for moment-independent measures based on the concept of dissimilarity between probability distributions is first introduced. By choosing the specific dissimilarity class of $f$-divergences, we then show that previously introduced sensitivity indices can be recovered easily. In particular, their estimation is discussed. This estimation happens to be difficult in practice and has not been studied thoroughly in previous works. Finally, a completely different dissimilarity is considered: the Cramér–von Mises distance, which possesses interesting theoretical and estimation properties.

In this section, we focus on the setting where the inputs are independent and the code output is scalar.

### 6.1.1 • New focus on Sobol' indices: General indices built on a dissimilarity measure

Let us first come back to the unnormalized first-order Sobol' sensitivity index for a given input variable $X_j$ and rewrite it as

$$V_j = \operatorname{Var}\mathbb{E}(Y|X_j) = \mathbb{E}_{X_j}\Big(\mathbb{E}Y - \mathbb{E}(Y|X_j)\Big)^2. \tag{6.1}$$

This alternative formulation provides a genuine intuition of what first-order Sobol' indices measure: They compare the output distribution $\mathrm{P}_Y$ and the output distribution conditionally to $X_j$, $\mathrm{P}_{Y|X_j}$, through their mean values and integrate the squared difference to get the unnormalized index.

A straightforward extension then readily consists in replacing the comparison of the means by some general dissimilarity measure $d(\cdot,\cdot)$ between probability distributions to propose a new sensitivity index $\mathcal{S}_j^d$ defined as

$$\mathcal{S}_j^d = \mathbb{E}_{X_j}\Big(d\left(\mathrm{P}_Y, \mathrm{P}_{Y|X_j}\right)\Big). \tag{6.2}$$

Here a dissimilarity measure is simply a function $d(\cdot,\cdot) : \mathcal{M}_1^+ \times \mathcal{M}_1^+ \to \mathbb{R}_+$ and should be considered in a broad sense. Such a general formulation has been discussed, for example, in Baucells and Borgonovo [21] and Da Veiga [99].

To provide some insight into how this sensitivity index operates, let us consider the Ishigami function (see Section C.4 with independent $X_j \sim \mathcal{U}(-\pi,\pi)$) with a fourth additional dummy variable $X_4$. Each input range is first discretized. Then, for each discretized value $x_j^{(l)}$, a kernel estimate of the conditional output distribution $\mathrm{P}_{Y|X_j=x_j^{(l)}}$ is computed. We can then compare each conditional distribution with an estimate of the output distribution $\mathrm{P}_Y$ through violin plots, as illustrated in Fig. 6.1. Each line represents a violin plot which compares $\mathrm{P}_Y$ (in pink) with the conditional distributions for each input fixed at a given discretized value (red for $X_1$, blue for $X_2$, green for $X_3$, and yellow for $X_4$). The mean of the distributions is also represented with a horizontal bar. When focusing on a given line, it is possible to investigate how the conditional distribution evolves with respect to the discretized value. For $X_1$, both the mean and the tails of the conditional distribution change, which implies that $X_1$ has a strong impact on the output. $X_2$ exhibits a similar behavior except that the tails do not change. On the contrary, for $X_3$ and $X_4$, the mean does not evolve, which explains why their first-order Sobol' index equals zero (recall that the alternative formulation of Sobol' illustrates that the probability distributions are only compared through their means). However, for $X_3$, the tails of the conditional distribution can significantly evolve for extreme discretized values: This means that $X_3$ has a more complex impact on the output than just a first-order effect.

Finally, the advantage of such a general formulation is that many choices for $d(\cdot,\cdot)$ are available: We see in what follows that using natural dissimilarity measures yields sensitivity indices related to well-known quantities, while others possess theoretical properties that make it possible to build efficient estimators as well as enjoyable additional characteristics.

### 6.1.2 • The $F$-divergence case

Coming back to the choice of the dissimilarity measure $d(\cdot,\cdot)$, a popular choice is the class of Csiszár $f$-divergences $d(\cdot,\cdot) = d_f(\cdot||\cdot)$ [93]. Assuming all considered random variables have an absolutely continuous distribution with respect to the Lebesgue measure on $\mathbb{R}$, the $f$-divergence

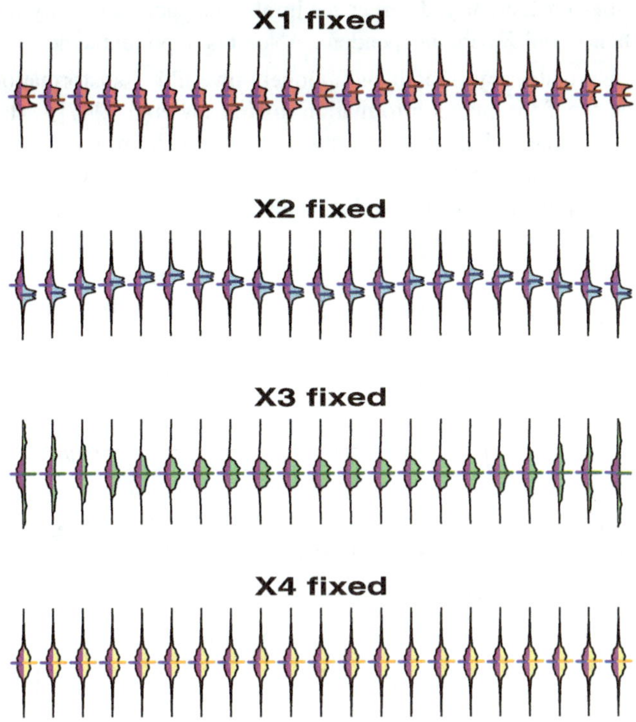


**Figure 6.1.** *Violin plots of conditional distributions for the Ishigami function with a fourth additional dummy variable* $X_4$.

between $\mathrm{P}_Y$ and $\mathrm{P}_{Y|X_j}$ is given by

$$d_f(\mathrm{P}_Y||\mathrm{P}_{Y|X_j}) = \int_{\mathbb{R}} f\left(\frac{p_Y(y)}{p_{Y|X_j}(y)}\right) f_{Y|X_j}(y)dy, \tag{6.3}$$

where $f$ is any convex function such that $f(1) = 0$. Standard choices for function $f$ include, for example,

- the Kullback–Leibler divergence $f(t) = -\ln(t)$ or $f(t) = t\ln(t)$;
- the Hellinger distance: $f(t) = \left(\sqrt{t} - 1\right)^2$;
- the total variation distance: $f(t) = |t - 1|$;
- the Pearson $\chi^2$ divergence: $f(t) = (t-1)^2$ or $f(t) = t^2 - 1$;
- the Neyman $\chi^2$ divergence: $f(t) = (t-1)^2/t$ or $f(t) = (1-t^2)/t$.

Plugging this dissimilarity measure into Eq. (6.2) and deconditioning with respect to $X_j$ yields the following sensitivity index:

$$S_j^{d_f} = \int_{\mathbb{R}^2} f\left(\frac{p_Y(y) f_{X_j}(x)}{p_{X_j,Y}(x,y)}\right) p_{X_j,Y}(x,y)dxdy. \tag{6.4}$$

Note that inequalities on Csiszár $f$-divergences imply that such sensitivity indices are positive and equal zero when $Y$ and $X_j$ are independent. Also, it is important to note that, given the form of $\mathcal{S}_j^{d_f}$, it is invariant under any smooth and uniquely invertible transformation of the variables $X_j$ and $Y$ (see the proof for mutual information in Kraskov et al. [247]). This is an enjoyable property over variance-based Sobol' sensitivity indices, which are only invariant under linear transformations for a scalar output.

It is easy to see that the total variation distance with $f(t) = |t-1|$ gives a sensitivity index equal to the one proposed by Borgonovo [42]:

$$\mathcal{S}_j^{d_f} = \int_{\mathbb{R}^2} |p_Y(y)p_{X_j}(x) - p_{X_j,Y}(x,y)| dxdy.$$

In addition, the Kullback–Leibler divergence with $f(t) = -\ln(t)$ yields

$$\mathcal{S}_j^{d_f} = \int_{\mathbb{R}^2} p_{X_j,Y}(x,y) \ln\left(\frac{p_{X_j,Y}(x,y)}{p_Y(y)p_{X_j}(x)}\right) dxdy,$$

that is, the *mutual information* between $X_j$ and $Y$, denoted $I(X_j;Y)$. A normalized version of this sensitivity index was extensively studied [358, 248, 247, 286, 13]. Similarly, the Neyman $\chi^2$ divergence with $f(t) = (1-t^2)/t$ leads to

$$\mathcal{S}_j^{d_f} = \int_{\mathbb{R}^2} \left(\frac{p_{X_j,Y}(x,y)}{p_Y(y)p_{X_j}(x)} - 1\right)^2 p_Y(y)p_{X_j}(x)dxdy,$$

which is the so-called *squared-loss mutual information* between $X_j$ and $Y$ (or *mean square contingency*). These results show that some previously proposed sensitivity indices are actually special cases of more general indices defined through Csiszár $f$-divergences. An extensive study of their properties is given in Rahman [389]. Moreover, the specific structure of Eq. (6.4) makes it possible to envision more efficient tools for the estimation of these sensitivity indices. Indeed, it only involves approximating a density ratio rather than full densities. This point is investigated below. But, more importantly, we see that special choices for $f$ define sensitivity indices that are actually well-known dependence measures, such as the mutual information. This paves the way for looking at new sensitivity indices relying on recent kernel-based dependence measures (see Section 6.2).

### Estimation

Coming back to Eq. (6.4), the goal is to estimate

$$\mathcal{S}_j^{d_f} = \int_{\mathbb{R}^2} f\left(\frac{1}{r(x,y)}\right) p_{X_j,Y}(x,y)dxdy = \mathbb{E}_{(X_j,Y)} f\left(\frac{1}{r(X_j,Y)}\right),$$

where $r(x,y) = p_{X_j,Y}(x,y)/(p_Y(y)p_{X_j}(x))$ is the ratio between the joint density of $(X_j,Y)$ and the marginals. Of course, straightforward estimation is possible if one estimates the densities $p_{X_j,Y}(x,y)$, $p_{X_j}(x)$, and $p_Y(y)$ with, e.g., kernel density estimators. Assuming that we have a sample $\boldsymbol{Z}^n = (\mathbf{X}_j^n, \mathbf{Y}^n) = \left(X_j^{(i)}, Y^{(i)}\right)_{i=1,\dots,n}$, the kernel density estimate of $p_{X_j,Y}(x,y)$ is given by

$$\hat{p}_{X_j,Y}(x,y) = \frac{1}{n}\sum_{i=1}^{n} K_H(z - Z_i)$$

for $z = (x,y)$ and $K_H(z) = |H|^{-1/2}K(H^{-1/2}z)$, where $K$ is a multivariate kernel and $H$ is the bandwidth matrix. We similarly get $\hat{p}_{X_j}(x)$ and $\hat{p}_Y(y)$, the kernel density estimates of $p_{X_j}(x)$

and $p_Y(y)$, respectively. We can deduce the following plug-in estimators for the ratio and the sensitivity index:

$$\hat{r}(x,y) = \frac{\hat{p}_{X_j,Y}(x,y)}{\hat{p}_Y(y)\hat{p}_{X_j}(x)}, \qquad \hat{S}^{d_f}_{X_j} = \frac{1}{n}\sum_{i=1}^{n} f\left(\frac{1}{\hat{r}\left(X_j^{(i)}, Y^{(i)}\right)}\right).$$

If we only consider the case of a scalar output and each input factor separately, this plug-in estimator can easily be computed. However, kernel density estimation suffers from the curse of dimensionality, and the estimator proposed above cannot be used when dealing with multidimensional outputs or group of inputs. For such a general setting, one can observe that only the ratio function $r(x,y)$ is needed: We can then expect more robust estimates by focusing only on it. Powerful estimation methods for such ratios include, e.g., maximum-likelihood estimation [470] and unconstrained least-squares importance fitting [231], among others [468]. A $k$-nearest-neighbors strategy dedicated to mutual information is also discussed in Kraskov et al. [247].

Unfortunately, all proposed estimation procedures lead to suboptimal estimators which exhibit a nonparametric rate of convergence. Potential progress toward the design of efficient estimators may come from adaptation of the work by Laurent [262] or Giné and Nickl [163].

#### Examples in R

We illustrate $f$-divergence sensitivity indices on the Ishigami function (see Section C.4 with independent $X_j \sim \mathcal{U}(-\pi,\pi)$) by choosing the total variation and Kullback–Leibler divergences:

```
> set.seed(12345)
> library(sensitivity)
> d <- 3
> n <- 200
> X <- data.frame(matrix(runif(d*n, min = -pi, max = pi), nrow = n))
> sa <- sensiFdiv(model=ishigami.fun, X=X, fdiv=c("TV","KL"), nboot=30)
> par(mfrow = c(1,2))
> library(ggplot2) ; ggplot(sa)
```

Figure 6.2 shows the results for both divergences. Here the conclusions are similar: The most important factor is $X_2$, followed by $X_1$. Interestingly, unlike first-order sensitivity indices, here the distributional indices detect that $X_3$ has an impact, even if it is small compared to the other factors.

### 6.1.3 ▪ The Cramér–von Mises distance case

In this section, a sensitivity index based on the so-called Cramér–von Mises distance is presented. Indeed, the general index defined in Eq. (6.2) is studied with the Cramér–von Mises distance. The frame considered here is the same as in Section 3.3.2. That is, we consider a multidimensional output-valued code ($Y \in \mathbb{R}^k$, $k \geq 1$). The material developed here comes mainly from Gamboa et al. [150]. Note that extensions about more complicated cases of manifold-valued codes and general metric spaces have been studied, respectively, in Fraiman et al. [140] and in Gamboa et al. [151] (also proposing U-statistics–based estimators).

To begin with, let $F_Y$ be the distribution function of $Y$, that is,

$$F_Y(t) = \mathbb{P}(Y \leqslant t) = \mathbb{E}\mathbf{1}_{\{Y \leqslant t\}} \text{ for } t = (t_1, \dots, t_k) \in \mathbb{R}^k.$$

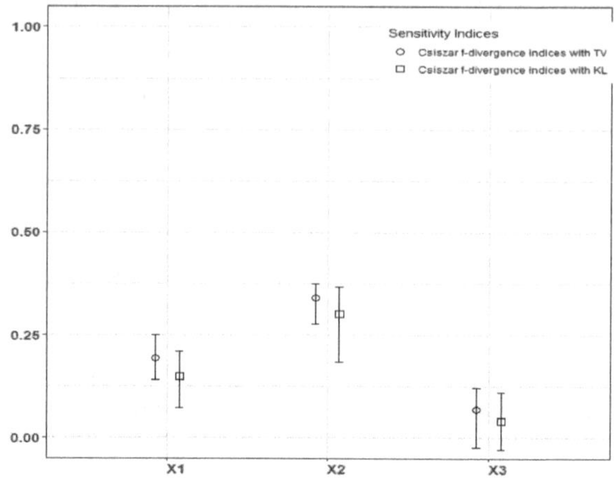


**Figure 6.2.** *$F$-divergence sensitivity measures for the Ishigami function.*

Here, $Y \leqslant t$ means that, for all $j = 1, \ldots, k$, $Y_i \leq t_i$. Let further $F_Y^A$ be the conditional distribution function of $Y$ conditionally on $X_A$:

$$F_Y^A(t) = \mathbb{P}(Y \leqslant t | \mathbf{X}_A) = \mathbb{E}\left[\mathbf{1}_{\{Y \leqslant t\}} | \mathbf{X}_A\right] \text{ for } t = (t_1, \ldots, t_k) \in \mathbb{R}^k.$$

Obviously, one has $\mathbb{E}F_Y^A(t) = F(t)$. Define $\check{Y}(t) = \mathbb{1}_{\{Y \leqslant t\}}$. Since for any fixed $t \in \mathbb{R}^k$, $\check{Y}(t)$ is a real-valued random variable, we may apply the framework presented in Chapter 3. More precisely, for any set $A \in \mathcal{P}_d$, we first perform the Hoeffding decomposition of $\check{Y}(t)$,

$$\begin{aligned}\check{Y}(t) = \mathbb{E}\check{Y}(t) &+ \left(\mathbb{E}[\check{Y}(t)|\mathbf{X}_A] - \mathbb{E}\check{Y}(t)\right) \\ &+ \left(\mathbb{E}[\check{Y}(t)|\mathbf{X}_{\bar{A}}] - \mathbb{E}\check{Y}(t)\right) + R(t, v),\end{aligned} \tag{6.5}$$

where

$$R(t, v) = \check{Y}(t) - \mathbb{E}\check{Y}(t) - \left(\mathbb{E}[\check{Y}(t)|\mathbf{X}_A] - \mathbb{E}\check{Y}(t)\right) - \left(\mathbb{E}[\check{Y}(t)|\mathbf{X}_{\bar{A}}] - \mathbb{E}\check{Y}(t)\right).$$

As usual, the variance of both sides of Eq. (6.5) is computed, leading to

$$\begin{aligned}\operatorname{Var}\check{Y}(t) &= F(t)(1 - F(t)) \\ &= \operatorname{Var}\left(\mathbb{E}[\check{Y}(t)|\mathbf{X}_A] - \mathbb{E}\check{Y}(t)\right) + \operatorname{Var}\left(\mathbb{E}[\check{Y}(t)|\mathbf{X}_{\bar{A}}] - \mathbb{E}\check{Y}(t)\right) + \operatorname{Var} R(t, v) \\ &= \operatorname{Var}\left(F_Y^A(t)\right) + \operatorname{Var}\left(F_Y^{\bar{A}}(t)\right) + \operatorname{Var} R(t, v) \\ &= \mathbb{E}\left(F_Y^A(t) - F_Y(t)\right)^2 + \mathbb{E}\left(F_Y^{\bar{A}}(t) - F_Y(t)\right)^2 + \operatorname{Var} R(t, v).\end{aligned} \tag{6.6}$$

Note that a straightforward application of the results of Chapter 3 gives, for any fixed $t \in \mathbb{R}^k$, a consistent and asymptotically normal procedure for the estimation of the following quantities:

$$\mathbb{E}\left(F_Y^A(t) - F_Y(t)\right)^2 = \operatorname{Var}\left(F_Y^A(t)\right) \quad \text{and} \quad \mathbb{E}\left(F_Y^{\bar{A}}(t) - F_Y(t)\right)^2 = \operatorname{Var}\left(F_Y^{\bar{A}}(t)\right).$$

Now use the distribution of $Y$ and integrate both sides of Eq. (6.6). This gives

$$\begin{aligned}&\int_{\mathbb{R}^k} F_Y(t)(1-F_Y(t))dF_Y(t)\\&=\int_{\mathbb{R}^k}\mathbb{E}\left(F_Y^A-F_Y(t)\right)^2 dF_Y(t)+\int_{\mathbb{R}^k}\mathbb{E}\left(F_Y^{\bar{A}}(t)-F_Y(t)\right)^2 dF_Y(t)\\&+\int_{\mathbb{R}^k}\operatorname{Var} R(t,v)dF_Y(t).\end{aligned} \tag{6.7}$$

This integration has to be understood in the Riemann–Stieltjes sense (see, e.g., [478]). Note that the first term in the right-hand side of Eq. (6.7) is the squared Cramér–von Mises distance of order 2 between the probability distributions $\mathrm{P}_Y$ of $Y$ and $\mathrm{P}_{Y|\mathbf{X}_A}$ of $Y$ given $\mathbf{X}_A$.

Following the construction of Sobol' indices, the previous equation is divided by

$$\int_{\mathbb{R}^k} F_Y(t)(1-F_Y(t))dF_Y(t).$$

This gives

$$\begin{aligned}1&=\frac{\int_{\mathbb{R}^k}\mathbb{E}\left(F_Y^A(t)-F_Y(t)\right)^2 dF_Y(t)}{\int_{\mathbb{R}^k} F_Y(t)(1-F_Y(t))dF_Y(t)}+\frac{\int_{\mathbb{R}^k}\mathbb{E}\left(F_Y^{\bar{A}}(t)-F_Y(t)\right)^2 dF_Y(t)}{\int_{\mathbb{R}^k} F_Y(t)(1-F_Y(t))dF_Y(t)}\\&+\frac{\int_{\mathbb{R}^k}\operatorname{Var} R(t,v)dF_Y(t)}{\int_{\mathbb{R}^k} F_Y(t)(1-F_Y(t))dF_Y(t)}.\end{aligned}$$

Then the Cramér–von Mises indices are defined with respect to $A$ and $\bar{A}$, respectively, by

$$\begin{aligned}S_{2,\mathrm{CvM}}^A&=\frac{\int_{\mathbb{R}^k}\mathbb{E}\left(F_Y(t)-F_Y^A(t)\right)^2 dF_Y(t)}{\int_{\mathbb{R}^k} F_Y(t)(1-F_Y(t))dF_Y(t)},\\S_{2,\mathrm{CvM}}^{\bar{A}}&=\frac{\int_{\mathbb{R}^k}\mathbb{E}\left(F_Y(t)-F_Y^{\bar{A}}(t)\right)^2 dF_Y(t)}{\int_{\mathbb{R}^k} F_Y(t)(1-F_Y(t))dF_Y(t)}.\end{aligned}$$

Let us now turn to the estimation of $S_{2,\mathrm{CvM}}^A$ (and $S_{2,\mathrm{CvM}}^{\bar{A}}$). We will see that one of the main advantages of the Cramér–von Mises index is that a pick-freeze estimation scheme is available. Note that one has to estimate both the numerator and the denominator defining the index. Nevertheless, when the output $Y$ has independent coordinates that are absolutely continuous with respect to the Lebesgue measure, we have

$$\int_{\mathbb{R}^k} F_Y(t)(1-F_Y(t))dF_Y(t)=\mathbb{E}[F_Y(Y)(1-F_Y(Y)]=\frac{1}{2^k}-\frac{1}{3^k}.$$

Thus, the normalizing factor reduces to $\frac{1}{2^k}-\frac{1}{3^k}$, and for the sake of simplicity, we only focus on the estimation of the numerator involved. Call by $N_{2,\mathrm{CvM}}^A$ the numerator of $S_{2,\mathrm{CvM}}^A$. Note first that this last quantity can be rewritten as

$$N_{2,\mathrm{CvM}}^A=\mathbb{E}\left(F_Y(Y'')-F_Y^A(Y'')\right)^2,$$

where $Y''$ is an independent copy of $Y$ and $Y^{\sim A}$ (defined in Section 3.2.1). Further, a double

Monte Carlo scheme for the estimation of $N^A_{2,\mathrm{CvM}}$ is applied using the empirical estimator given by

$$\widehat{N}^A_{2,\mathrm{CvM}} = \frac{1}{n}\sum_{k=1}^{n}\left\{\frac{1}{n}\sum_{j=1}^{n}\mathbf{1}_{\{Y^{(j)}\leq Y''^{(k)}\}}\mathbf{1}_{\{Y^{\sim A,(j)}\leq Y''^{(k)}\}} - \left[\frac{1}{2n}\sum_{j=1}^{n}\left(\mathbf{1}_{\{Y^{(j)}\leq Y''^{(k)}\}} + \mathbf{1}_{\{Y^{\sim A,(j)}\leq Y''^{(k)}\}}\right)\right]^2\right\}.$$

Some asymptotic results on the empirical estimator of $N^A_{2,\mathrm{CvM}}$ are given in the following theorem. Roughly speaking, the proofs of these results follow along the lines of the ones developed in Chapter 3 with a technical complication due to the Monte Carlo integration on the third sample $Y''$. The interested reader can find the detailed proofs in Gamboa et al. [150].

**Theorem 6.1.** *We have the following properties:*

- $\widehat{N}^A_{2,\mathrm{CvM}}$ *converges almost surely to* $N^A_{2,\mathrm{CvM}}$ *as* $n$ *goes to infinity so that the sequence of estimators is convergent.*
- $\sqrt{n}(\widehat{N}^A_{2,\mathrm{CvM}} - N^A_{2,\mathrm{CvM}})$ *converges in distribution toward the centered Gaussian distribution with a limiting variance* $\xi^2$ *whose explicit expression can be found in Gamboa et al. [150].*

## 6.2 ▪ Kernel-based sensitivity analysis

The dissimilarity point of view introduced above, though appealing (especially with the Cramér–von Mises distance properties), still only provides a complementary perspective to Sobol' indices. In particular, none of the importance measures introduced so far enjoys the highly desirable FANOVA decomposition property. Such a decomposition not only provides a proper normalization constant for sensitivity indices but also appropriately defines the concept of higher-order effects.

In this section, starting from another general class of dissimilarities between probability distributions, we introduce recent advances in distances based on kernel representations of probability distributions. This framework can be seen as a straightforward generalization of Sobol' indices with profitable characteristics: Estimation is far less prone to the curse of dimensionality, and a FANOVA-like decomposition is readily available.

Unless stated otherwise, we consider here the case of independent inputs and scalar output. Extensions to vectorial and functional outputs are briefly discussed in Section 6.2.5, and kernelized versions of Shapley effects for handling dependent inputs are introduced in Section 6.2.3.

### 6.2.1 ▪ Integral probability metrics and the RKHS case

*Integral probability metrics* (IPM) [335] is a popular family of distance measures on probabilities given by

$$\gamma_{\mathcal{F}}(\mathrm{P},\mathrm{Q}) = \sup_{f\in\mathcal{F}}\left|\int_S f d\mathrm{P} - \int_S f d\mathrm{Q}\right| \tag{6.8}$$

for two probability measures P and Q defined on a measurable space $S$ and where $\mathcal{F}$ is a class of real-valued bounded measurable functions on $S$. Just as the choice of function $F$ in Csiszár $F$-divergences gives rise to different measures, the choice of $\mathcal{F}$ generates different IPM, e.g., the Wasserstein distance, the Dudley metric, or the total variation distance. It is interesting to note that Csiszár $F$-divergences and IPM are very distinct classes of dissimilarity measures since they only intersect at the total variation distance [452]. Notable examples of $\mathcal{F}$ and associated distances include

- the space of bounded continuous functions for the Dudley metric;
- the space of functions with bounded variations for the Kolmogorov metric;
- the space of Lipschitz bounded functions for the Wasserstein distance;
- the space of characteristic functions on Borel sets for the total variation distance.

Except for the total variance case, plugging the general expression (6.8) of an IPM into Eq. (6.2) no longer yields a closed-form expression for a sensitivity index as in Eq. (6.4).

Recently, in a rich sequence of papers, several authors focused on another function class $\mathcal{F}$, namely, the space of functions in the unit ball of a characteristic reproducing kernel Hilbert space (RKHS), which gives rise to the maximum mean discrepancy (MMD) distance [439].

In the following, we assume the reader is familiar with RKHS, and we discuss this specific kernel-based distance, which possesses interesting properties. A reminder of definitions and basic properties of RKHS is given in Appendix B.

#### MMD distance via kernel embedding of distributions

When particularizing $\mathcal{F}$ as the unit ball of a RKHS $\mathcal{H}$ in Eq. (6.8), we can write

$$\begin{aligned}
\gamma_{\mathcal{F}}(\mathrm{P},\mathrm{Q})^2 &= \left(\sup_{f\in\mathcal{H},\ \|f\|\leqslant 1} \int_S f d\mathrm{P} - \int_S f d\mathrm{Q}\right)^2 \\
&= \left(\sup_{f\in\mathcal{H},\ \|f\|\leqslant 1} \mathbb{E}_{\xi\sim\mathrm{P}} f(\xi) - \mathbb{E}_{\zeta\sim\mathrm{Q}} f(\zeta)\right)^2 \\
&= \left(\sup_{f\in\mathcal{H},\ \|f\|\leqslant 1} \mathbb{E}_{\xi\sim\mathrm{P}}\langle f, k(\xi,\cdot)\rangle - \mathbb{E}_{\zeta\sim\mathrm{Q}}\langle f, k(\zeta,\cdot)\rangle\right)^2 \quad \begin{pmatrix}\text{reproducing}\\ \text{property}\end{pmatrix} \\
&= \left(\sup_{f\in\mathcal{H},\ \|f\|\leqslant 1} \langle f, \mathbb{E}_{\xi\sim\mathrm{P}} k(\xi,\cdot) - \mathbb{E}_{\zeta\sim\mathrm{Q}} k(\zeta,\cdot)\rangle\right)^2 \quad \begin{pmatrix}\text{linearity of}\\ \text{inner product}\end{pmatrix} \\
&= \left(\sup_{f\in\mathcal{H},\ \|f\|\leqslant 1} \langle f, \mu_{\mathrm{P}} - \mu_{\mathrm{Q}}\rangle\right)^2 \quad \begin{pmatrix}\text{kernel embedding}\\ \text{definition}\end{pmatrix} \\
&= \|\mu_{\mathrm{P}} - \mu_{\mathrm{Q}}\|^2 \ \text{(definition of the norm).}
\end{aligned}$$

Here, $\|\cdot\|$ is the Hilbertian norm (under $\mathcal{H}$) and $\mu_{\mathrm{P}} = \mathbb{E}_{\xi\sim\mathrm{P}} k(\xi,\cdot)$ is the kernel embeddings of P. In this particular case, the squared IPM $\gamma_{\mathcal{F}}(\mathrm{P},\mathrm{Q})^2$ is denoted by $\mathrm{MMD}^2(\mathrm{P},\mathrm{Q};\mathcal{H})$ and is the square of the so-called MMD distance. Using again the reproducing property, it is possible to obtain an explicit formulation involving the kernel only. Indeed, $\xi, \xi' \sim \mathrm{P}$, meaning that $\xi$ and $\xi'$ are independent copies of law P, we may write

$$\begin{aligned}
\mathrm{MMD}^2(\mathrm{P},\mathrm{Q};\mathcal{H}) &= \|\mu_{\mathrm{P}} - \mu_{\mathrm{Q}}\|^2 \\
&= \mathbb{E}_{\xi,\xi'\sim\mathrm{P}} k(\xi,\xi') + \mathbb{E}_{\zeta,\zeta'\sim\mathrm{Q}} k(\zeta,\zeta') \\
&\quad -2\mathbb{E}_{\xi\sim\mathrm{P},\zeta\sim\mathrm{Q}} k(\xi,\zeta),
\end{aligned} \tag{6.9}$$

assuming $\mathbb{E}_{\xi,\xi'\sim \mathrm{P}} k(\xi,\xi') < \infty$ and $\mathbb{E}_{\zeta,\zeta'\sim \mathrm{Q}} k(\zeta,\zeta') < \infty$. The MMD is a distance if and only if the kernel $k$ is characteristic (i.e., the kernel embedding is injective) in order to ensure that two different probability distributions are not mapped to the same function in the RKHS. Most kernels used in the machine learning community, such as the Gaussian kernel, the exponential kernel, or the class of Matérn kernels, are characteristic. However, polynomial kernels $k(y,y') = (yy'+\alpha)^d$ are not characteristic since probability distributions with the same moments up to order $d$ but with different higher-order ones are mapped to the same function. In particular, the vanilla linear kernel $k(y,y') = yy'$ is worth mentioning since in this case the MMD boils down to a simple comparison of the means, i.e., $\mathrm{MMD}^2(\mathrm{P},\mathrm{Q};\mathcal{H}) = \left(\mathbb{E}_{\xi\sim\mathrm{P}}\xi - \mathbb{E}_{\zeta\sim\mathrm{Q}}\zeta\right)^2$, which obviously does not define a distance on probability distributions.

### MMD estimation

The kernel formulation (6.9) is easily used to define an estimator for $\mathrm{MMD}^2(\mathrm{P},\mathrm{Q};\mathcal{H})$ given by

$$\widehat{\mathrm{MMD}}^2(\mathrm{P},\mathrm{Q};\mathcal{H}) = \frac{1}{n(n-1)}\sum_{i=1}^{n}\sum_{j\neq i}^{n}\left[k(\xi_i,\xi_j) - k(\xi_i,\zeta_j) - k(\zeta_i,\xi_j) + k(\zeta_i,\zeta_j)\right],$$

where $\{\xi_i\}_{i=1,\dots,n} \sim \mathrm{P}$ and $\{\zeta_i\}_{i=1,\dots,n} \sim \mathrm{Q}$. This is an unbiased one-sample U-statistic estimate, but a biased V-statistic has also been proposed. Both converge in probability at rate $\mathcal{O}(n^{-\frac{1}{2}})$ and have been used, for example, in the context of two-sample tests [171].

### Sensitivity index based on the MMD

Now as a kernel-based dissimilarity has been introduced, we may plug its definition just like we did for the $F$-divergences into Eq. (6.2) to obtain an *unnormalized* kernel-based sensitivity index [100]

$$\mathcal{S}_j^{\mathrm{MMD,unnorm}} = \mathbb{E}_{X_j}\mathrm{MMD}^2(\mathrm{P}_Y,\mathrm{P}_{Y|X_j};\mathcal{H}),$$

provided the following assumption holds.

**Assumption 1.** *For all $A \subseteq \mathcal{P}_d$ and $\mathrm{P}_{\mathbf{X}_A}$-almost all $\mathbf{x}_A \in \mathcal{X}_A$, $\mathbb{E}_{\xi\sim\mathrm{P}_{Y|\mathbf{X}_A=\mathbf{x}_A}} k_{\mathcal{Y}}(\xi,\xi) < \infty$ with the convention $\mathrm{P}_{Y|\mathbf{X}_A} = \mathrm{P}_Y$ if $A = \emptyset$.*

Notice that this index is not normalized (just as all indices defined through Eq. (6.2)). Nevertheless, we will see later on that we may obtain a proper normalization via a FANOVA-like decomposition; see Section 6.2.2.

An interesting property of $\mathcal{S}_j^{\mathrm{MMD,unnorm}}$ holds considering again the linear kernel $k(y,y') = yy'$. Indeed, in this case, we get

$$\begin{aligned}\mathcal{S}_j^{\mathrm{MMD,unnorm}} &= \mathbb{E}_{X_j}\left(\mathbb{E}_{\xi\sim\mathrm{P}_Y}(\xi) - \mathbb{E}_{\zeta\sim\mathrm{P}_{Y|X_j}}(\zeta)\right)^2\\ &= \mathbb{E}_{X_j}\left[\mathbb{E}Y - \mathbb{E}(Y|X_j)\right]^2\\ &= \mathrm{Var}\,\mathbb{E}(Y|X_j);\end{aligned}$$

that is, we recover the unnormalized Sobol' first-order index.

More generally, we can also consider the unnormalized MMD-based sensitivity index for a group of variables $\mathbf{X}_A$:

$$\mathcal{S}_A^{\mathrm{MMD,unnorm}} = \mathbb{E}_{\mathbf{X}_A}\left(\mathrm{MMD}^2(\mathrm{P}_Y,\mathrm{P}_{Y|\mathbf{X}_A};\mathcal{H})\right).$$

To make another connection with Sobol' indices, let us assume that Mercer's theorem holds; i.e., we assume that the output $Y$ lies in the compact set $\mathcal{Y}$ so that $k$ admits the representation

$$k(y,y') = \sum_{i=1}^{\infty} \phi_i(y)\phi_i(y'),$$

where $\phi_i(y) = \sqrt{\lambda_i} e_i(y)$ are orthogonal functions in $\mathbb{L}^2(\mathcal{Y})$; see Appendix B. We first notice that in this setting, we can write

$$\begin{aligned}
\mathbb{E}_{\mathbf{X}_A}\left(\mathrm{MMD}^2(\mathrm{P}_Y, \mathrm{P}_{Y|\mathbf{X}_A}; \mathcal{H})\right) &= \mathbb{E}_{\mathbf{X}_A}\left[\|\mu_{\mathrm{P}_{Y|\mathbf{X}_A}}\|^2 - \|\mu_{\mathrm{P}_Y}\|^2\right] \\
&= \mathbb{E}_{\mathbf{X}_A}\mathbb{E}_{\xi,\xi'\sim \mathrm{P}_{Y|\mathbf{X}_A}} k(\xi,\xi') - \mathbb{E}_{\zeta,\zeta'\sim \mathrm{P}} k(\zeta,\zeta') \\
&= \mathbb{E}_{\mathbf{X}_A}\mathbb{E}_{\xi,\xi'\sim \mathrm{P}_{Y|\mathbf{X}_A}} \left(\sum_{i=1}^{\infty}\phi_i(\xi)\phi_i(\xi')\right) \\
&\quad -\mathbb{E}_{\zeta,\zeta'\sim \mathrm{P}}\left(\sum_{i=1}^{\infty}\phi_i(\zeta)\phi_i(\zeta')\right).
\end{aligned}$$

Now, since the convergence of the series is absolute, we may interchange the expectations and the summations to get

$$\begin{aligned}
\mathbb{E}_{\mathbf{X}_A}\left(\mathrm{MMD}^2(\mathrm{P}_Y, \mathrm{P}_{Y|\mathbf{X}_A}; \mathcal{H})\right) &= \sum_{i=1}^{\infty} \mathbb{E}_{\mathbf{X}_A}\mathbb{E}_{\xi,\xi'\sim \mathrm{P}_{Y|\mathbf{X}_A}}\left(\phi_i(\xi)\phi_i(\xi')\right) \\
&\quad -\mathbb{E}_{\zeta,\zeta'\sim \mathrm{P}}\left(\phi_i(\zeta)\phi_i(\zeta')\right) \\
&= \sum_{i=1}^{\infty} \mathbb{E}_{\mathbf{X}_A}\mathbb{E}\left(\phi_i(Y)|\mathbf{X}_A\right)^2 - \mathbb{E}\left(\phi_i(Y)\right)^2 \\
&= \sum_{i=1}^{\infty} \mathrm{Var}\,\mathbb{E}\left[\phi_i(Y)|\mathbf{X}_A\right]. 
\end{aligned} \tag{6.10}$$

In other words, the kernel-based sensitivity index $\mathbb{E}_{\mathbf{X}_A}\left(\mathrm{MMD}^2(\mathrm{P}_Y, \mathrm{P}_{Y|\mathbf{X}_A}; \mathcal{H})\right)$ generalizes the Sobol' one in the sense that we measure the impact of the inputs not only on the conditional expectation of the output but also on a possibly infinite number of transformations of the output, given by the eigenfunctions of the kernel. We will see that this connection is essential when establishing a general decomposition later.

Given their close links to Sobol' indices, MMD-based indices can be estimated with the same class of algorithms: double-loop Monte Carlo, pick-freeze from Chapter 3, Section 3.2.1, or even ranking and nearest neighbors as discussed in Chapter 3, Section 3.4. Details and convergence properties of such estimators can be found in Da Veiga [100].

### Going further with the HSIC dependence measure

Coming back to moment-independent sensitivity indices, we have seen in Section 6.1 that some special cases of $f$-divergences lead to sensitivity indices which are dependence measures, such as the mutual information

$$\mathcal{S}_j^{MI} = I(X_j;Y) = \int_{\mathbb{R}^2} p_{X_j,Y}(x,y)\ln\left(\frac{p_{X_j,Y}(x,y)}{p_Y(y)p_{X_j}(x)}\right)dxdy$$

or the squared-loss mutual information

$$\mathcal{S}_j^{SMI} = SMI(X_j;Y) = \int_{\mathbb{R}^2}\left(\frac{p_{X_j,Y}(x,y)}{p_Y(y)p_{X_j}(x)} - 1\right)^2 p_Y(y)p_{X_j}(x)dxdy.$$

To measure the dependence between any two random variables $(U,V) \sim \mathrm{P}_{UV}$ with marginals $U \sim \mathrm{P}_U$ and $V \sim \mathrm{P}_V$, it is common practice to measure the dissimilarity between the joint distribution $\mathrm{P}_{UV}$ and the product of the marginals $\mathrm{P}_U \times \mathrm{P}_V$. In the examples above, the dissimilarity measures are the Kullback–Leibler divergence and the $\chi^2$ divergence, respectively. However, the kernel-based dissimilarity measure MMD has been introduced in the previous section: It is then possible to define a dependence measure equal to the MMD distance between $\mathrm{P}_{UV}$ and $\mathrm{P}_U \times \mathrm{P}_V$. This measure is the so-called *Hilbert–Schmidt independence criterion* (HSIC) and is defined below.

**Definition 6.2.** *Consider an RKHS $\mathcal{H}_{\mathcal{X}}$ of functions $\mathcal{X} \to \mathbb{R}$ with kernel $k_{\mathcal{X}}$ and a second RKHS $\mathcal{H}_{\mathcal{Y}}$ of functions $\mathcal{Y} \to \mathbb{R}$ with kernel $k_{\mathcal{Y}}$. The Hilbert–Schmidt independence criterion* HSIC$(U,V)$ *between two random vectors $(U,V)$ with joint distribution $\mathrm{P}_{UV}$ on $\mathcal{X} \times \mathcal{Y}$ and marginal distributions $\mathrm{P}_U$ on $\mathcal{X}$ and $\mathrm{P}_V$ on $\mathcal{Y}$ is given by*

$$\mathrm{HSIC}(U,V) = \|\mu_{\mathrm{P}_{UV}} - \mu_{\mathrm{P}_U \times \mathrm{P}_V}\|^2_{\mathcal{H}_{\mathcal{X}} \times \mathcal{H}_{\mathcal{Y}}},$$

*where $\mu_{\mathrm{P}_{UV}} = \int_{\mathcal{X}\times\mathcal{Y}} k_{\mathcal{X}}(\cdot,u)k_{\mathcal{Y}}(\cdot,v)d\mathrm{P}_{UV}(u,v)$ is the kernel embedding of the joint distribution and $\mu_{\mathrm{P}_U\times\mathrm{P}_V} = \int_{\mathcal{X}\times\mathcal{Y}} k_{\mathcal{X}}(\cdot,u)k_{\mathcal{Y}}(\cdot,v)d\mathrm{P}_U(u)d\mathrm{P}_V(v)$ is the kernel embedding of the product of the marginals. Furthermore, we have the following formulation with kernels:*

$$\begin{aligned}\mathrm{HSIC}(U,V) = {} & \mathbb{E}_{U,U',V,V'} k_{\mathcal{X}}(U,U')k_{\mathcal{Y}}(V,V') \\ & + \mathbb{E}_{U,U'} k_{\mathcal{X}}(U,U')\mathbb{E}_{V,V'} k_{\mathcal{Y}}(V,V') \\ & - 2\mathbb{E}_{U,V}\left[\mathbb{E}_{U'} k_{\mathcal{X}}(U,U')\mathbb{E}_{V'} k_{\mathcal{Y}}(V,V')\right],\end{aligned}$$

*where $(U,V)$ and $(U',V')$ are independent copies distributed as $\mathrm{P}_{UV}$, provided $\mathbb{E}_{U,U'} k_{\mathcal{X}}(U,U') < \infty$ and $\mathbb{E}_{V,V'} k_{\mathcal{Y}}(V,V') < \infty$.*

If both kernels $k_{\mathcal{X}}$ and $k_{\mathcal{Y}}$ are characteristic, the HSIC dependence measure equals 0 if and only if the variables are independent (just like for the mutual information). Concerning estimation, assume that $(U_i,V_i)_{i=1,\dots,n}$ is a sample of the random vector $(U,V)$, and denote $K_{\mathcal{X}}$ and $K_{\mathcal{Y}}$ the Gram matrices with entries $K_{\mathcal{X}}(i,j) = k_{\mathcal{X}}(U_i,U_j)$ and $K_{\mathcal{Y}}(i,j) = k_{\mathcal{Y}}(V_i,V_j)$ and $H$ the centering matrix such that $H(i,j) = \delta_{ij} - \frac{1}{n}$. Gretton et al. [172] propose the following consistent estimator for HSIC$(U,V)$:

$$\begin{aligned}\widehat{\mathrm{HSIC}}(U,V) &= \frac{1}{n^2}\mathrm{Tr}\left(K_{\mathcal{X}} H K_{\mathcal{Y}} H\right) \qquad (6.11) \\ &= \frac{1}{n^2}\sum_{i,j=1}^{n} k_{\mathcal{X}}(U_i,U_j)k_{\mathcal{Y}}(V_i,V_j) \\ &\quad + \frac{1}{n^2}\sum_{i,j=1}^{n} k_{\mathcal{X}}(U_i,U_j)\frac{1}{n^2}\sum_{i,j=1}^{n} k_{\mathcal{Y}}(V_i,V_j) \\ &\quad - \frac{2}{n}\sum_{i=1}^{n}\left[\frac{1}{n}\sum_{j=1}^{n} k_{\mathcal{X}}(U_i,U_j)\frac{1}{n}\sum_{j=1}^{n} k_{\mathcal{Y}}(V_i,V_j)\right].\end{aligned}$$

Notice that an unbiased estimator is also introduced by Song et al. [448]. HSIC has been widely used in a large panel of applications, such as feature selection [448] and independence testing [173].

### Sensitivity index based on HSIC

Similarly to the $F$-divergence case with mutual information, we can now define an unnormalized first-order HSIC index as

$$\mathcal{S}_j^{\text{HSIC,unnorm}} = \text{HSIC}(X_j, Y)$$

relying on a kernel $k_{\mathcal{X}_j}$ for the input $X_j$ and $k_{\mathcal{Y}}$ for the output $Y$. More generally, considering a product kernel $k_{\mathcal{X}}(\mathbf{x}, \mathbf{x}') = \prod_{j=1}^d k_{\mathcal{X}_j}(x_j, x'_j)$, the HSIC-based index for a group of inputs $\mathbf{X}_A$ writes

$$\mathcal{S}_A^{\text{HSIC,unnorm}} = \text{HSIC}(\mathbf{X}_A, Y)$$

using the kernel $k_A(\mathbf{x}_A, \mathbf{x}'_A) = \prod_{j\in A} k_{\mathcal{X}_j}(x_j, x'_j)$ for the inputs $\mathbf{X}_A$. For this, we have to assume that the following assumption holds.

**Assumption 2.** *For all $A \subseteq \mathcal{P}_d$, $\mathbb{E}_{\xi\sim \mathrm{P}_{\mathbf{X}_A}} k_{\mathcal{X}_A}(\xi, \xi) < \infty$ and $\mathbb{E}_{\xi\sim \mathrm{P}_Y} k_{\mathcal{Y}}(\xi, \xi) < \infty$.*

Again, this index can be seen as a further extension of the previous one since when the input kernel $k_{\mathcal{X}}$ degenerates to a Dirac kernel, the HSIC-based index is equal to the MMD based one, as stated in the following proposition from Da Veiga [100].

**Proposition 6.3.** *For every subset $A \subseteq \mathcal{P}_d$, assume that for all $\mathbf{x}_A \in \mathcal{X}_A$, $p_{\mathbf{X}_A}(\mathbf{x}_A) > 0$ and that*

$$k_A(\mathbf{x}_A, \mathbf{x}'_A) = \frac{1}{\sqrt{p_{\mathbf{X}_A}(\mathbf{x}_A)}\sqrt{p_{\mathbf{X}_A}(\mathbf{x}'_A)}} \prod_{l\in A} \frac{1}{h} K\left(\frac{x_l - x'_l}{h}\right), \tag{6.12}$$

*where $K : \mathbb{R} \to \mathbb{R}$ is a symmetric kernel function satisfying $\int_u K(u)du = 1$ and $h > 0$. Then we have, for all $A \subseteq \mathcal{P}_d$,*

$$\lim_{h\to 0} \text{HSIC}(\mathbf{X}_A, Y) = \mathbb{E}_{\mathbf{X}_A}\left(\text{MMD}^2(\mathrm{P}_Y, \mathrm{P}_{Y|\mathbf{X}_A})\right).$$

Normalized versions of the first-order index have been proposed and investigated in Da Veiga [99]. In particular, it is a powerful tool in the context of screening. Indeed, the estimation of all first-order HSIC indices is independent of the number of variables. Further, Gretton et al. [172] also give the rate of convergence of their estimator. In practice, this means that with a small number of observations, it is possible to compute a robust estimate of the index. For example, Da Veiga [99] considers several test cases with $n = 50$ observations only where screening with HSIC is powerful.

Another interesting screening property relies on the HSIC-based independence test proposed in Gretton et al. [173], which can be used as a robust way to detect noninfluential inputs among a large number. These HSIC-based independence tests have been further developed for screening (i.e., *factor fixing* setting) in SA by De Lozzo and Marrel [106], who also proposed non-asymptotic test versions, adapted to small-size sample cases. Chabridon et al. [72] have adapted this tool for an SA industrial application dealing with 100 inputs.

**Remark 37.** *In case of dependent inputs, HSIC indices cannot be used for the factor fixing setting. Indeed, in a dependence framework, it is not obvious to understand the impact of fixing the input with the lowest HSIC on the dependence structure of the remaining inputs or on the conditional distribution of the output with respect to the value of the fixed input. The setting of dependent inputs remains an open problem where adequate backward-forward selection procedures have to be adopted as in Song et al. [448] or variants based on the lasso [510]. Another point of view is to consider kernelized versions of Shapley effects that are introduced in Section* 6.2.3.

### 6.2.2 ▪ FANOVA-like decomposition for kernel-based indices

As mentioned earlier, the two previously introduced kernel-based indices can be thought as more general kernelized version of Sobol' indices. But they also admit a FANOVA-like decomposition. This makes them the only moment-independent indices with this property. In particular, such decomposition not only provides their natural normalization constant but also yields a systematic procedure to define higher-order interactions.

Recall that the variance decomposition states that the variance of the output can be decomposed as $\operatorname{Var} Y = \sum_{A \in \mathcal{P}_d} V_A$, where each term is given by

$$V_A = \sum_{B \subset A} (-1)^{|A|-|B|} \operatorname{Var} \mathbb{E}(Y|\mathbf{X}_B).$$

We start with the MMD-based equivalent decomposition. It is described in the following theorem [100].

**Theorem 6.4.** *Assume the same assumptions as the one assumed in Theorem* 3.3 *(in particular, the random vector* $\mathbf{X}$ *has independent components). Assume further that Assumption* 1 *and Mercer's theorem hold. Denote* $\mathrm{MMD}^2_{tot} = \mathbb{E}k(Y,Y) - \mathbb{E}k(Y,Y')$*, where* $Y'$ *is an independent copy of* $Y$*. Then the total MMD can be decomposed as*

$$\mathrm{MMD}^2_{tot} = \sum_{A \in \mathcal{P}_d} \mathrm{MMD}^2_A,$$

*where each term is given by*

$$\mathrm{MMD}^2_A = \sum_{B \subset A} (-1)^{|A|-|B|} \mathbb{E}_{\mathbf{X}_B} \left( \mathrm{MMD}^2(\mathrm{P}_Y, \mathrm{P}_{Y|\mathbf{X}_B}; \mathcal{H}) \right).$$

***Proof.*** The theorem is proved in the case where Mercer's theorem holds; i.e., the output is assumed to be such that $Y \in \mathcal{Y}$ with $\mathcal{Y}$ a compact set and $k$ has the representation

$$k(y, y') = \sum_{i=1}^{\infty} \phi_i(y)\phi_i(y')$$

as in Eq. (6.10). Consider now the random variable $W = \sum_{i=1}^{\infty} G^{[i]}(\mathbf{X})$, where $G^{[i]}(\mathbf{X}) = \phi_i(Y) = \phi_i(G(\mathbf{X}))$. To prove the theorem, we will provide two expansions of $\operatorname{Var} W$. On the one hand, since the functions $\phi_i$ are orthogonal in $\mathbb{L}^2(\mathcal{Y})$ and using the absolute convergence of the series, we have

$$\begin{aligned}
\operatorname{Var} W &= \sum_{i=1}^{\infty} \operatorname{Var} \phi_i(Y) \\
&= \sum_{i=1}^{\infty} \mathbb{E}\left(\phi_i(Y)\phi_i(Y)\right) - \sum_{i=1}^{\infty} \mathbb{E}\left(\phi_i(Y)\phi_i(Y')\right) \\
&= \mathbb{E}\left( \sum_{i=1}^{\infty} \phi_i(Y)\phi_i(Y) \right) - \mathbb{E}\left( \sum_{i=1}^{\infty} \phi_i(Y)\phi_i(Y') \right) \\
&= \mathbb{E}k(Y,Y) - \mathbb{E}k(Y,Y').
\end{aligned}$$

On the other hand, using the variance decomposition for each $G^{[i]}(\mathbf{X}) = \phi_i(G(\mathbf{X}))$, we get

$$\begin{aligned}
\operatorname{Var} W &= \sum_{i=1}^{\infty} \operatorname{Var} G^{[i]}(\mathbf{X}) \\
&= \sum_{i=1}^{\infty} \sum_{A \in \mathcal{P}_d} \sum_{B \subset A} (-1)^{|A|-|B|} \operatorname{Var} \mathbb{E}\left(G^{[i]}(\mathbf{X}) | \mathbf{X}_B\right) \\
&= \sum_{A \in \mathcal{P}_d} \sum_{B \subset A} (-1)^{|A|-|B|} \sum_{i=1}^{\infty} \operatorname{Var} \mathbb{E}\left(\phi_i(Y) | \mathbf{X}_B\right) \\
&= \sum_{A \in \mathcal{P}_d} \sum_{B \subset A} (-1)^{|A|-|B|} \mathbb{E}_{\mathbf{X}_B}\left(\operatorname{MMD}^2(\mathrm{P}_Y, \mathrm{P}_{Y|\mathbf{X}_B}; \mathcal{H})\right)
\end{aligned}$$

using again the absolute continuity and the expansion of $\mathbb{E}_{\mathbf{X}_B}\left(\operatorname{MMD}^2(\mathrm{P}_Y, \mathrm{P}_{Y|\mathbf{X}_B}; \mathcal{H})\right)$ obtained above in Eq. (6.10). The theorem follows by equating both expansions of $\operatorname{Var} W$. □

Theorem 6.4 is very similar to the FANOVA one (Corollary 3.5, page 52). One can note that the total variance of the output is replaced by a generalized variance $\operatorname{MMD}^2_{tot}$ defined by the kernel and that each subset effect is obtained by removing lesser-order ones in the MMD distance of the conditional distributions (instead of the variance of the conditional expectations in the FANOVA). The following corollary states that these two decompositions coincide when using the linear kernel (Sobol' decomposition).

**Corollary 6.5.** *When $k(y, y') = yy'$ in Theorem 6.4, the decomposition is the same as the one provided in Corollary 3.5 (page 52). In other words,*

$$\operatorname{MMD}^2_{tot} = V \quad \text{and} \quad \forall B \in \mathcal{P}_d,\ \mathbb{E}_{\mathbf{X}_B}\left(\operatorname{MMD}^2(\mathrm{P}_Y, \mathrm{P}_{Y|\mathbf{X}_B}; \mathcal{H})\right) = \operatorname{Var} \mathbb{E}(Y | \mathbf{X}_B).$$

*It further implies that for all $A \in \mathcal{P}_d$, $\operatorname{MMD}^2_A = V_A$.*

Thanks to Theorem 6.4, we can now define properly normalized MMD-based indices.

**Definition 6.6.** *In the frame of Theorem 6.4, let $A \in \mathcal{P}_d$ and $j \in [1:d]$:*

- *The normalized MMD-based sensitivity index associated to $A$ is defined as*

$$S_A^{\mathrm{MMD}} = \frac{\operatorname{MMD}^2_A}{\operatorname{MMD}^2_{tot}}.$$

- *$S_j^{\mathrm{MMD}} = S_{\{j\}}^{\mathrm{MMD}}$ is the MMD-based index associated to the singleton $\{j\}$. It is called the first-order MMD-based index for the input variable $X_j$. More generally, if $l = |A|$, $S_A^{\mathrm{MMD}}$ is called the MMD-based index of order $l$ associated to $X_A$.*

- *The total MMD-based index associated to $X_A$ is defined as $S_A^{T,\mathrm{MMD}} = 1 - S_{\bar{A}}^{\mathrm{MMD}}$. In particular,*

$$S_j^{T,\mathrm{MMD}} = S_{\{j\}}^{T,\mathrm{MMD}} = 1 - \frac{\operatorname{MMD}^2_{\overline{\{j\}}}}{\operatorname{MMD}^2_{tot}}.$$

From Theorem 6.4, we have the fundamental normalization identity

$$\sum_{A \in \mathcal{P}_d} S_A^{\mathrm{MMD}} = 1.$$

Now, HSIC-based indices also enjoy a FANOVA-like decomposition. Nevertheless, it requires an additional strong assumption on the kernel structure assigned to the inputs.

**Assumption 3.** *The reproducing kernel $k_{\mathcal{X}}$ is of the form*

$$k_{\mathcal{X}}(\mathbf{x},\mathbf{x}') = \prod_{j=1}^{d} \left(1 + k_j(x_j, x_j')\right), \tag{6.13}$$

*where, for each $j = 1,\ldots,d$, $k_j(\cdot,\cdot)$ is the reproducing kernel of a RKHS $\mathcal{H}_j$ of real functions depending only on variable $x_j$ and such that $1 \notin \mathcal{H}_j$.*

*Furthermore, for all $j = 1,\ldots,d$ and for all $x_j \in \mathcal{X}_j$, we have*

$$\int_{\mathcal{X}_j} k_j(x_j, x_j') d\mathrm{P}_{X_j}(x_j') = 0. \tag{6.14}$$

Observe that this assumption involves using univariate kernels associated to an RKHS, which does not include constant functions. This may seem to be an original setting, but this is actually what is usually done in standard support vector machines where a bias is explicitly introduced when using RBF kernels. Indeed, it is a well-known fact that they define an RKHS, which does not contain constant functions [456]. With Assumption 3, we can now state a decomposition for HSIC-based sensitivity indices.

**Theorem 6.7.** *Under the same assumptions of Theorem* 3.3 *(in particular, the random vector* $\mathbf{X}$ *has independent components), with Assumptions* 2 *and* 3 *and assuming Mercer's theorem holds, the HSIC dependence measure between $\mathbf{X} = (X_1,\ldots,X_d)$ and $Y$ can be decomposed as*

$$\mathrm{HSIC}\,(\mathbf{X}, Y) = \sum_{A \subseteq \mathcal{P}_d} \mathrm{HSIC}_A,$$

*where each term is given by*

$$\mathrm{HSIC}_A = \sum_{B \subset A} (-1)^{|A|-|B|} \mathrm{HSIC}\,(\mathbf{X}_B, Y)$$

*and* HSIC $(\mathbf{X}_B, Y)$ *is defined with kernel $k_B(\mathbf{x}_B, \mathbf{x}_B') = \prod_{j\in B}\left(1 + k_j(x_j, x_j')\right)$ on the inputs as in* (6.13).

The proof relies on Theorem 4.1 from Kuo et al. [254] and is given in Da Veiga [100]. Normalized HSIC-based indices can finally be defined.

**Definition 6.8.** *In the framework of Theorem 6.7, let $A \subseteq \mathcal{P}_d$. The normalized HSIC-based sensitivity index associated to a subset $A$ of input variables is defined as*

$$S_A^{\mathrm{HSIC}} = \frac{\mathrm{HSIC}_A}{\mathrm{HSIC}\,(\mathbf{X}, Y)},$$

*while the total HSIC-based index associated to $A$ is*

$$S_A^{T,\mathrm{HSIC}} = \sum_{B \subseteq \mathcal{P}_d,\, B\cap A \neq \emptyset} S_B^{\mathrm{HSIC}} = 1 - \frac{\mathrm{HSIC}(\mathbf{X}_{-A}, Y)}{\mathrm{HSIC}\,(\mathbf{X}, Y)}.$$

*From Theorem 6.7, we have the fundamental normalizing identity providing the interpretation of HSIC-based indices as a percentage of the explained HSIC dependence measure between $\mathbf{X} = (X_1, \ldots, X_d)$ and $Y$:*

$$\sum_{A \subseteq \mathcal{P}_d} S_A^{\text{HSIC}} = 1.$$

Notice that Assumption 3 requires the use of kernels defining a so-called RKHS of *zero-mean functions* [492]. A general procedure to build such kernels was introduced by [120]: A zero-mean kernel $k_0^D(\cdot,\cdot)$ is given by

$$k_0^D(x, x') = k(x, x') - \frac{\int k(x,t) d\mathrm{P}(t) \int k(x',t) d\mathrm{P}(t)}{\iint k(s,t) d\mathrm{P}(s) d\mathrm{P}(t)}$$

from an arbitrary univariate $k(\cdot,\cdot)$, where $k_0^D(\cdot,\cdot)$ satisfies, for all $x$, $\int k_0^D(x,t) d\mathrm{P}(t) = 0$.

### 6.2.3 ▪ Kernel-embedding Shapley effects for dependent inputs

When the inputs are no longer independent, we discussed in Chapter 5, Section 5.5, the Shapley effects defined with a value function $\text{val}(A) = S_A^{\text{clos}} = \frac{\operatorname{Var} \mathbb{E}[G(\mathbf{X})|\mathbf{X}_A]}{V}$ and showed how they can be valuable for sensitivity analysis with dependent inputs. Nevertheless, notice that Shapley effects with such a value function are still variance-based indices. Using the previously introduced kernel-based sensitivity indices as building blocks for new value functions, we define *kernel-embedding Shapley effects*.

**Definition 6.9.** *For any $j = 1, \ldots, d$, we define*

(a) *the MMD-Shapley effect*

$$\begin{aligned} Sh_j^{\text{MMD}} = \frac{1}{\text{MMD}_{\text{tot}}^2} \frac{1}{p} \sum_{A \subset -j} \binom{p-1}{|A|}^{-1} \Big\{ & \mathbb{E}_{\mathbf{X}_{A \cup \{j\}}} \left( \text{MMD}^2(\mathrm{P}_Y, \mathrm{P}_{Y|\mathbf{X}_{A\cup\{j\}}}) \right) \\ & - \mathbb{E}_{\mathbf{X}_A} \left( \text{MMD}^2(\mathrm{P}_Y, \mathrm{P}_{Y|\mathbf{X}_A}) \right) \Big\} \end{aligned} \tag{6.15}$$

*provided Assumption 1 holds;*

(b) *the HSIC-Shapley effect*

$$Sh_j^{\text{HSIC}} = \frac{1}{\text{HSIC}(\mathbf{X}, Y)} \frac{1}{p} \sum_{A \subset -j} \binom{p-1}{|A|}^{-1} \Big\{ \text{HSIC}(\mathbf{X}_{A \cup \{j\}}, Y) - \text{HSIC}(\mathbf{X}_A, Y) \Big\} \tag{6.16}$$

*provided Assumptions 2 and 3 hold.*

*We further have the decompositions*

$$\sum_{j=1}^{d} Sh_j^{\text{MMD}} = \sum_{j=1}^{d} Sh_j^{\text{HSIC}} = 1.$$

As in the independent setting, kernel-embedding Shapley effects (6.15) and (6.16) are general kernelized versions of Shapley effects. More details on these indices and estimators are discussed in Da Veiga [100].

### 6.2.4 ▪ Examples in R

We use the Sobol' $g$-function (see Section C.7 with independent $X_j \sim \mathcal{U}(0,1)$) as an illustration for the HSIC indices by using standard Gaussian kernels for both the inputs and the output:

```
> set.seed(12345)
> library(sensitivity)
> d <- 8
> n <- 200
> X <- data.frame(matrix(runif(d * n), nrow = n))
> sa <- sensiHSIC(model=sobol.fun, X, kernelX="rbf", kernelY="rbf",
+      estimator.type = "V-stat", test.method = "Permutation",
+      B = 3000, nboot = 100)
> library(ggplot2) ; ggplot(sa)
```

The HISC indices estimates are given in Figure 6.3. Confidence intervals are computed by bootstrap. $X_1$ is easily detected as influential, while $X_3$ to $X_8$ appear to have no impact on the output. The effect of $X_2$ is very small. We can also investigate the p-values to make a decision whether or not it is influential:

```
> print(sa$Pvalue)

       original
X1 0.000000e+00
X2 1.447751e-06
X3 9.682952e-02
X4 1.566254e-01
X5 6.877311e-01
X6 5.075607e-01
X7 6.526675e-01
X8 6.765713e-02
```

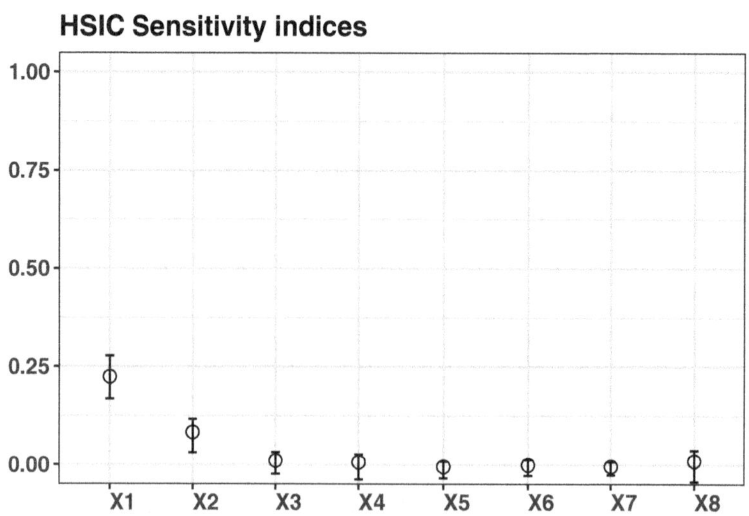


**Figure 6.3.** *Normalized Hilbert–Schmidt independence criteria for the Sobol g-function.*

Here, the test of independence rejects the hypothesis that both $X_1$ and $X_2$ are independent from the output.

### 6.2.5 ▪ More general kernels

The kernel point of view also provides an elegant and powerful framework for dealing with categorical outputs as well as functional ones.

#### Categorical variables

The categorical case is common practice in applications for both inputs and outputs. For example, if we focus on different regions of interest for a model output, the target output can be represented as labels. Appropriate kernels include, for example, $k(y, y') = \delta_{yy'}/n_y$, where $n_y$ is the number of samples with label $y$; see, e.g., [448, 510].

#### Functional inputs and outputs

It is also possible to design dedicated semimetrics for functional outputs which can be incorporated in the definition of the kernels; see, e.g., [134]. For example, let $\Delta(\cdot, \cdot)$ be such a semimetric defined on $\mathcal{Y} \times \mathcal{Y}$ if the output variable is of functional type $Y \in \mathcal{Y}$. Usually, we may take $\Delta(y, y') = \|\phi(y) - \phi(y')\|_{\mathcal{Z}}$, where $\phi(\cdot)$ is a function from $\mathcal{Y}$ to $\mathcal{Z}$ and $\|\cdot\|_{\mathcal{Z}}$ is a norm on $\mathcal{Z}$. This is a semimetric since $\phi(\cdot)$ is not necessarily injective. The kernel associated to $\mathcal{Y}$ is then given by $k_{\mathcal{Y}}(y, y') = k_{\mathcal{Z}}(\phi(y), \phi(y')) = k(\Delta(y, y'))$, where $k_{\mathcal{Z}}$ is a stationary kernel on $\mathcal{Z} \times \mathcal{Z}$. However, a theoretical shortcoming lies in our current inability to ensure that such semimetric kernels are characteristic, which implies they are not very often. Despite this deficiency, from a practical perspective, the use of a semimetric based on principal components can efficiently deal with a functional output given as a curve or a 2D map, for example. The same also directly applies when considering inputs, which makes the kernel-based approach well suited for handling both functional inputs and outputs.

#### Exotic kernels

Finally, the design of appropriate kernels for many sorts of objects has been extensively studied in the machine learning community. Of particular interest are kernels for curves via dynamic time warping or variants [98], kernels for images via graphs [179], and many others. Depending on the SA task and the practical knowledge of problem invariance, the use of such kernels may help to prevent a misleading sensitivity study. Invariant kernels have been considered, for example, in Walder and Chapelle [493] and Hamsici and Martinez [178].

Among all these kernels, let us mention a specific one proposed for probability distributions, that is, for $Y \in \mathcal{M}_1^+$,

$$\forall y, y' \in \mathcal{M}_1^+, \; k(y, y') = \exp\left(-\lambda \mathrm{MMD}^2(y, y'; \mathcal{H})\right),$$

which is a valid symmetric positive definite kernel for any $\lambda > 0$ [453]. From a practical perspective, such a situation may arise when, for a given value of the inputs, the code output is a probability distribution obtained through an internal Monte Carlo sampling. Typical examples include predictive maintenance (where the code simulates Monte Carlo samples of component failures), molecular dynamics, or codes reproducing nondestructive testing.

Other kernels of interest include the ones developed for rankings, which means that the output $Y$ is a permutation of a fixed number of items. If we denote all items by $\{1, \ldots, n\}$ and $\Pi_n$ the set of all the permutations of this set, a possible valid kernel is given by

$$\forall y, y' \in \Pi_n,\ k(y, y') = \exp\left(-\lambda K(y, y')\right),$$

where $K(\cdot,\cdot)$ is the Kendall tau rank distance [221]. In practice, the code output can be a ranking list of scenarios or a ranking list of objectives. Both kernels on probability distributions and rankings will be used later when discussing Bayesian perturbations for robustness in Section 6.4.2.

## 6.3 ▪ Goal-oriented sensitivity analysis

In all the previous chapters and the preceding sections of this chapter, SA has been applied by considering the overall variation domain of the output. However, in UQ studies, one might often be interested in some specific events on the output, for instance, a critical event, such as the exceedence of a threshold occurring in the right tail of its distribution. Such events being related to specific QoI, goal-oriented sensitivity analysis (GOSA) aims at analyzing the influence of the inputs with respect to this QoI. For instance, in simulation-based system optimization [136], the QoI is the optimization result [450], while in simulation-based risk analysis [110, 328], it is the considered risk measure, such as a probability of failure (probability that the output exceeds a critical threshold), a quantile, or a superquantile [397]. This section first describes simple adaptations of sensitivity measures (Sobol', density-based, and HSIC ones) to this context. Then a bibliographical analysis is made on sensitivity measures closely related to particular rare event estimation techniques. Finally, the mathematical generalization of Sobol' indices to different QoI is introduced.

### 6.3.1 ▪ Target SA via Monte Carlo thresholding

Let us define $\mathcal{R} \subset \mathcal{Y}$ a region of interest of the output domain $\mathcal{Y}$. From an SA perspective, our objective here is to measure the impact of each input on the fact that the output reaches $\mathcal{R}$. By a simple change of variable

$$Z = \mathbb{1}_{\{Y \in \mathcal{R}\}} \tag{6.17}$$

(for example, $Z = \mathbb{1}_{\{Y>t\}}$ for a given threshold $t$), this implies that we have to evaluate the impact of the inputs on level sets of the output. Such an objective has been called *target SA* by Raguet and Marrel [387] and Marrel and Chabridon [306]. Note that other changes of variable than Eq. (6.17) are possible in order to smooth the discontinuity of the indicator function.

**Remark 38.** *Another objective (that is not discussed here), called "conditional SA," aims at evaluating the influence of inputs within the region of interest only, ignoring what happens outside [387, 306].*

The first straightforward target SA approach just consists in recycling a first-order Sobol' index with the new output $Z$,

$$V_j = \operatorname{Var} \mathbb{E}(Z|X_j), \tag{6.18}$$

where the unnormalized version of the Sobol' index is used (as in Eq. (6.1) of Section 6.1.1) in order to be consistent with Sections 6.1 and 6.2 of this chapter. From Eq. (6.18), it is easy to see that [94, 279]

$$V_j = \mathbb{E}(p_j - p)^2, \tag{6.19}$$

where $p = \mathbb{E}Z = \mathbb{P}(Y \in \mathcal{R})$ (resp., $p_j = \mathbb{E}(Z|X_j) = \mathbb{P}(Y \in \mathcal{R}|X_j)$) is called the probability of failure (resp., the conditional probability of failure). Different estimation schemes of this quantity have been considered, for instance, in Wei et al. [497].

Interestingly, by developing Eq. (6.19) via the Bayes' theorem, it can also be shown that we have [363, 449]

$$V_j = \mathbb{E}\left(\frac{p f_{X_j|Y\in\mathcal{R}}(X_j)}{f_{X_j}(X_j)} - p\right)^2$$
$$= p^2\mathbb{E}\left(\frac{f_{X_j|Y\in\mathcal{R}}(X_j)}{f_{X_j}(X_j)} - 1\right)^2,$$

which is proportional (up to the factor $p^2$) to the Pearson $\chi^2$ divergence between $f_{X_j|Y\in\mathcal{R}}$ and $f_{X_j}$. This perspective demonstrates that this simple sensitivity index on an indicator function operates as a measure comparing the initial input distribution and its conditional counterpart when the output is forced to lie in the region of interest. In this context, several research works (see, e.g., [423]) try to develop some given-data estimation procedures.

In a similar way but without this connection in mind, Spear and Hornberger [451] proposed performing the same comparison between the probability distributions via the Kolmogorov distance. More precisely, they define a sensitivity index as

$$\mathcal{S}_j = \sup_{x\in\mathcal{X}} |F_{X_j|Y\in\mathcal{R}}(x) - F_{X_j}(x)|.$$

Saltelli et al. [414] have called it the Monte Carlo filtering method (see also [142]).

Finally, we can also define the following HSIC-based sensitivity index [449]:

$$\mathcal{S}_j^{\text{HSIC,unnorm}} = \text{HSIC}(X_j, \mathbb{1}_{\{Y\in\mathcal{R}\}}) = \text{HSIC}(X_j, Z),$$

where $Z = \mathbb{1}_{\{Y\in\mathcal{R}\}}$. If we set $k(z, z') = \delta_{zz'}$ and choose any kernel $k_\mathcal{X}$ for the inputs, it can be shown that

$$\mathcal{S}_j^{\text{HSIC,unnorm}} = \mathbb{P}(Y \in \mathcal{R})^2 \,\text{MMD}^2\left(\text{P}_{X_j}, \text{P}_{X_j|Y\in\mathcal{R}}; \mathcal{H}\right),$$

where $\mathcal{H}$ is the RKHS with kernel $k_\mathcal{X}$; see Spagnol et al. [450] for a proof in a goal-oriented context related to optimization. In other words, such a sensitivity index detects an influential variable if, when conditioned on the fact that the output reaches the region of interest, its resulting conditional distribution varies from the original one. This is the same idea as in the Monte Carlo filtering approach presented in Section 6.3.1, except that the distributions are compared with the MMD instead of the Kolmogorov distance.

### Example in R

Target SA is illustrated in Fig. 6.4 on the Sobol' $g$-function (see Section C.7 with independent $X_j \sim \mathcal{U}(0,1)$). The region of interest is $\mathcal{R} = \{y \in \mathcal{Y} : y < 0.4\}$, meaning that we want to identify the inputs which have an influence on the fact that the output is less than 0.4. We first generate a sample of inputs, compute the corresponding output values, and transform them with the indicator function to define $Z = \mathbb{1}_{\{Y<0.4\}}$:

```
> set.seed(12345)
> library(sensitivity)
> d <- 8
> n <- 1000
> X1 <- data.frame(matrix(runif(d * n), nrow = n))
> X2 <- data.frame(matrix(runif(d * n), nrow = n))
> sa <- sobolSalt(model = NULL, X1, X2, scheme="A")
> Y <- sobol.fun(sa$X)
```

```
> threshold <- 0.4
> Z <- as.numeric(Y<threshold)
> tell(sa,Z)

> library(ggplot2) ; ggplot(sa, choice=1)
```

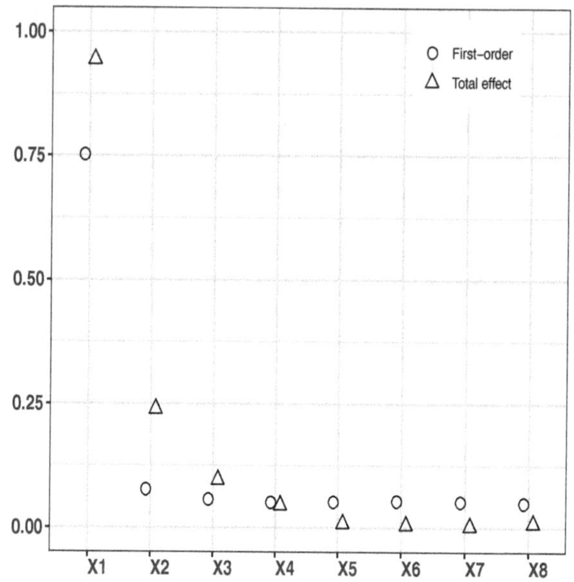


**Figure 6.4.** *Target SA on Sobol' g-function with a region of interest* $\mathcal{R} = \{y \in \mathcal{Y} : y < 0.4\}$.

In a standard global SA setting, $X_1$ is the most influential input on the $g$-function. But here, its influence is even larger if we focus on reaching output values below the defined threshold.

## 6.3.2 ▪ Reliability-oriented sensitivity analysis

In a general context, structural reliability aims at determining the failure probability (or a quantile) of the numerical model output $Y = G(\mathbf{X})$ by considering the input variables as random. To solve this problem, several authors from the structural reliability community (see, e.g., [297]) have largely contributed to the development and promotion of Monte Carlo techniques and geometrically based approximation approaches, such as the *first-/second-order reliability methods* (FORM/SORM). These techniques are now widely used in other physical domains, from hydraulics to aerospace engineering [110]. A large number of more sophisticated methods have also been developed to compute a failure probability and, more generally, to infer rare events with a small number of runs of the model $G(\cdot)$. The two main classes of advanced techniques are the variance-reduction Monte Carlo methods [270, 328] and the metamodel-based approaches [27, 28] (see Section 4.2, page 105, for a reminder on metamodel concepts). From a practical point of view, it appears that importance measures of the inputs are particularly useful for a post hoc analysis of rare event algorithm results or even as help for accelerating the rare event inference algorithm (see, e.g., [210]).

The reliability-oriented SA aims at studying the dependence of a rare event–related QoI to the model input variables [71]. One major difficulty is that the QoI can be very restrictive on the output (due to the rareness of the event). It is then necessary to find methods able to analyze correctly the influence of the input uncertainties on this critical domain of the output. In Section 6.3.1, several importance measures have been introduced, but their computations require the use of a Monte Carlo (i.i.d.) sample with a sufficient data number in the two classes of $Z$ in order to compute significant statistics. In the structural reliability community, sensitivity indices embedded within different failure probability estimation methods have been proposed (see Chabridon [69] for a complete overview). As an example, one can mention the following:

- the local indices based on partial derivatives of the failure probability with respect to the parameters of the pdf of the inputs [40], which can easily deduced from a sample via the so-called score function [405];
- the FORM/SORM and reliability index–based importance factors [297, 296, 239, 356];
- the global indices based on conditional failure probability (as in Section 6.3.1) and obtained by conditional samples, which are by-products of rare event simulation techniques (e.g., importance sampling or subset simulation) [9, 516].

### 6.3.3 ▪ Contrast function–based indices

In this section, a general methodology is introduced and discussed to build sensitivity indices. For the sake of clarity, let us consider that the output of interest $Y$ is scalar. Let us emphasize that, in the probabilistic framework of SA, the output $Y$ is random. Indeed, the uncertainty on the inputs is propagated to the output via the model. The output is thus described by a probability distribution whose support is included in $\mathbb{R}$. Let us now come back to the definition of first-order Sobol' indices

$$S_j = \frac{\mathrm{Var}\,\mathbb{E}(Y|X_j)}{\mathrm{Var}\,Y} = 1 - \frac{\mathbb{E}\,\mathrm{Var}\,(Y|X_j)}{\mathrm{Var}\,Y}\,.$$

It is well known that

$$\begin{aligned}\mathrm{Var}\,Y &= \inf_{\theta\in\mathbb{R}} \mathbb{E}(Y-\theta)^2,\\ \mathrm{Var}\,(Y|X_j) &= \inf_{\theta\in\mathbb{R}} \mathbb{E}[(Y-\theta)^2|X_j].\end{aligned}$$

Thus, we can write

$$S_j = 1 - \frac{\mathbb{E}\inf_{\theta\in\mathbb{R}}\mathbb{E}[(Y-\theta)^2|X_j]}{\inf_{\theta\in\mathbb{R}}\mathbb{E}(Y-\theta)^2} = 1 - \frac{\mathbb{E}\inf_{\theta\in\mathbb{R}}\mathbb{E}[l_2(Y;\theta)|X_j]}{\inf_{\theta\in\mathbb{R}}\mathbb{E}l_2(Y;\theta)},$$

with $l_2(y;\theta) = (y-\theta)^2$ the quadratic loss function. Thus $S_j$ appears to compare the optimal value of the function $\mathbb{E}(Y-\theta)^2$ to the expected optimal value of the conditional function $\mathbb{E}[(Y-\theta)^2|X_j]$. A function $\theta \mapsto \mathbb{E}\Psi(Y;\theta)$ is a contrast function if it admits a unique minimizer $\theta^*$. The quadratic loss function is a contrast function, as $\theta \mapsto l_2(Y;\theta) = \mathbb{E}\,(Y-\theta)^2$ admits a unique minimizer, namely, $\mathbb{E}Y$.

As noticed in Fort et al. [139], a contrast function is a very useful object in *statistical learning theory* (see, e.g., Massart [314]), where it defines estimation procedures of some feature $\theta^* \in \Theta$ (scalar or functional) associated to a random variable $Y$. For instance, when observing an $n$-sample $Y^{(1)},\dots,Y^{(n)}$ of the random variable $Y$, an estimator of $\theta^*$ is given by the minimizer of $\theta \mapsto \Psi_n(\theta)$, where $\Psi_n$ is obtained by substituting the expectation with respect to the variable $Y$ by the expectation with respect to the empirical measure of the sample. For example, by considering the quadratic loss function $\theta \mapsto l_2(Y;\theta) = \mathbb{E}(Y-\theta)^2$, one gets $\hat{\theta}_n = \frac{1}{n}\sum_{i=1}^n Y^{(j)}$, namely, the empirical mean to estimate $\theta^* = \mathbb{E}Y$.

Thus, it seems that first-order Sobol' indices are well tailored when one is interested in the influence of input parameters for the estimation of the mean behavior of the output $Y$. If one is interested in estimating other QoI (i.e., characteristics of the probability distribution of $Y$), such as quantiles, one has to consider different loss functions leading to different contrast functions. It leads to new sensitivity indices called goal-oriented sensitivity indices in Fort et al. [139]. More precisely, let us state Definition 6.10 below.

**Definition 6.10.** *Let $\theta \mapsto \mathbb{E}\Psi(Y;\theta)$ be a contrast function such that $\mathbb{E}\min_\theta \Psi(Y;\theta)$ is finite. The sensitivity indices associated to $\Psi$ are then defined as*

$$S_j^{loss} = 1 - \frac{\mathbb{E}\inf_{\theta\in\mathbb{R}} \mathbb{E}[l(Y,\theta)|X_j]}{\inf_{\theta\in\mathbb{R}} \mathbb{E}l(Y,\theta)}, \; j = 1,\dots,d.$$

### Examples

1. If the QoI is the mean of $Y$, one considers

$$\Psi(y;\theta) = l_2(y;\theta) = (y-\theta)^2.$$

2. If the QoI is the median of $Y$, one considers

$$\Psi(y;\theta) = l_1(y;\theta) = |y-\theta|.$$

3. If the QoI is the quantile of order $\alpha \in (0,1)$ of $Y$, one considers

$$\Psi(y;\theta) = l_\alpha(y,\theta) = \begin{cases} (1-\alpha)(\theta-y) & \text{if } y \leqslant \theta, \\ \alpha(y-\theta) & \text{if } y > \theta. \end{cases}$$

   In this case, the literature refers to QOSA for quantile-oriented SA. We refer the reader to Maume-Deschamps et al. [316] and Browne et al. [58] for a complete study including the statistical estimation of the corresponding indices. Another line of research explores the use of random forest given-data estimation procedures [123].

Let us also remark that Kala [230] has introduced second-order and total QOSA indices in the same way as that for Sobol' indices. Finally, Kucherenko et al. [252] have proposed similar quantile-based global sensitivity measures based on averaging metrics between unconditional and conditional quantiles at any order.

### Example in R

Focusing on the QoI defined by a quantile, we use as an illustration the test case of Fort et al. [139] involving the difference of two exponential distributions and show the results in Fig. 6.5:

```
> set.seed(12345)
> n <- 10000
> X <- data.frame(matrix(rexp(2 * n,1), nrow = n))
> Y <- X[,1] - X[,2]
> sa <- qosa(model = NULL, X1 = X, type = "quantile", alpha = 0.7)
> tell(sa,Y)

> library(ggplot2) ; ggplot(sa)
```

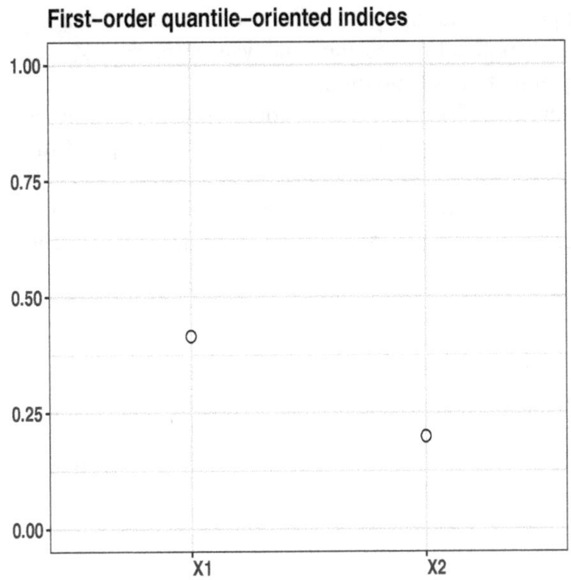


**Figure 6.5.** *Quantile-oriented sensitivity indices for the difference of two exponential distributions.*

### 6.3.4 ▪ Further references

Other GOSA techniques develop higher-order indices than the second-order ones of Sobol' by considering higher-order moments decomposition [352, 156, 505]. They aim at analyzing the tails of the output distribution by using the skewness and kurtosis of the output distribution. Indeed, a skewed distribution or a large-kurtosis distribution generates more extreme values than a symmetric distribution with the same variance.

## 6.4 ▪ Robustness analysis in uncertainty quantification

Up to this point, we only considered an SA framework where the inputs joint probability distribution $P_{\mathbf{X}}$ is fixed (whether the inputs are independent or not). However, in practice, it is common to encounter industrial applications where engineers are not able to assign a given probability distribution to some of the inputs. This happens, for example, for inputs corresponding to physical parameters for which no data are available. If experts in the field provide information in the form of an interval, we could choose (assuming the equiprobability hypothesis is relevant) a uniform distribution on this interval. Yet a customary but tricky question can arise: Would the results of SA change if we misspecified the bounds? Similarly, if they also yield a most probable value in addition to the interval, should we prefer a triangular distribution or a truncated normal distribution? Does our choice matter with respect to SA? Such concerns fall into the area of so-called *robustness sensitivity analysis*, where the objective is to assess the impact of a misspecification in the inputs probability distribution.

### 6.4.1 ▪ Perturbed law–based indices

The PLI (perturbed law–based indices) method is devoted to quantify the impact of a change in input distribution on the QoI. It was first developed by Lemaître [271] and Lemaître et al. [272] (under the acronym DMBRSI for "Density Modification Based Reliability Sensitivity Indices") to perform SA of inputs on a failure probability estimation. In Delage et al. [112], the method is clearly designed and applied for robustness analysis purposes instead of a standard SA setting (e.g., screening or quantitative partitioning).

The practical interest of PLI is that it does not require any new model evaluation to estimate the indices. Further, it has been adapted to quantile [466, 467] and superquantile (distribution mean beyond a quantile) [215] as the QoI (instead of a failure probability). However, it can be used for any QoI as the model output mean or the model output variance.

In the following, we use $Y = G(\mathbf{X})$ as usual, consider independent inputs $\mathbf{X}$ of pdf $f$, and denote $\phi = \phi(Y)$ the QoI. Given an input variable $X_j$ with pdf $f_j$, let us call $X_{j\delta} \sim f_{j\delta}$ the corresponding perturbed random input. This perturbed input takes the place of the real random input $X_j$ in the sense of modeling error: What if the real distribution were $f_{j\delta}$ instead of $f_j$?

More precisely, Lemaître et al. [272] define a perturbed input density $f_{j\delta}$ as the closest distribution to the original $f_j$ in the entropy sense and under some constraints of perturbation by using the Kullback–Leibler divergence (see Eq. (6.3)) to quantify the distance between the reference model input density and the perturbed density. The perturbed density is then found as the solution to the following constrained optimization problem:

$$\begin{aligned} &\min_{f_{\text{mod}}}\{\text{KL}(f_{\text{mod}}, f_j)\} \\ \text{s.t.} \quad &\int g_k(x_j) f_{\text{mod}}(x_j)dx_j = \delta_{k,j} \;\; (k = 1, \ldots, K)\,, \end{aligned} \tag{6.20}$$

where $\text{KL}(f_{mod}, f_j)$ is the Kullback–Leibler divergence between the original and the perturbed model input density, $g_k$ are assigned univariate functions, and $\delta_{k,j}$ are real numbers for $j = 1, \ldots, d$ and $k = 1, \ldots, K$.

A unique vector $\boldsymbol{\lambda}^* \in \mathbb{R}^K$ solves the minimization problem in Eq. (6.20), and the explicit solution is [272]

$$f_{j\delta}(x_j) = f_j(x_j)\exp\left[\sum_{k=1}^{K} \lambda_k^* g_k(x_j) - \psi_j(\boldsymbol{\lambda}^*)\right]$$

with

$$\psi_j(\boldsymbol{\lambda}) = \log \int f_j(x)\exp\left[\sum_{k=1}^{K} \lambda_k g_k(x)\right] dx.$$

Different types of perturbations can be considered [271, 272]. The most natural is a *mean shifting*, i.e., a perturbation in the mean:

$$\int x_j f_{\text{mod}}(x_j)dx_j = \delta_j.$$

Another kind of perturbation can be based on a *variance shifting*, i.e., when the analyst is perturbing the variance of $X_j$:

$$\begin{cases} \int x_j f_{\text{mod}}(x_j)dx_j = \mathbb{E}X_j, \\ \int x_j^2 f_{\text{mod}}(x_j)dx_j = \delta_j + \mathbb{E}X_j^2. \end{cases}$$

**Remark 39.** *As in Perrin and Defaux [363], an iso-probabilistic operator (e.g., the Rosenblatt transform; see Section* 5.4.1*) can be applied to transform all the input random variables into standard Gaussian ones. Indeed, the problem greatly simplifies for Gaussian pdf: For a mean perturbation, it gives* $f_{j\delta}(x_j) = f_j(x_j) + \delta$.

For a specific QoI, the perturbed QoI (denoted $\phi_{j\delta}$) is naturally defined. If we consider the failure probability, denoted by $p$, which is written as in Section 6.3.1 ($\mathcal{R}$ being a region of interest),

$$\phi = p = \mathbb{P}(G(\mathbf{X}) \in \mathcal{R}) = \int \mathbb{1}_{\{G(\mathbf{x})\in\mathcal{R}\}} f(\mathbf{x})d\mathbf{x},$$

we then have

$$\phi_{j\delta} = p_{j\delta} = \int \mathbb{1}_{\{G(\mathbf{x})\in\mathcal{R}\}} \frac{f_{j\delta}(x_j)}{f_j(x_j)} f(\mathbf{x})d\mathbf{x}.$$

If we consider the QoI $q^\alpha$, the quantile of order $\alpha$ of $Y$, which is written as

$$\phi = q^\alpha = \inf\{t \in \mathbb{R}, F_Y(t) \geq \alpha\},$$

we have

$$\phi_{j\delta} = q^\alpha_{j\delta} = \inf\{t \in \mathbb{R}, F_{Y,j\delta}(t) \geq \alpha\},$$

where $F_{Y,j\delta}$ is the cumulative distribution function corresponding to the input variable $X_j$ sampled from $f_{j\delta}$.

A first PLI formulation is then defined as follows:

$$\mathrm{PLI}_{j\delta} = \left(\frac{\phi_{j\delta}}{\phi} - 1\right) \mathbb{1}_{\{\phi_{j\delta}\geq\phi\}} + \left(1 - \frac{\phi}{\phi_{j\delta}}\right) \mathbb{1}_{\{\phi_{j\delta}<\phi\}}.$$

It is easy to see that $\mathrm{PLI}_{j\delta} = 0$ if $\phi_{j\delta} = \phi$, as expected if $X_j$ is a noninfluential variable or if $\delta$ expresses a negligible perturbation. The sign of $\mathrm{PLI}_{j\delta}$ indicates how the perturbation impacts the QoI qualitatively. The formulation of Eq. (6.4.1) provides a ratio of increase or decrease of the QoI with respect to the reference QoI (without perturbing the inputs). Another PLI formulation that can be interpreted all the time in percentage of variation is the following [153, 215]:

$$\mathrm{PLI}^p_{j\delta} = \frac{\phi_{j\delta}}{\phi} - 1. \tag{6.21}$$

The PLI can be estimated directly from the set of simulations used to compute the reference QoI by using the reverse importance sampling technique (see, e.g., [192]), thus limiting the number of model evaluations. For example, for a failure probability, a consistent estimator of $p_{j\delta}$ with a Monte Carlo sample $(\mathbf{x}^1, \ldots, \mathbf{x}^n)$ of size $n$ is given by

$$\hat{p}_{j\delta n} = \frac{1}{n} \sum_{i=1}^{n} \mathbb{1}_{\{G(\mathbf{x}^{(i)})\in\mathcal{R}\}} \frac{f_{j\delta}(x_j^{(i)})}{f_j(x_j^{(i)})}.$$

Asymptotic properties of the PLI probability (i.e., PLI for a failure probability) are derived in Lemaître et al. [272], including a CLT for the plug-in estimator of $S_{j\delta}$. In practice, the asymptotic variance in this CLT is used to compute confidence intervals of the PLI estimates (see the example and Fig. 6.6 below).

The PLI approach has been applied on the computation of a probability of failure [208, 363]. For the PLI quantile, formulas have been first developed in Sueur et al. [466, 467], while the asymptotic properties are given in Gauchy et al. [153]. The asymptotic variance of the PLI quantile remains difficult to compute, and the bootstrap method is preferred in practice to compute, confidence intervals.

**Remark 40.** *The perturbation method presented above is not fully satisfactory. Indeed, no straightforward equivalence can be drawn between perturbations of two different inputs or even two different parameters of the same density. Moreover, some distributions do not have defined moments. Dealing with inputs belonging in the standard Gaussian space (via a Rosenblatt transform; see Remark* 39*) allows one to make perturbations comparable. However, it remains difficult to translate this interpretation in the initial physical space, which is the one of interest for the practitioners. Gauchy et al. [153] have proposed a more rigorous way of perturbing densities via the use of the Fisher distance on manifolds of probability distributions.*

### Examples in R

We illustrate the principles of the PLI quantile (at the 95% order) on the Ishigami function (see Section C.4 with independent $X_j \sim \mathcal{U}(-\pi, \pi)$). Our QoI is then $q^{0.95}$ (empirical estimation at approximately 9.25). We choose to move the distribution mean of each input by a perturbation on the interval $[-2, 2]$. The commands are given below, and the PLI graph is shown in Fig. 6.6:

```
> library(sensitivity)
> set.seed(12345)
> distrIshigami <- list()
> for (i in 1:3){
+   distrIshigami[[i]] <- list("unif", c(-pi,pi))
+   distrIshigami[[i]]$r <- ("runif")
+ }
> #
> # Monte Carlo sampling
> n <- 10^4
> X <- matrix(runif(n * 3, -pi, pi), ncol = 3)
> Y <- ishigami.fun(X)
> #
> # PLI with perturbation of the mean
> v_delta = seq(-2,2,1/10)
> PLI = PLIquantile(0.95, X, Y, deltasvector = v_delta,
+   InputDistributions = distrIshigami, type = "MOY", samedelta = TRUE,
+         percentage = TRUE, nboot = 200, bootsample = FALSE)
> #
> # Plotting the PLI
> par(mar=c(4,5,1,1))
> plot(v_delta,PLI$PLI[,2],ylim=c(-0.3,0.3),xlab=expression(delta),
+      ylab=expression(hat(PLI[i*delta])),pch=19,cex=1.)
> points(v_delta,PLI$PLI[,1],col="darkgreen",pch=15,cex=1.)
> points(v_delta,PLI$PLI[,3],col="red",pch=17,cex=1.)
> lines(v_delta,PLI$PLICIinf[,2],col="black")
> lines(v_delta,PLI$PLICIsup[,2],col="black")
> lines(v_delta,PLI$PLICIinf[,1],col="darkgreen")
> lines(v_delta,PLI$PLICIsup[,1],col="darkgreen")
> lines(v_delta,PLI$PLICIinf[,3],col="red")
> lines(v_delta,PLI$PLICIsup[,3],col="red")
> abline(h=0,lty=2)
> legend(-1,0.3,legend=c("X1","X2","X3"),
+         col=c("darkgreen","black","red"),pch=c(15,19,17),cex=1.5)
```

## 6.4.2 ▪ Bayesian perturbation

Here, we adopt a somewhat higher-level approach in the sense that the considered perturbations on distributions are no longer parametric and can encompass any type of probabilistic uncertainty on the distributions. For illustration purposes, imagine that for some inputs we know from expert experience that the input distribution should be unimodal and within a given interval, with a consensus on the most probable value. However, among the experts panel, some suggest a triangular distribution, while others vouch for a truncated normal distribution. Depending on

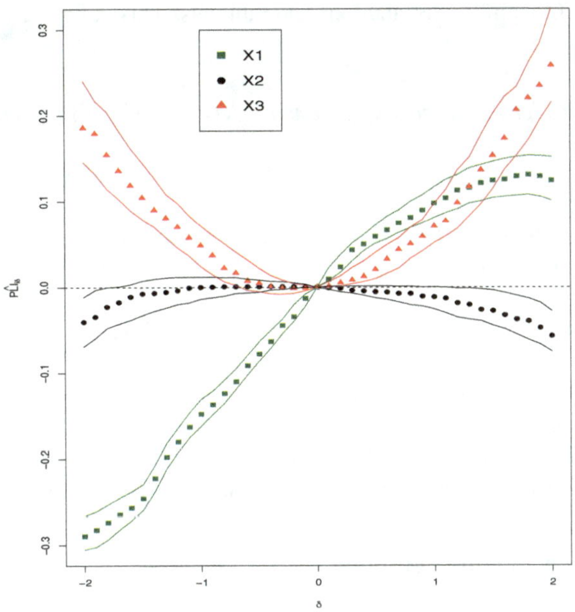


**Figure 6.6.** *Illustration of the PLI quantile applied on the Ishigami function. The lines surrounding the points represent the* 95% *confidence intervals.*

the choices made for each input, the result of SA may of course vary. The objective here is to identify which input choice of distribution has the most impact on the variability of SA results. Note that Meynaoui et al. call this approach a second-level sensitivity analysis [323].

### Theory

Formally, this problem can be written as a higher-level SA study where the inputs and the output definition differ. Until now, we considered the setting of a code $G(\cdot)$ with inputs $\mathbf{X}$ with distribution $\mathrm{P}_{\mathbf{X}}$. Instead, we investigate now an *all-in-one code* which takes as inputs the $d$ probability distributions $\mathrm{P}_{X_1}, \mathrm{P}_{X_2}, \ldots, \mathrm{P}_{X_d}$ and outputs the result of a sensitivity study carried out with this choice of input distributions. Note that we assume for simplicity that the inputs are independent. We also need to specify the uncertainty on each input probability distribution $\mathrm{P}_{X_j}$ as a higher-level probability distribution $\mathrm{P}_{\mathrm{P}_{X_j}}$ for all $j = 1, \ldots, d$ and define what type of object is a *result of a sensitivity study*. Here, we deliberately focus on rankings; i.e., the all-in-one code provides, as an output, a ranking of the inputs with respect to a given sensitivity index. For example, it may be the ranking of the inputs after the estimation of their Sobol' total sensitivity index. In this framework, we thus consider that the result of an SA changes with the input distribution if the ranking of the inputs varies. Finally, one denotes by $Y \in \Pi_d$ the output of such an all-in-one code. However, note that it is straightforward to envision other sensitivity results, such as the vector of all total indices, for example.

Coming back to our objective, we now have an all-in-one code consisting of inputs $\mathrm{P}_{X_1}, \mathrm{P}_{X_2}, \ldots, \mathrm{P}_{X_d} \in \mathcal{M}_1^+$ and output $Y \in \Pi_d$. In order to quantify the impact of the uncertainty attached to each input $\mathrm{P}_{X_j}$ on $Y$, a simple yet powerful approach hinges upon the use of the HSIC sensitivity index with adapted kernels, just as mentioned in Section 6.2.1. Indeed, the HSIC approach is generic in the sense that it can evaluate the dependence between inputs and outputs as long as a kernel adapted to their type is provided. Here, we thus take advantage of the previously introduced kernels for probability distributions and rankings. This defines a robustness sensitivity index as follows.

**Definition 6.11.** *Consider the kernel $k_{\text{MMD}}(u,u') : \mathcal{M}_1^+ \times \mathcal{M}_1^+ \to \mathbb{R}$ given by*

$$k_{\text{MMD}}(u,u') = \exp\left(-\lambda_1 \text{MMD}^2(u,u';\mathcal{H})\right)$$

*for $\lambda_1 > 0$ and where $\mathcal{H}$ is an RKHS with kernel $k_{\mathcal{X}}$ used to compute the kernel embeddings of $u$ and $u'$. Further define the kernel $k_{\text{rank}}(y,y') : \Pi_d \times \Pi_d \to \mathbb{R}$ as*

$$k_{\text{rank}}(y,y') = \exp\left(-\lambda_2 K(y,y')\right),$$

*where $K(\cdot,\cdot)$ is the Kendall tau rank distance and $\lambda_2 > 0$. Then, for all $j = 1,\dots,d$, $\text{HSIC}\left(\mathrm{P}_{X_j}, Y\right)$ with kernels $k_{\text{MMD}}$ and $k_{\text{rank}}$ is a robustness sensitivity measure quantifying the impact of the uncertainty attached to $\mathrm{P}_{X_j}$ on the result of an SA $Y$ given as a ranking. This property holds no matter what influence measure is used to obtain the input ranking.*

### Examples and practical considerations

Let us consider first a parametric setting where the input distributions are Gaussian but with an uncertainty on their mean. We thus have $\mathrm{P}_{X_j|\mu_j} = \mathcal{N}(\mu_j, \sigma_j^2)$. We further model the uncertainty on each distribution by specifying $\mathrm{P}_{\mu_j} = \mathcal{N}(\nu_j, \tau_j^2)$. The computation of $\text{HSIC}\left(\mathrm{P}_{X_j}, Y\right)$ can then be performed as follows:

1. Choose the RKHS $\mathcal{H}$ and the parameters $\lambda_1 > 0$, $\lambda_2 > 0$. Decide on the sensitivity measure used to rank the inputs, and choose also the number of different sensitivity studies $n_{\mathrm{P}}$ that will be carried out (with modified inputs distributions) and $n_S$ the sample size dedicated to each sensitivity study.

2. For all $i = 1,\dots,n_{\mathrm{P}}$,

   (a) sample a mean $\mu_j^{(i)}$ for all inputs according to $\mathrm{P}_{\mu_j}$ to determine the current input distributions $\mathrm{P}_{X_j}^{(i)} = \mathrm{P}_{X_j|\{\mu_j=\mu_j^{(i)}\}}$ allocated to the inputs;

   (b) perform a sensitivity study and rank the inputs according to the chosen measure and collect them in the vector $Y^{(i)}$.

3. To compute an estimate of $\text{HSIC}\left(\mathrm{P}_{X_j}, Y\right)$, use the formula in Eq. (6.11):

$$\forall i_1, i_2 = 1,\dots,n_{\mathrm{P}},\ K_{\text{MMD}}(i_1,i_2) = k_{\text{MMD}}(\mathrm{P}_{X_j}^{(i_1)}, \mathrm{P}_{X_j}^{(i_2)}),$$

$$\forall i_1, i_2 = 1,\dots,n_{\mathrm{P}},\ K_{\text{rank}}(i_1,i_2) = k_{\text{rank}}(Y^{(i_1)}, Y^{(i_2)}),$$

$$\widehat{\text{HSIC}}\left(\mathrm{P}_{X_j}, Y\right) = \frac{1}{n^2}\text{Tr}\left(K_{\text{MMD}} H K_{\text{rank}} H\right).$$

The impact of a misspecification on each inputs distribution is finally given by $\widehat{\text{HSIC}}\left(\mathrm{P}_{X_j}, Y\right)$. Note that we may be tempted to initially include the uncertainty on the input distribution by following a standard Bayesian prior/posterior point of view where we would set $\mathrm{P}_{X_j} = \mathcal{N}\left(\nu_j, \sigma_j^2 + \tau_j^2\right)$ and compute directly sensitivity indices. But this approach is nothing like what we proposed above with HSIC, since it does not yield a measure of the impact of the misspecification of the inputs distribution.

In a nonparametric framework, the estimation procedure is similar. If we come back to our example where the uncertainty on the distributions stems from an indetermination between a triangular and a truncated normal distribution, step 2(a) is modified as follows: For each new sensitivity study, we randomly choose a triangular or a truncated normal for each input and carry out the sensitivity study. If the number of inputs is small, we do not even need to use a random procedure and can investigate instead all the possible combinations of distributions. It is also straightforward to incorporate weights in the indetermination, which may serve as a proxy to the confidence assigned to each expert.

Finally, although this robustness measure based on HSIC is appealing from a theoretical perspective, the computational cost of the estimation is substantial. Indeed, for each new combination of inputs distribution, an SA is carried out: The number of calls to the code is then $n_P n_S$. Even if HSIC indices usually necessitate a moderate sample size $n_P$ around 50–100 for a relatively accurate estimation, this means that at least 50 sensitivity studies are required. Recently, Meynaoui et al. [323] proposed an approach related to importance sampling to greatly reduce this cost.

### Examples in R

We illustrate the principles of uncertainty on the distributions to be assigned on the inputs of the Ishigami function (see Section C.4). In the original test case, the inputs uncertainty is given by a uniform distribution on $[-\pi, \pi]$. Here, we consider that the experts propose three different distributions: the original uniform one, a triangular one, and a Gaussian one. They are given in Fig. 6.7:

```
> library(truncnorm)
> library(triangle)
> xplot <- seq(-pi, pi, length.out = 200)
> plot(xplot, dtruncnorm(xplot,a=-pi,b=pi,mean=0,sd=2*pi/10),type='l')
> lines(xplot, dtriangle(xplot, a = -pi, b = pi, c = 0))
> lines(xplot, dunif(xplot, min = -pi, max = pi))
```

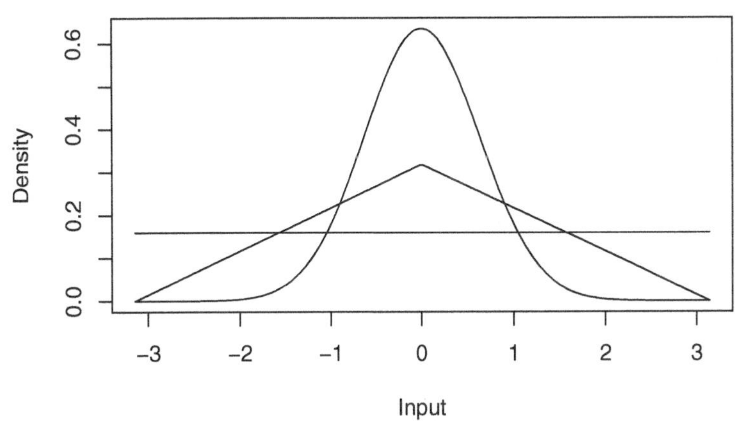


**Figure 6.7.** *The three inputs distributions proposed by the experts.*

We also define a generic sampling function:

```
> SamplingInputs <- function(id,n){
+   if (id == 1){ # Uniform
+     S <- runif(n, min = -pi, max = pi)
+   }
+   if (id == 2){ # Triangular
+     S <- rtriangle(n, a = -pi, b = pi, c = 0)
+   }
+   if (id == 3){ # Truncated Gaussian+
+     S <- rtruncnorm(n, a = -pi, b = pi, mean = 0, sd = 2*pi/10)
+   }
+   return(S)
+ }
```

Our objective is then to detect if SA results vary when one assigns different distributions to the inputs. The first step is to build all the possible assignment combinations (27 total, with three possibles choices for each of the three inputs):

```
> DOEdistrd <- expand.grid(P1 = 1:3, P2 = 1:3, P3 = 1:3)
> head(DOEdistrd)

  P1 P2 P3
1  1  1  1
2  2  1  1
3  3  1  1
4  1  2  1
5  2  2  1
6  3  2  1
```

For each assignment, we then perform an SA and store the results (total sensitivity indices and ranking of the inputs):

```
> set.seed(12345)
> library(sensitivity)
> d <- 3
> sensiDOE <- matrix(NA, 27, d)
> rankDOE <- matrix(NA, 27, d)
> nsensi <- 1000 # Sample size to compute sensitivity indices
> for (i in 1:27){
+   iddistr <- as.matrix(DOEdistrd[i,])
+   # Get a sample from the inputs distributions for this assignment
+   X1 <- matrix(NA, nrow = nsensi, ncol = d)
+   X2 <- matrix(NA, nrow = nsensi, ncol = d)
+   for (j in 1:d){
+     X1[,j] <- SamplingInputs(id = iddistr[j], n = nsensi)
+     X2[,j] <- SamplingInputs(id = iddistr[j], n = nsensi)
+   }
+   # Compute total sensitivity indices
+   Stemp <- sobolSalt(model = ishigami.fun,scheme="A",
+                      X1 = data.frame(X1), X2 = data.frame(X2))$T
+   # Store indices and the inputs ranking
+   sensiDOE[i,] <- as.matrix(Stemp)
```

```
+   rankDOE[i,] <- order(Stemp, decreasing = TRUE)
+ }
> head(sensiDOE)

          [,1]      [,2]      [,3]
[1,] 0.4964976 0.4418931 0.2245031
[2,] 0.5371562 0.4115294 0.2167618
[3,] 0.3955123 0.5704655 0.1622479
[4,] 0.5407160 0.4459649 0.2314891
[5,] 0.5699053 0.4626211 0.2514510
[6,] 0.4110439 0.6125830 0.1653234

> head(rankDOE)

     [,1] [,2] [,3]
[1,]    1    2    3
[2,]    1    2    3
[3,]    2    1    3
[4,]    1    2    3
[5,]    1    2    3
[6,]    2    1    3
```

Both the values of the total indices and the input rankings change according to the assignment, which shows that a misspecification of the distributions impacts the prioritization study. Our goal now is to identify which misspecification has the largest impact. We can start a visual exploration of the SA results by looking at how the values of the sensitivity indices change when we fix the distribution of each input. For example, in Figure 6.8, we can compare the summary of SA results when the first input distribution is fixed to one of the possible choices. It is clear that changing the distribution of the first input has a small impact on SA results:

```
> par(mfrow = c(1,3))
> figtitles <- c("Uniform", "Triangular", "Gaussian")
> for (i in 1:3){
+   # Get assignments where the first input
+   # is fixed to the i-th possible choice
+   idfixed <- (DOEdistrd$P1 == i)
+   # Summarize the SA results for this case
+   boxplot(sensiDOE[idfixed,], ylim = c(0,1), main = figtitles[i],
+           xlab = "Inputs", ylab = "Total index")
+ }
```

But by repeating this analysis for the third input, we can see in Figure 6.9 that now the impact is large:

```
> par(mfrow = c(1,3))
> for (i in 1:3){
+   # Get assignments where the third input
+   # is fixed to the i-th possible choice
+   idfixed <- (DOEdistrd$P3 == i)
+   # Summarize the SA results for this case
+   boxplot(sensiDOE[idfixed,], ylim = c(0,1), main = figtitles[i],
+           xlab = "Inputs", ylab = "Total index")
+ }
```

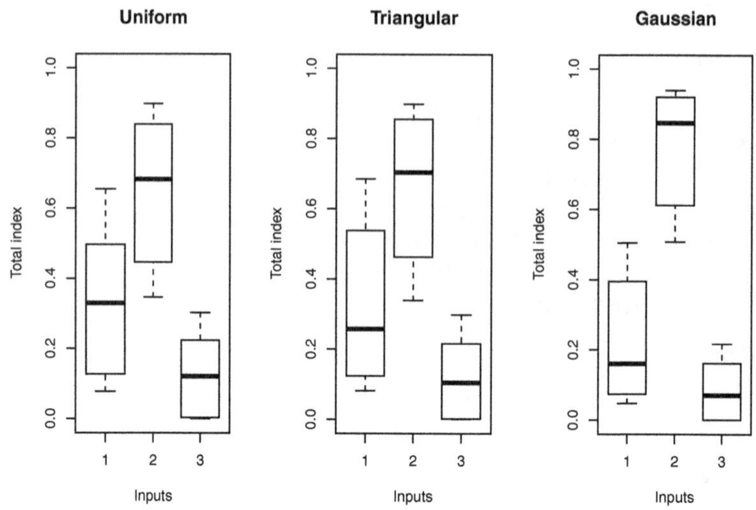


**Figure 6.8.** *Sensitivity analysis results when the first input distribution changes.*

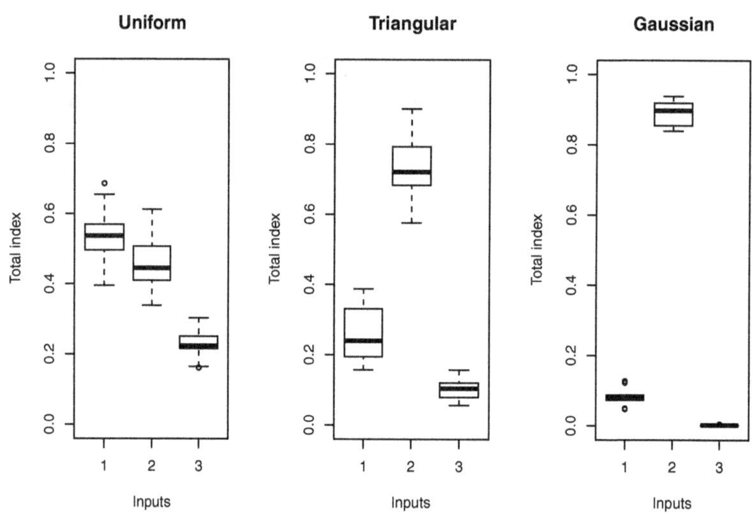


**Figure 6.9.** *Sensitivity analysis results when the third input distribution changes.*

In order to measure quantitatively such impacts, we compute as in (6.11) the HSIC measure between the input distribution and the SA ranking for each input. The first step is to compute the $K_{\text{rank}}$ matrix:

```
> library("TopKLists")
> Drank <- matrix(0, nrow = 27, ncol = 27)
> for (i in 1:26){
+   for (j in (i+1):27){
+     Drank[i,j] <- Kendall2Lists(rankDOE[i,], rankDOE[j,], d, d, d)
+   }
+ }
```

```
> s <- mean(c(Drank)) # Simple normalization
> Drank <- Drank + t(Drank)
> Krank <- exp(-0.5 * (Drank/s)^2)
```

Then, for each input the $K_{\text{MMD}}$ matrix must be assembled. Since for all possible assignments there are only three possible MMDs to compute (distance between two possible distributions out of three), we first compute and store these three possible distances. Let us define the Gaussian kernel used in the MMD and the empirical estimator of the MMD:

```
> library(flexclust)
> rbfkernel <- function(x, y = NULL, param){
+   if (is.null(y)){
+     d <- as.matrix(dist(x))
+   }else{
+     d <- as.matrix(dist2(x, y))
+   }
+   return(exp(-0.5 * (d/param)^2))
+ }
> MMD <- function(samplex, sampley){
+   param <- sqrt(sd(samplex) * sd(sampley))
+   dxx <- rbfkernel(samplex, param = param)
+   diag(dxx) <- 0
+   dyy <- rbfkernel(sampley, param = param)
+   diag(dyy) <- 0
+   dxy <- as.matrix(dist2(samplex, sampley))
+   dxy <- rbfkernel(samplex, sampley, param = param)
+   diag(dxy) <- 0
+   m <- length(samplex)
+   res <- (sum(dxx) + sum(dyy) - 2 * sum(dxy)) / m / (m-1)
+ }
```

We then compute the three possible MMDs:

```
> nsampleinit <- 1000 # Sample size to compute MMD
> # Generate only one sample for each distribution
> sampleinit <- matrix(NA, nsampleinit, 3)
> for (i in 1:3){
+   sampleinit[,i] <- SamplingInputs(id = i, n = nsampleinit)
+ }
> # Compute the MMD between each distribution
> MMDinit <- matrix(NA, nrow = 3, ncol = 3)
> for (i in 1:3){
+   for (j in 2:3){
+     MMDinit[i,j] <- MMD(sampleinit[,i], sampleinit[,j])
+   }
+ }
> tt <- lower.tri(MMDinit, diag = TRUE)
> MMDinit[tt] <- 0
> MMDinit <- MMDinit + t(MMDinit)
> print(MMDinit)

           [,1]       [,2]      [,3]
[1,] 0.00000000 0.03921671 0.2883083
```

```
[2,] 0.03921671 0.00000000 0.1198590
[3,] 0.28830831 0.11985895 0.0000000
```

The values computed above are the distance between a uniform and a triangular, a uniform and a truncated Gaussian, and a triangular and a Gaussian. Now we can assemble $K_{\text{MMD}}$ for each possible assignment:

```
> KMMD <- array(0, dim = c(27, 27, d))
> for (k in 1:d){
+   MMDtemp <- matrix(NA, nrow = 27, ncol = 27)
+   for (i in 1:26){
+     for (j in (i+1):27){
+       MMDtemp[i,j] <- MMDinit[DOEdistrd[i,k], DOEdistrd[j,k]]
+     }
+   }
+   s2 <- median(MMDtemp, na.rm = TRUE)
+   tt <- lower.tri(MMDtemp, diag = TRUE)
+   MMDtemp[tt] <- 0
+   MMDtemp <- MMDtemp + t(MMDtemp)
+   KMMD[,,k] <- exp(-0.5 * MMDtemp / s2)
+ }
```

The HSIC measure is finally obtained as in Eq. (6.11) with $K_{\text{MMD}}$ and $K_{\text{rank}}$ for each input:

```
> HSIC <- matrix(0, nrow = 1, ncol = d)
> H <- diag(27) - 1/27 * matrix(1, 27, 27)
> for (k in 1:d){
+   HSIC[k] <- sum(diag(KMMD[,,k] %*% H %*% Krank %*% H)) / 27^2
+ }
```

For interpretation purposes, we finally normalize the HSIC measures and rank the inputs misspecification:

```
> HSIC <- HSIC / sum(HSIC)
> print(HSIC)

          [,1]         [,2]      [,3]
[1,] 0.2855092 2.186797e-31 0.7144908
```

The HSIC measure indicates that a misspecification on the third input distribution has a far larger impact on the input ranking which would be obtained via total indices. This is consistent with what was deduced from the preliminary visual exploration. This result could have been anticipated if one understands the peculiarity of the Ishigami function. Indeed, in the original setting (i.e., all inputs are uniform), the third input has a nonnegligible total index via an interaction when it has very low or very large values. If we misspecify its distribution with, for example, a Gaussian one with light tails just as in our example above, this interaction vanishes, and the third input now is completely negligible.

### 6.4.3 ▪ Further references

In UQ, the branch of robustness analysis has emerged during recent years. Borgonovo et al. [47] have developed the FANOVA expansion when a multiplicity of distributions is assigned to the inputs. For global SA, Hart and Gremaud [180] use "optimal perturbations" of the probability density functions to analyze the robustness of the Sobol' indices. In a rare event estimation context, Chabridon et al. [70, 71] also propose approaches to dealing with the so-called second-level uncertainty, i.e., uncertainty on the parameters of the input distributions. Another approach, called optimal uncertainty quantification, avoids specifying the input probability distributions, turning the problem to the definition of constraints on moments [354, 458]. It allows one to evaluate bounds on the Sobol' indices over certain classes of probability measures, as in Stenger et al. [457]. Considering the robustness of a quantile (as the QoI) with dependent inputs, Benoumechiara et al. [32] have also proposed a greedy algorithm to provide the "worst" statistical dependency structure reaching a minimum (or maximum) quantile .

In the same spirit as the approach presented in Section 6.4.1, another type of robustness analysis has been proposed in quantitative finance by Cont et al. [88]. The authors investigate whether the estimated QoI is sensitive to a small perturbation of the empirical distribution function. For this purpose, they define the robustness of a QoI as its continuity with respect to the Prokhorov distance on the set of integrable random variables. Pesenti et al. [365] have proposed other ideas where the computation of the QoI's derivative comes from a certain kind of input density perturbation (called a stress).

## 6.5 ▪ Summary

Moment-independent sensitivity analysis has proven useful in order to grasp further types of input influence on a given model, thus providing an effective complementary tool to Sobol' indices. Since the seminal work of Borgonovo [42], moment-independent indices can now be considered as mature approaches in the traditional global SA toolkit. On the other hand, the kernel point of view for sensitivity analysis presented in this chapter is still under development but has already provided a major breakthrough for screening purposes via the introduction of HSIC indices. Despite their appealing properties and clear potential for handling a large variety of complex output types, such indices still need to be extensively examined and validated on more industrial applications.

The goal-oriented perspective is also common nowadays since many sensitivity studies are dedicated to the investigation of critical states in a system. In particular, target and quantile-oriented sensitivity indices are now popular and insightful tools for such settings.

Finally, assessing the impact of a misspecification in the input probability distribution with robustness sensitivity analysis is a rather new topic. But interestingly, this research area has been heavily driven by operational considerations in the sense that global SA studies may be challenged by engineers and regulatory authorities regarding the sometimes arbitrary choice of input distributions. We expect that robustness considerations in global SA will expand in the near future, similarly to the current challenge of *training data robustness* in the machine learning community.

## Chapter 7

# A case study in R: COVID-19 epidemic model

The aim of this last chapter is to provide a case study using the statistical SA methods developed in this book, associated to UQ issues. In the same spirit as in the book's other chapters, all the R code commands written for the statistical analysis are also provided. Users are then able to reproduce the entire analysis. The sensitivity R package is mainly used, associated to other useful packages.

The study of a compartmental epidemiological model emerged as a natural choice, as we finished the book during the COVID-19 outbreak. Let us mention that we have been inspired mainly by the early work of the mathematician Josselin Garnier on the uncertainty quantification and Bayesian calibration of the COVID-19 epidemiological model.[7] We must also acknowledge the impactful work of the Imperial College Team [133].

The COVID-19 epidemic outbreak involves a new human coronavirus which occurred in Wuhan, China. The first cases were detected in early December 2019. Within a few weeks, it took most of the world by surprise, starting with Asia, then Europe, and finally North and South America and Africa.

The use of mathematical models in epidemiology is not new, even if the literature on the topic has blown up since early January 2020. We refer the reader to Brauer [53] for a review of the development of mathematical models in epidemiology through the centuries. Mathematical models help in simulating and clarifying biological processes, quantifying sensitivities to changes in parameter values, and calibrating parameters from data in the absence and presence of preventive and therapeutic interventions (vaccination, lockdown, medical treatment, etc.). The basic compartmental models were introduced in a sequence of three papers by Kermack and McKendrick [236, 237, 238]. There exist both deterministic compartmental models (based on ODEs) and stochastic compartmental models, which are probably more natural but whose analysis is often more complex (see, e.g., [4, 54]). Stochastic models are based on the definition of the probability of transmission between two individuals, while deterministic models describe the mean behavior and rely on the law of large numbers. More recently, agent-based approaches were also developed to pursue the progression of a disease through each individual. For these models, all the rules for individual agent movements and contacts with and transmissions to other people are explicit (see, e.g., [124]).

In this chapter, we do not pretend to predict the evolution of COVID-19, which remains a scientific challenge, but rather to demonstrate how the numerous SA tools developed in the chapters of this book can help in characterizing and propagating the model parameter uncertainties. In the

[7] Presentation in French on April 2, 2020, `https://www.youtube.com/watch?v=Rwhco42urkA`.

recent comment published by Saltelli et al. [413], the authors recommend uncertainty and sensitivity analyses as good practices toward responsible modeling by considering uncertainties on mathematical models and boundary conditions. Among interesting papers considering sensitivity analysis for epidemiological models, let us cite the paper by Alexanderian et al. [3], in which the authors illustrate sensitivity analysis for time-dependent processes in the context of a cholera model proposed in Hartley et al. [183]. The very recent paper by Lu and Borgonovo [293] is also interesting in proposing sensitivity analysis to measure the sensitivity of functional model outputs, such as quarantined, recovered, or dead people, to parameters driving intervention policies, such as the intervention time in the context of COVID-19 in Italy. In our study, we focus on two SA settings (see Section 1.2): the *factor fixing* in a first step, then the *factor prioritization* in a second step. We emphasize that this chapter will illustrate, on one possible model, the techniques we have developed in the previous chapters. In particular, some modelization choices (calibration and priors, for instance) are partially arbitrary but we think meaningful and easily adaptable to other practitioner choices. But we strongly believe that the strategy discussed here should be used for the study of many other models in an advantageous manner.

This chapter is organized as follows. Section 7.1 describes the epidemiological model under study. Then Section 7.2 provides the preliminary UQ and SA (screening via HSIC and quantitative SA via Sobol' indices) of this model. Finally, Section 7.3 presents the necessary calibration of the model using official data, helped by the preliminary SA results. Once the model is calibrated, a final SA (via the use of Shapley effects) is provided in order to clearly identify the most influential model inputs.

## 7.1 ▪ A compartmental model for COVID-19

We start by defining the compartmental model which will be used throughout this chapter. However, we insist that the aim of this chapter is not to demonstrate the superiority of one model over others, or to provide short- to long-term predictions for the evolution of the pandemic, but rather to conduct a complete UQ (see Section 1.1; see also Smith [438]) involving SA on a complex model by using the tools presented in the previous chapters with an R implementation. In order to be representative of this epidemic, some specific characteristics of the COVID-19 disease have to be taken into account: the important role of asymptomatic individuals, testing strategies, hospitalized individuals with severe symptoms going to the intensive care unit (ICU), etc. Therefore, we build upon the models proposed in Charpentier et al. [74] and Di Domenico et al. [114] as well as on the parametric form for the transmission rate from Liu et al. [288, 289]. The model we present is a deterministic compartmental model, which can be considered as an approximation of the mean behavior of a stochastic compartmental model if the infectious periods are assumed to be i.i.d. with an exponential distribution and if the population is assumed to be homogeneous. These are strong assumptions which are not discussed in this chapter. We do not consider regional or age stratification (see, e.g., [135, 114, 407]) but instead a global model at the French national scale.

An overview of the model and its compartments, which are highly inspired by Charpentier et al. [74], is given in Fig. 7.1. The considered compartments are defined as follows:

- $S$: susceptible individuals who are not infected. During the epidemic spread, they can either stay uninfected or get infected and move to compartment $I^-$.

- $I^-$: infected individuals who are not detected. They will typically be not tested or isolated so that they can contaminate others. If they are tested with type-1 tests (virological tests like nasal ones) while still being infected, they move to compartment $I^+$. They can also get sicker, in which case they can go to the hospital ($H$) or directly to ICU ($U$). If they

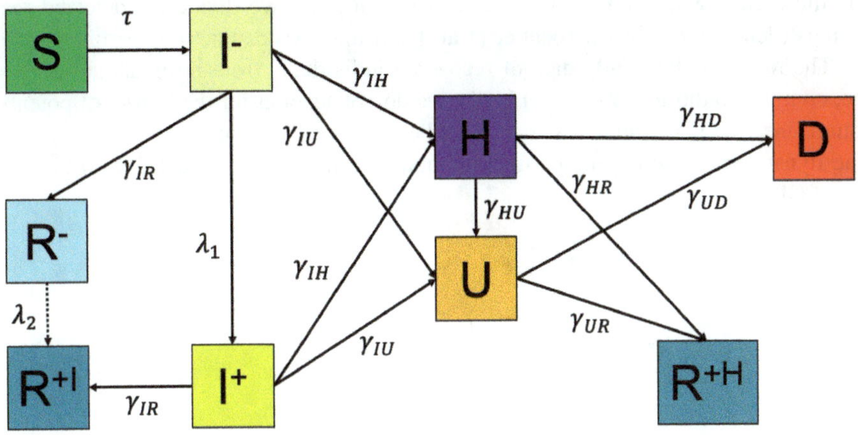


**Figure 7.1.** *Overview of the COVID-19 compartmental model.*

recover without having been tested, they get immune but are still undetected (compartment $R^-$). The daily number of individuals leaving $I^-$ is reported each day in France except for the undetected who move to $R^-$.

- $I^+$: infected individuals who have been detected with type-1 tests but did not need hospitalization. They are isolated so that they cannot contaminate others. They can either get sicker and go to the hospital ($H$ or $U$) or recover (compartment $R^{+I}$).
- $R^-$: recovered individuals who have not been detected. If tested with type-2 tests (serological tests), they move to $R^{+I}$. Infected people who died at a nursing home or in their own home but were never detected (and consequently never went to the hospital) also end up in this compartment.
- $R^{+I}$: recovered detected individuals who never needed hospitalization. Deaths from nursing homes or at home are also included in this compartment if they were detected and never went to the hospital.
- $H$: hospitalized individuals with severe symptoms who are all detected and isolated. They can get even sicker and go to the ICU ($U$), recover ($R^{+H}$), or die ($D$). The number of cases in $H$ is reported every day in France.
- $U$: hospitalized people in ICU who need special medical care. They can come directly from infected individuals ($I^-$ and $I^+$) or already hospitalized ones ($H$) whose state deteriorates. They can recover and move to $R^{+H}$ or die ($D$). The number of cases in $U$ is also reported every day in France.
- $R^{+H}$: recovered detected individuals from hospitals ($H$ and $U$). They are reported every day in France.
- $D$: people who have died from COVID-19 in hospitals. We do not consider here deaths from nursing homes or people's homes (which are accounted for in $R^-$ or $R^{+I}$).

This model differs from the one introduced in Charpentier et al. [74] in three points. First, infected individuals are allowed to go directly to ICU as proposed in Di Domenico et al. [114].

This is more consistent with daily reported data. Second, we divide the detected recovered individuals (denoted $R^+$ in Charpentier et al. [74]) into two distinct compartments, $R^{+I}$ and $R^{+H}$. The reason is that daily data of recovered individuals from hospitals are accessible in France, contrary to those from $R^{+I}$. Finally, we do not account for the impact of potential ICU saturation since the ICU limit capacity has not been reached in France.

The dynamics of such a compartmental model is then easily modeled with the following system of ODEs:

$$
\begin{aligned}
\frac{dS}{dt} &= -\tau S I^-, \\
\frac{dI^-}{dt} &= \tau S I^- - \lambda_1 I^- - (\gamma_{IR} + \gamma_{IH} + \gamma_{IU}) I^-, \\
\frac{dI^+}{dt} &= \lambda_1 I^- - (\gamma_{IR} + \gamma_{IH} + \gamma_{IU}) I^+, \\
\frac{dR^-}{dt} &= \gamma_{IR} I^- - \lambda_2 R^-, \\
\frac{dR^{+I}}{dt} &= \gamma_{IR} I^+ + \lambda_2 R^-, \\
\frac{dH}{dt} &= \gamma_{IH}(I^- + I^+) - (\gamma_{HR} + \gamma_{HD} + \gamma_{HU}) H, \\
\frac{dU}{dt} &= \gamma_{IU}(I^- + I^+) + \gamma_{HU} H - (\gamma_{UR} + \gamma_{UD}) U, \\
\frac{dR^{+H}}{dt} &= \gamma_{HR} H + \gamma_{UR} U, \\
\frac{dD}{dt} &= \gamma_{UD} U + \gamma_{HD} H, \\
\frac{dDR}{dt} &= (\lambda_1 + \gamma_{IH} + \gamma_{IU}) I^- - DR.
\end{aligned}
$$

We also consider an additional compartment $DR$ consisting of the daily reported infected individuals. This compartment is not involved in the dynamics and is virtual in the sense that it is only added to the model for calibration purposes.

The different rate parameters involved in this ODE system have a straightforward interpretation, as some of them are related to more interpretable quantities:

- $\tau$: the transmission rate. We use a parametric form proposed by Liu et al. [288] and validated on early data in China given by

  $$\tau(t) = \tau_0 \exp(-\mu \max(t - N, 0)).$$

  The assumption is that, before the epidemic outbreak, the transmission rate is constant ($\tau_0$) and decreases exponentially at a rate $\mu$ when social distancing and lockdowns start to have an effect after $N$ days. Of course, other parameterizations of $\tau(t)$ may be more consistent with data from other countries, but for our illustration purpose focusing on UQ, such a choice is representative enough.

- $\gamma_{IR}$: rate at which infected individuals from $I^-$ or $I^+$ recover. They have been either asymptomatic or with mild/severe symptoms but without needing hospitalization or intensive care. Let $p_a$ be the probability of being asymptomatic (conditioned on being infected) and $N_a$ the number of days until recovery if asymptomatic. Then the recovery rate for asymptomatic individuals can be written as $p_a/N_a$. Furthermore, if $p_H$ denotes the probability of needing hospitalization ($H$ or $U$) conditioned on being mild/severely ill, then the recovery rate for mild/severe can be written as $(1-p_a)(1-p_H)/N_s$. Here, without hospitalization or ICU $(1-p_H)$, mild/severe individuals $(1-p_a)$ recover in $N_s$ days. Finally, the recovery rate is thus

$$\gamma_{IR} = \frac{p_a}{N_a} + \frac{(1-p_a)(1-p_H)}{N_s}.$$

- $\gamma_{IH}$: rate at which infected individuals move to the hospital. They have mild/severe symptoms $(1-p_a)$ but do not need ICU $(1-p_U)$, where $p_U$ is the probability to go directly to ICU knowing that the individual needs hospitalization), so that this rate is given by

$$\gamma_{IH} = \frac{(1-p_a)p_H(1-p_U)}{N_{IH}},$$

  where $N_{IH}$ is the number of days until hospitalization if severely symptomatic.

- $\gamma_{IU}$: rate at which infected individuals move to ICU given by

$$\gamma_{IU} = \frac{(1-p_a)p_H p_U}{N_{IH}}.$$

  Here we assume the number of days until hospitalization is the same for $H$ and $U$, hence a number of days until hospitalization for people in ICU equal to $N_{IH}$ in the denominator.

- $\gamma_{HD}$: rate at which hospitalized individuals in $H$ die. If $p_{HD}$ is the probability of dying conditioned on being in $H$, we write this as

$$\gamma_{HD} = \frac{p_{HD}}{N_{HD}},$$

  where $N_{HD}$ is the number of days until death for hospitalized in $H$.

- $\gamma_{HR}$: rate at which hospitalized individuals in $H$ recover given by

$$\gamma_{HR} = \frac{1-p_{HD}}{N_{HR}},$$

  where $N_{HR}$ is the number of days until recovery for hospitalized in $H$.

- $\gamma_{UD}$: rate at which hospitalized individuals in ICU die. If $p_{UD}$ is the probability of dying conditioned on being in ICU, it is written as

$$\gamma_{UD} = \frac{p_{UD}}{N_{UD}},$$

  where $N_{UD}$ is the number of days until death for hospitalized in ICU.

- $\gamma_{UR}$: rate at which hospitalized individuals in ICU recover given by

$$\gamma_{UR} = \frac{1-p_{UD}}{N_{UR}},$$

  where $N_{UR}$ is the number of days until recovery for hospitalized in ICU.

- $\gamma_{HU}$: rate at which hospitalized individuals who did not go directly to ICU finally need it. If $p_{HU}$ is the probability of needing ICU conditioned on having already been hospitalized in $H$, we write this as

$$\gamma_{HU} = \frac{p_{HU}}{N_{HU}},$$

  where $N_{HU}$ is the number of days until ICU if already in $H$.

- $\lambda_1$: rate at which infected individuals are detected with type-1 tests.
- $\lambda_2$: rate at which recovered individuals are detected with type-2 tests.

The final model is written in R as follows, using the package deSolve for solving the ODE system:

```
> library(deSolve)
> siduhr.modified.model <- function(t,x,parms){
+   with(as.list(c(parms, x)), {
+
+     # ************
+     # Compartments
+     # ************
+     S <- x[1]
+     Iminus <- x[2]
+     Iplus <- x[3]
+     Rminus <- x[4]
+     RplusI <- x[5]
+     RplusHU <- x[6]
+     H <- x[7]
+     U <- x[8]
+     D <- x[9]
+     DailyRep <- x[10]
+
+     # ****************************************
+     # Rates as function of time and parameters
+     # ****************************************
+     gammaIR <- (1-pa)*(1-pH)/Ns + pa/Na
+     gammaIH <- (1-pa)*pH*(1-pU)/NIH
+     gammaIU <- (1-pa)*pH*pU/NIH
+     gammaHR <-  (1-pHD)/NHR
+     gammaHD <- pHD/NHD
+     gammaHU <- pHU/NHU
+     gammaUR <- (1-pUD)/NUR
+     gammaUD <- pUD/NUD
+
+     tau <- tau0*exp(-mu*max(t-N,0))
+
+     # ****************************************
+     # ODE
+     # ****************************************
+     dS <- -tau*S*Iminus
+     dIminus <- tau*S*Iminus - lambda1*Iminus -
+       (gammaIR+gammaIH+gammaIU)*Iminus
+     dIplus <- lambda1*Iminus - (gammaIR+gammaIH+gammaIU)*Iplus
+     dRminus <- gammaIR*Iminus - lambda2*Rminus
+     dRplusHU <- gammaHR*H + gammaUR*U
+     dRplusI <- gammaIR*Iplus + lambda2*Rminus
+     dH <- gammaIH*(Iminus+Iplus) - (gammaHR+gammaHD+gammaHU)*H
+     dU <- gammaIU*(Iminus+Iplus) + gammaHU*H - (gammaUR+gammaUD)*U
```

```
+       dD <- gammaUD*U + gammaHD*H
+       dDailyRep <- (lambda1+gammaIH+gammaIU)*Iminus - DailyRep
+       list(c(dS, dIminus, dIplus, dRminus, dRplusI, dRplusHU,
+              dH, dU, dD, dDailyRep))
+   })
+ }
```

In addition to the above model parameters, the initial conditions for the ODE system must also be given: First the initial date $t_0$ at which the epidemic starts, then the initial number $I_0^-$ of undetected infected. The number of susceptible $S_0$ at the start of the epidemic is equal to the whole population, and all other compartments are empty ($I_0^+ = R_0^- = R_0^{+I} = R_0^{+H} = H_0 = U_0 = D_0 = DR_0 = 0$). The final date for our simulation is May 11, corresponding to the end of the strict lockdown in France. Finally, instead of considering $\tau_0$ as a parameter, we work with the reproducing number $R_0$ since the two are related through the equation

$$\tau_0 = \frac{R_0(\lambda_1 + \gamma_{IR} + \gamma_{IH} + \gamma_{IU})}{S_0};$$

see [74].

For convenience, the ODE is solved inside a function simu.model which takes as arguments all the parameters. The initial date is parameterized with a continuous variable between 0 and 1 (which is then transformed into a date), and the type-2 testing strategy is not available in France ($\lambda_2 = 0$):

```
> # Simulation times (common for all studies in France,
> #   from January to end of lockdown)
> times.date <- seq(as.Date("15/01/20",format="%d/%m/%y"),
+                   as.Date("11/05/20",format="%d/%m/%y"),by=1)
> ntimes.date <- length(times.date)
> # Possible starting date
> times.date.sampling <- seq(as.Date("25/01/20",format="%d/%m/%y"),
+                            as.Date("24/02/20",format="%d/%m/%y"),by=1)
> ntimes.date.sampling <- length(times.date.sampling)
> simu.model <- function(x){
+   # For a given value of the parameters x, it solves the ODE system
+
+   Iminus <- Iplus <- Rminus <- RplusI <- RplusHU <-
+     H <- U <- D <- DR <- matrix(0,1,ntimes.date)
+
+   S0 <- 66.99e6
+   pa <- x[1]
+   pH <- x[2]
+   pU <- x[3]
+   pHD <- x[4]
+   pUD <- x[5]
+   Na <- x[6]
+   Ns <- x[7]
+   NIH <- x[8]
+   NHD <- x[9]
+   NUD <- x[10]
+   NHR <- x[11]
```

```
+   NUR <- x[12]
+   basic.reprod.nb <- x[13]
+   t0 <- times.date.sampling[round(1+(ntimes.date.sampling-1)*x[14])]
+   mu <- x[15]
+   N <- x[16]
+
+   lambda1 <- x[17]
+   pHU <- x[18]
+   NHU <- x[19]
+   Iminus0 <- x[20]
+   lambda2 <- 0 # No type-2 test strategy
+   gammaIR <- (1-pa)*(1-pH)/Ns + pa/Na
+   gammaIH <- (1-pa)*pH*(1-pU)/NIH
+   gammaIU <- (1-pa)*pH*pU/NIH
+   tau0 <- basic.reprod.nb * (lambda1+gammaIR+gammaIH+gammaIU)/S0
+
+   # Simulation of the model
+   times.simu.date <- seq(t0,as.Date("11/05/20",format="%d/%m/%y"),by=1)
+   times.simu <- seq(from = 0, to = length(times.simu.date)-1, by = 1)
+   parms <- c(pa = pa, pH = pH, pU = pU, pHD = pHD, pUD = pUD,
+              pHU = pHU, Na = Na, Ns = Ns, NIH = NIH, NHD = NHD,
+              NUD = NUD, NHR = NHR, NUR = NUR, NHU = NHU,
+              tau0 = tau0, mu = mu, N = N,
+              lambda1 = lambda1, lambda2 = lambda2)
+   xini  <- c(S = S0, Iminus = Iminus0, Iplus = 0, Rminus = 0,
+              RplusI = 0, RplusHU = 0, H = 0, U = 0, D = 0, DailyRep = 0)
+   out <- ode(y = xini, times = times.simu,
+              func = siduhr.modified.model, parms)
+   id <- seq(which(t0==times.date),ntimes.date,by=1)
+
+   Iminus[id] <- out[,"Iminus"]
+   Iplus[id] <- out[,"Iplus"]
+   Rminus[id] <- out[,"Rminus"]
+   RplusI[id] <- out[,"RplusI"]
+   RplusHU[id] <- out[,"RplusHU"]
+   H[id] <- out[,"H"]
+   U[id] <- out[,"U"]
+   D[id] <- out[,"D"]
+   DR[id] <- out[,"DailyRep"]
+
+   id.maxU <- which.max(out[,"U"])
+   time.maxU <- times.simu.date[id.maxU]
+   value.maxU <- out[id.maxU,"U"]
+   cumul.DR <- sum(out[,"DailyRep"])
+
+   return(list(Iminus=Iminus, Iplus=Iplus, Rminus=Rminus,
+               RplusI=RplusI, RplusHU=RplusHU, H=H, U=U, D=D, DR=DR,
+               time.maxU=time.maxU, value.maxU=value.maxU,
+               cumul.DR=cumul.DR))
+ }
```

Note that some of the compartments are posttreated to extract additional relevant information for the study, such as the maximum number of individuals in ICU, the date at which this maximum value is reached (the "peak"), and the total number of reported cases.

## 7.2 ▪ Prior uncertainty quantification and sensitivity analysis

### 7.2.1 ▪ Uncertainty quantification

The compartmental model introduced in the previous section exhibits a large number of input parameters (20 if we neglect type-2 tests) which are not known exactly (or estimated using statistical tools) and are thus uncertain. Some of the values chosen in several studies are discussed in Charpentier et al. [74]. Here, in order to deal with these uncertainties, we consider that each parameter is uniformly distributed with bounds consistent with typical reported values. Using uniform distributions for the model input variables, despite the assumptions made, is a standard practice for a first exploratory study using a numerical design of experiments and a preliminary global sensitivity analysis [414, 127, 110]. These input distributions will be refined a second time.

We turn then to probabilistic SA, which allows us to illustrate the potential of quantitative partitioning techniques (see Section 1.4.3) for investigating numerical models. The distributions assigned to each parameter are summarized in Table 7.1. Since we have not yet calibrated the parameter values on available data, we denote such probability distributions as "prior" and further assume that all parameters are independent.

The parameter names, lower and upper bounds of their associated uniform distributions, and their nominal values (simply equal to the mean value here) are defined in the following code:

```
> parms.name <- c("pa","pH","pU","pHD","pUD","Na","Ns",
+                 "NIH","NHD","NUD","NHR","NUR","basic.reprod.nb",
+                 "t0","Mu","N","lambda1","pHU","NHU","Iminus0")
> nparms <- length(parms.name)
> lb.all <- c(0.5,0.15,0.15,0.15,0.2,8,8,8,15,8,15,15,
+             3,0.4,0.03,20,1e-4,0.001,1,1) # Parameter lower bounds
> ub.all <- c(0.9,0.2,0.2,0.25,0.3,12,12,12,20,12,25,20,3.5,
+             0.6,0.08,50,1e-3,0.01,10,100) # Parameter upper bounds
> nom.all <- (lb.all+ub.all)/2 # Nominal value for each parameter
```

Before diving into an SA, we first have to define R functions to generate a sample of the parameter distributions. First of all, since some parameters may be fixed to their nominal value in a second SA process, we use an argument id.active to control the set of parameters which are randomly generated. The generate.sample.UQ() function takes as input a sample of random variables which are uniformly distributed between 0 and 1 and rescales them according to the parameter bounds. To generate this initial uniform sample between 0 and 1, we can either use a crude Monte Carlo approach (generate.basic.sample.MC() function) or a quasi-Monte Carlo one (generate.basic.sample.QMC() function); see Appendix A. The command lines are the following:

```
> library(numbers)
> library(randtoolbox)
> generate.basic.sample.MC <- function(nsample,id.active){
+   # Distributions: U[0,1]
+   p <- length(id.active)
```

**Table 7.1.** *Prior probability distributions of the compartmental model parameters.*

| Parameter | Description | Distribution |
|---|---|---|
| $p_a$ | Conditioned on being infected, the probability of having light symptoms or no symptoms | $\mathcal{U}(0.5, 0.9)$ |
| $p_H$ | Conditioned on being mildly/severely ill, the probability of needing hospitalization ($H$ or $U$) | $\mathcal{U}(0.15, 0.2)$ |
| $p_U$ | Conditioned on going to hospital, the probability of needing ICU | $\mathcal{U}(0.15, 0.2)$ |
| $p_{HD}$ | Conditioned on being hospitalized but not in ICU, the probability of dying | $\mathcal{U}(0.15, 0.25)$ |
| $p_{UD}$ | Conditioned on being admitted to ICU, the probability of dying | $\mathcal{U}(0.2, 0.3)$ |
| $N_a$ | If asymptomatic, number of days until recovery | $\mathcal{U}(8, 12)$ |
| $N_s$ | If symptomatic, number of days until recovery without hospitalization | $\mathcal{U}(8, 12)$ |
| $N_{IH}$ | If severe symptomatic, number of days until hospitalization | $\mathcal{U}(8, 12)$ |
| $N_{HD}$ | If in $H$, number of days until death | $\mathcal{U}(15, 20)$ |
| $N_{UD}$ | If in ICU, number of days until death | $\mathcal{U}(8, 12)$ |
| $N_{HR}$ | If hospitalized but not in ICU, the number of days until recovery | $\mathcal{U}(15, 25)$ |
| $N_{UR}$ | If in ICU, number of days until recovery | $\mathcal{U}(15, 20)$ |
| $R_0$ | Basic reproducing number (basic.reprod.nb) | $\mathcal{U}(3, 3.5)$ |
| $t_0$ | Starting date of epidemics (in 2020) | $\mathcal{U}(01/25, 02/24)$ |
| $\mu$ | Decaying rate for transmission (after social distancing and lockdown) | $\mathcal{U}(0.03, 0.08)$ |
| $N$ | Date of effect of social distancing and lockdown | $\mathcal{U}(20, 50)$ |
| $\lambda_1$ | Type-1 testing rate | $\mathcal{U}(1e{-}4, 1e{-}3)$ |
| $p_{HU}$ | Conditioned on being hospitalized in $H$, the probability of needing ICU | $\mathcal{U}(0.15, 0.2)$ |
| $N_{HU}$ | If in $H$, number of days until ICU | $\mathcal{U}(1, 10)$ |
| $I_0^-$ | Number of infected undetected at the start of epidemics | $\mathcal{U}(1, 100)$ |

```
+    sample <- matrix(runif(nsample*p),nsample,p)
+    return(sample)
+ }
> generate.basic.sample.QMC <- function(nsample,id.active){
+    # Distributions: U[0,1]
+    sample <- randtoolbox::sobol(nsample,length(id.active))
+    return(sample)
+ }
> generate.sample.UQ <- function(X,lb,ub,nom,id.active){
+    n <- nrow(X)
+    p <- length(id.active)
+    sample <- matrix(NA,n,p)
+    # Active variables
```

```
+    for (i in 1:p){
+      sample[,i] <- lb[id.active[i]] +
+        (ub[id.active[i]]-lb[id.active[i]])*X[,i]
+    }
+    return(sample)
+ }
```

With these functions, we can now perform a prior UQ (via a classical uncertainty propagation) on the model outputs of interest (all the compartment dynamics as well as the peak and the total number of reported cases):

```
> set.seed(12345)
> nsample <- 5000
> UQ.sample <-
+   generate.sample.UQ(generate.basic.sample.QMC(nsample,1:nparms),
+                      lb.all,ub.all,nom.all,1:nparms)
> Iminus.sample <- Iplus.sample <- Rminus.sample <- RplusI.sample <-
+   RplusHU.sample <- H.sample <- U.sample <- D.sample <- DR.sample <-
+   matrix(0,nsample,ntimes.date)
> time.maxU.sample <- value.maxU.sample <- cumul.DR.sample <-
+   matrix(0,nsample,1)
> for (i in 1:nsample){
+   if (!as.logical(mod(i,100))) print(paste0("Sample ",i,"/",nsample))
+   res.simu <- simu.model(UQ.sample[i,])
+   Iminus.sample[i,] <- res.simu$Iminus
+   Iplus.sample[i,] <- res.simu$Iplus
+   Rminus.sample[i,] <- res.simu$Rminus
+   RplusI.sample[i,] <- res.simu$RplusI
+   RplusHU.sample[i,] <- res.simu$RplusHU
+   H.sample[i,] <- res.simu$H
+   U.sample[i,] <- res.simu$U
+   D.sample[i,] <- res.simu$D
+   DR.sample[i,] <- res.simu$DR
+   time.maxU.sample[i,] <- res.simu$time.maxU
+   value.maxU.sample[i,] <- res.simu$value.maxU
+   cumul.DR.sample[i,] <- res.simu$cumul.DR
+ }
```

Note that we have used a quasi-Monte Carlo sample of 5000 simulations. For readability, we first compute the pointwise quartiles of the output curves:

```
> p.low <- 0.25
> p.up <- 0.75
> Iminus.med <- apply(Iminus.sample,2,median,na.rm=T)
> Iplus.med <- apply(Iplus.sample,2,median,na.rm=T)
> Rminus.med <- apply(Rminus.sample,2,median,na.rm=T)
> RplusI.med <- apply(RplusI.sample,2,median,na.rm=T)
> RplusHU.med <- apply(RplusHU.sample,2,median,na.rm=T)
> H.med <- apply(H.sample,2,median,na.rm=T)
> U.med <- apply(U.sample,2,median,na.rm=T)
> D.med <- apply(D.sample,2,median,na.rm=T)
```

```
> DR.med <- apply(DR.sample,2,median,na.rm=T)
> Iminus.qlow <- apply(Iminus.sample,2,quantile,probs=p.low,na.rm=T)
> Iplus.qlow <- apply(Iplus.sample,2,quantile,probs=p.low,na.rm=T)
> Rminus.qlow <- apply(Rminus.sample,2,quantile,probs=p.low,na.rm=T)
> RplusI.qlow <- apply(RplusI.sample,2,quantile,probs=p.low,na.rm=T)
> RplusHU.qlow <- apply(RplusHU.sample,2,quantile,probs=p.low,na.rm=T)
> H.qlow <- apply(H.sample,2,quantile,probs=p.low,na.rm=T)
> U.qlow <- apply(U.sample,2,quantile,probs=p.low,na.rm=T)
> D.qlow <- apply(D.sample,2,quantile,probs=p.low,na.rm=T)
> DR.qlow <- apply(DR.sample,2,quantile,probs=p.low,na.rm=T)
> Iminus.qup <- apply(Iminus.sample,2,quantile,probs=p.up,na.rm=T)
> Iplus.qup <- apply(Iplus.sample,2,quantile,probs=p.up,na.rm=T)
> Rminus.qup <- apply(Rminus.sample,2,quantile,probs=p.up,na.rm=T)
> RplusI.qup <- apply(RplusI.sample,2,quantile,probs=p.up,na.rm=T)
> RplusHU.qup <- apply(RplusHU.sample,2,quantile,probs=p.up,na.rm=T)
> H.qup <- apply(H.sample,2,quantile,probs=p.up,na.rm=T)
> U.qup <- apply(U.sample,2,quantile,probs=p.up,na.rm=T)
> D.qup <- apply(D.sample,2,quantile,probs=p.up,na.rm=T)
> DR.qup <- apply(DR.sample,2,quantile,probs=p.up,na.rm=T)
```

For example, we can visualize in Fig. 7.2 the uncertainty on the outputs related to hospitals, which has been obtained by the following commands:

```
> library(ggplot2)
> dfUQ <- data.frame(Min=c(H.qlow,U.qlow,D.qlow,RplusHU.qlow),
+                    Med=c(H.med,U.med,D.med,RplusHU.med),
+                    Max=c(H.qup,U.qup,D.qup,RplusHU.qup),
+                    Compartment=rep(c("H","U","D","R+H"),each=ntimes.date),
+                    Date=rep(times.date,4))
> ggplot(dfUQ,aes(x=Date,y=Med,colour=Compartment)) + geom_line(size=1) +
+   geom_ribbon(aes(ymin=Min,ymax=Max,fill=Compartment),
+               alpha=0.2,size=0) +
+   scale_color_manual(values =
+                        c("H"=rgb(147,112,219,maxColorValue = 255),
+                          "D"=rgb(220,20,60,maxColorValue = 255),
+                          "U"=rgb(205,133,63,maxColorValue = 255),
+                          "R+H"=rgb(70,130,180,maxColorValue = 255))) +
+   scale_fill_manual(values =
+                       c("H"=rgb(147,112,219,maxColorValue = 255),
+                       "D"=rgb(220,20,60,maxColorValue = 255),
+                       "U"=rgb(205,133,63,maxColorValue = 255),
+                       "R+H"=rgb(70,130,180,maxColorValue = 255))) +
+   ylab("Number of cases")  + theme_bw() + theme(legend.position="top")
```

Figure 7.2 shows that the prior uncertainty is pretty large since, for example, the highest 75% quantile for the number of people in ICU is around 40,000 and for the number of hospitalized individuals around 200,000.

We can do the same for a more precise study of the uncertainty on other quantities of interest: the maximum number of people in ICU, the date at which this value is attained, and the total number of reported cases; see Fig. 7.3:

```
> library(gridExtra) # for grid.arrange
```

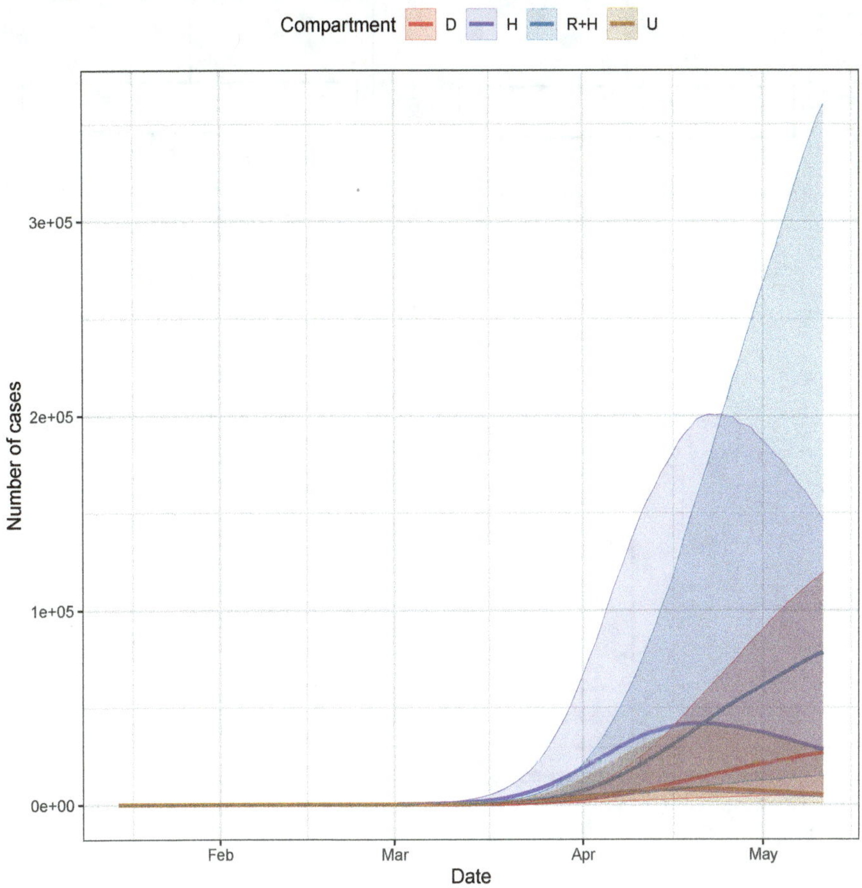


**Figure 7.2.** *Prior uncertainty quantification for compartments H, U, D, and* $R^{+H}$.

```
> gmaxU <- ggplot(as.data.frame(log(value.maxU.sample))) +
+   geom_boxplot(aes(y=V1)) + ylab("Maximum U (log)") + theme_bw()
> gDR <- ggplot(as.data.frame(log(cumul.DR.sample))) +
+   geom_boxplot(aes(y=V1)) + ylab("Cumul Daily Cases (log)") + theme_bw()
> gtimeU <- ggplot(as.data.frame(time.maxU.sample)) +
+   geom_boxplot(aes(x=V1)) + xlab("Date Max. U") + theme_bw() +
+   scale_x_continuous(
+     labels = as.character(times.date[seq(1,ntimes.date,2)]),
+     breaks = as.numeric(times.date)[seq(1,ntimes.date,2)]) +
+   theme(axis.text.x = element_text(angle = 90, hjust = 1))
> grid.arrange(gmaxU, gDR, gtimeU, nrow=2, ncol=2)
```

We see, for example, that the a priori median estimate of the "peak" date is around April 16.

### 7.2.2 ▪ Screening

The number of parameters in the compartmental model is quite large (20), which motivates a first screening step (*factor fixing* setting) before moving to a more quantitative SA analysis. We resort

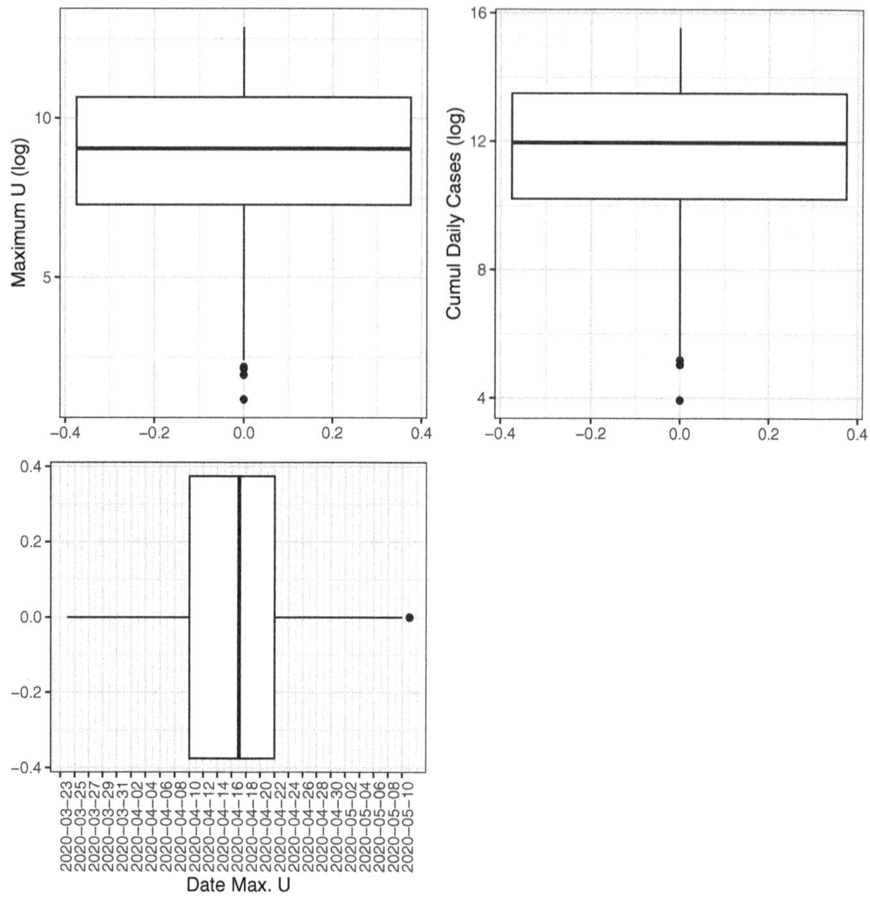


**Figure 7.3.** *Prior uncertainty quantification for maximum value of U, total reported cases, and time where maximum value of U is reached.*

to an SA sampling-based approach which requires the availability of an input-output sample. We focus here on the compartments for which we can acquire data for later calibration, namely, $DR$, $H$, $U$, $D$, and $R^{+H}$:

```
> library(sensitivity)
> GSAfunc <- function(UQ.sample, nom, id.active){
+   n <- nrow(UQ.sample)
+   resout <- matrix(0,n,5*ntimes.date+3)
+   for (i in 1:n){
+     parms.simu <- matrix(nom,nrow=1)
+     parms.simu[id.active] <- as.matrix(UQ.sample[i,])
+     res.simu <- simu.model(parms.simu)
+     resout[i,1:ntimes.date] <- res.simu$DR
+     resout[i,(1+ntimes.date):(2*ntimes.date)] <- res.simu$H
+     resout[i,(1+2*ntimes.date):(3*ntimes.date)] <- res.simu$U
+     resout[i,(1+3*ntimes.date):(4*ntimes.date)] <- res.simu$D
+     resout[i,(1+4*ntimes.date):(5*ntimes.date)] <- res.simu$RplusHU
```

```
+       resout[i,(5*ntimes.date)+1] <- res.simu$time.maxU
+       resout[i,(5*ntimes.date)+2] <- res.simu$value.maxU
+       resout[i,(5*ntimes.date)+3] <- res.simu$cumul.DR
+     }
+     return(resout)
+ }
```

The HSIC sensitivity indices, defined in Section 6.2.1, are useful for such a goal of screening, and we compute them from a 2000-size sample. Each of the compartments is given as a curve, which implies that we need to use the multidimensional version of the HSIC indices with a principal component analysis step as proposed in Da Veiga [99]. Briefly, we first collect a sample of all the calibrated compartments (five curves and three scalar values). Since they are all strongly correlated, we use principal component analysis to compress them into a small number of new variables, representative of the output variability. The output kernel in HSIC is finally given as a product of 1D kernels on each of these principal components:

```
> n.screening <- 2000
> X.screening <- generate.sample.UQ(
+     generate.basic.sample.MC(n.screening,1:nparms),
+     lb.all,ub.all,nom.all,1:nparms)
> Y.screening <- GSAfunc(X.screening,nom.all,1:nparms)
> x <- sensiHSIC(model = NULL, X.screening,
+                   kernelX = "dcov", kernelY = "dcov",
+                   expl.var.PCA = 0.95, test.method = "Permutation")
> tell(x,Y.screening[,1:(5*ntimes.date)]) # Warning: time consuming step
> S.HSIC <- data.frame(parms = 1:nparms, SH = unname(x$S))
> ggplot(S.HSIC, aes(x=parms, y=SH)) + geom_point() +
+     ylab("HSIC sensitivity index") + xlab("Parameters") +
+     scale_x_continuous(breaks=1:nparms,labels=parms.name) + theme_bw() +
+     theme(axis.text.x = element_text(angle = 45, hjust = 1))
```

Figure 7.4 is then obtained and clearly illustrates that a small number of parameters have a much larger influence on the considered outputs than the others.

However, the final selection of the influential parameters is not entirely straightforward: Should we select $N_s$, $N_{IH}$, and $p_{HU}$? To perform a more qualitative selection, it is possible to check the p-values of the independence test based on HSIC:

```
> pval.HSIC <- data.frame(parms = 1:nparms, pval = unname(x$Pvalue))
> ggplot(pval.HSIC, aes(x=parms, y=pval)) + geom_point() +
+     ylab("HSIC P-value") + xlab("Parameters") +
+     scale_x_continuous(breaks=1:nparms,labels=parms.name) + theme_bw() +
+     theme(axis.text.x = element_text(angle = 45, hjust = 1))
```

Figure 7.5 shows that $N_{IH}$ and $p_{HU}$ should not be selected. By using a p-value limit with a Bonferroni correction, we see that $N_s$ is also not selected, which leads us to select only eight parameters out of the 20 initial ones:

```
> pval.lim <- 0.05/nparms # Bonferroni
> id.reduced <- which(pval.HSIC$pval < pval.lim)
> parms.name.reduced <- parms.name[id.reduced]
> nparms.reduced <- length(id.reduced)
```

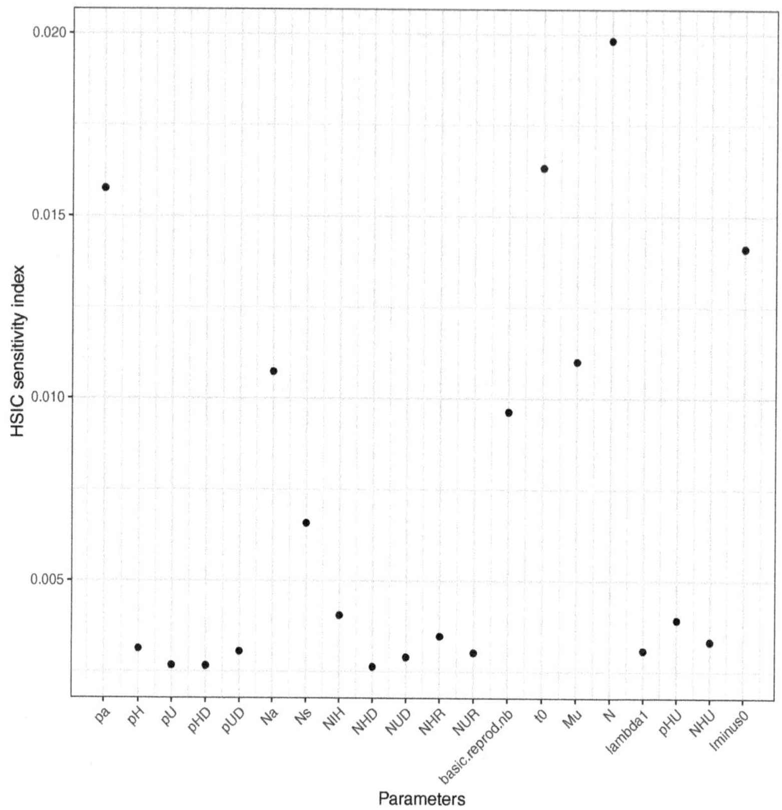


**Figure 7.4.** *Normalized HSIC indices for daily reported, H, U, D, and* $R^{+H}$.

## 7.2.3 ▪ Sobol' sensitivity analysis

Now that the number of parameters has been greatly reduced with the screening step, we can gain more precise insight into the impact of each parameter (*factor prioritization* setting). In particular, since the outputs of interest are curves, we can rely on the ubiquitous sensitivity indices defined in Section 3.3.1 to see which parameters have an impact on each period of the epidemic spread. The sobolMultOut() function of the sensitivity package is used for this purpose:

```
> nGSA <- 5000
> X1 <- generate.sample.UQ(generate.basic.sample.MC(nGSA,id.reduced),
+                          lb.all,ub.all,nom.all,id.reduced)
> X2 <- generate.sample.UQ(generate.basic.sample.MC(nGSA,id.reduced),
+                          lb.all,ub.all,nom.all,id.reduced)
> # First-order sensitivity indices
> ss1 <- sobolMultOut(model = GSAfunc, q = 5*ntimes.date,
+                     X1=as.data.frame(X1), X2=as.data.frame(X2),
+                     MCmethod = "sobol2007", ubiquitous = T,
+                     nom=nom.all,id.active=id.reduced)
> # Total sensitivity indices
> sst <- sobolMultOut(model = GSAfunc, q = 5*ntimes.date,
+                     X1=as.data.frame(X1), X2=as.data.frame(X2),
```

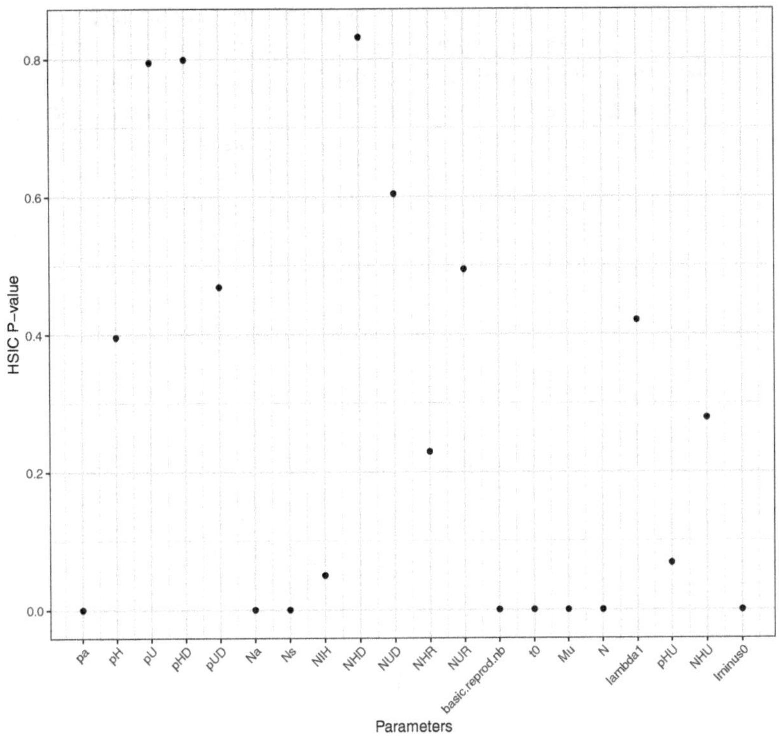


**Figure 7.5.** *p-values from independence tests with HSIC indices for daily reported, H, U, D, and* $R^{+H}$.

```
+                          MCmethod = "soboljansen", ubiquitous = T,
+                          nom=nom.all,id.active=id.reduced)
> print(ss1$S)
> print(sst$T)
> S1.DR <- as.data.frame(ss1$Sfct[1:ntimes.date,])
> ST.DR <- as.data.frame(sst$Tfct[1:ntimes.date,])
> S1.H <- as.data.frame(ss1$Sfct[(ntimes.date+1):(2*ntimes.date),])
> ST.H <- as.data.frame(sst$Tfct[(ntimes.date+1):(2*ntimes.date),])
> S1.U <- as.data.frame(ss1$Sfct[(2*ntimes.date+1):(3*ntimes.date),])
> ST.U <- as.data.frame(sst$Tfct[(2*ntimes.date+1):(3*ntimes.date),])
> S1.D <- as.data.frame(ss1$Sfct[(3*ntimes.date+1):(4*ntimes.date),])
> ST.D <- as.data.frame(sst$Tfct[(3*ntimes.date+1):(4*ntimes.date),])
> S1.R <- as.data.frame(ss1$Sfct[(4*ntimes.date+1):(5*ntimes.date),])
> ST.R <- as.data.frame(sst$Tfct[(4*ntimes.date+1):(5*ntimes.date),])
> colnames(S1.DR) <- colnames(ST.DR) <- colnames(S1.H) <-
+   colnames(ST.H) <- colnames(S1.U) <- colnames(ST.U) <-
+   colnames(S1.D) <- colnames(ST.D) <- colnames(S1.R) <-
+   colnames(ST.R) <- parms.name.reduced
> library(reshape2)
> dfS1.DR <- cbind(melt(S1.DR,measure.vars=1:nparms.reduced),
+                       data.frame(times=rep(times.date,nparms.reduced)))
> dfST.DR <- cbind(melt(ST.DR,measure.vars=1:nparms.reduced),
+                       data.frame(times=rep(times.date,nparms.reduced)))
```

```
> dfS1.H <- cbind(melt(S1.H,measure.vars=1:nparms.reduced),
+                 data.frame(times=rep(times.date,nparms.reduced)))
> dfST.H <- cbind(melt(ST.H,measure.vars=1:nparms.reduced),
+                 data.frame(times=rep(times.date,nparms.reduced)))
> dfS1.U <- cbind(melt(S1.U,measure.vars=1:nparms.reduced),
+                 data.frame(times=rep(times.date,nparms.reduced)))
> dfST.U <- cbind(melt(ST.U,measure.vars=1:nparms.reduced),
+                 data.frame(times=rep(times.date,nparms.reduced)))
> dfS1.D <- cbind(melt(S1.D,measure.vars=1:nparms.reduced),
+                 data.frame(times=rep(times.date,nparms.reduced)))
> dfST.D <- cbind(melt(ST.D,measure.vars=1:nparms.reduced),
+                 data.frame(times=rep(times.date,nparms.reduced)))
> dfS1.R <- cbind(melt(S1.R,measure.vars=1:nparms.reduced),
+                 data.frame(times=rep(times.date,nparms.reduced)))
> dfST.R <- cbind(melt(ST.R,measure.vars=1:nparms.reduced),
+                 data.frame(times=rep(times.date,nparms.reduced)))
> g1DR <- ggplot(dfS1.DR,aes(x=times,y=value,colour=variable)) +
+   geom_line(size=1) + coord_cartesian(ylim=c(0,1)) +
+   ylab("1st order index for DR") + xlab("Date") + theme_bw() +
+   theme(legend.position="right",
+         legend.text = element_text(size=6),
+         legend.title = element_blank())
> gTDR <- ggplot(dfST.DR,aes(x=times,y=value,colour=variable)) +
+   geom_line(size=1) + coord_cartesian(ylim=c(0,1)) +
+   ylab("Total index for DR") + xlab("Date") + theme_bw() +
+   theme(legend.position="right",
+         legend.text = element_text(size=6),
+         legend.title = element_blank())
> g1H <- ggplot(dfS1.H,aes(x=times,y=value,colour=variable)) +
+   geom_line(size=1) + coord_cartesian(ylim=c(0,1)) +
+   ylab("1st order index for H") + xlab("Date") + theme_bw()+
+   theme(legend.position="right",
+         legend.text = element_text(size=6),
+         legend.title = element_blank())
> gTH <- ggplot(dfST.H,aes(x=times,y=value,colour=variable)) +
+   geom_line(size=1) + coord_cartesian(ylim=c(0,1)) +
+   ylab("Total index for H") + xlab("Date") + theme_bw() +
+   theme(legend.position="right",
+         legend.text = element_text(size=6),
+         legend.title = element_blank())
> g1U <- ggplot(dfS1.U,aes(x=times,y=value,colour=variable)) +
+   geom_line(size=1) + coord_cartesian(ylim=c(0,1)) +
+   ylab("1st order index for U") + xlab("Date") + theme_bw() +
+   theme(legend.position="right",
+         legend.text = element_text(size=6),
+         legend.title = element_blank())
> gTU <- ggplot(dfST.U,aes(x=times,y=value,colour=variable)) +
+   geom_line(size=1) + coord_cartesian(ylim=c(0,1)) +
+   ylab("Total index for U") + xlab("Date") + theme_bw()+
+   theme(legend.position="right",
```

```
+          legend.text = element_text(size=6),
+          legend.title = element_blank())
> g1D <- ggplot(dfS1.D,aes(x=times,y=value,colour=variable)) +
+    geom_line(size=1) + coord_cartesian(ylim=c(0,1)) +
+    ylab("1st order index for D") + xlab("Date") + theme_bw() +
+    theme(legend.position="right",
+          legend.text = element_text(size=6),
+          legend.title = element_blank())
> gTD <- ggplot(dfST.D,aes(x=times,y=value,colour=variable)) +
+    geom_line(size=1) + coord_cartesian(ylim=c(0,1)) +
+    ylab("Total index for D") + xlab("Date") + theme_bw() +
+    theme(legend.position="right",
+          legend.text = element_text(size=6),
+          legend.title = element_blank())
> g1R <- ggplot(dfS1.R,aes(x=times,y=value,colour=variable)) +
+    geom_line(size=1) + coord_cartesian(ylim=c(0,1)) +
+    ylab("1st order index for R+H") + xlab("Date") + theme_bw() +
+    theme(legend.position="right",
+          legend.text = element_text(size=6),
+          legend.title = element_blank())
> gTR <- ggplot(dfST.R,aes(x=times,y=value,colour=variable)) +
+    geom_line(size=1) + coord_cartesian(ylim=c(0,1)) +
+    ylab("Total index for R+H") + xlab("Date") + theme_bw() +
+    theme(legend.position="right",
+          legend.text = element_text(size=6),
+          legend.title = element_blank())
> grid.arrange(g1DR, gTDR, g1H, gTH, g1U, gTU, g1D, gTD, g1R, gTR,
+               nrow=5, ncol=2)
```

Unsurprisingly, Fig. 7.6 shows that at the beginning of the simulation, the most influential parameters are $t_0$ (time at which the epidemic starts) and $I_0^-$ and $p_a$ (parameters related to the first compartments in the model). The number of days before recovering without going to the hospital $N_a$ and $N_s$ then arise in the middle of the spread, while the starting time of the lockdown effect $N$ dominates at the end of epidemic.

We can also compute first-order and total indices for the scalar outputs of interest (results are given in Fig. 7.7):

```
> GSA.scalars <- function(UQ.sample, nom, id.active){
+    n <- nrow(UQ.sample)
+    resout <- matrix(0,n,3)
+    for (i in 1:n){
+      if (!as.logical(mod(i,1000))) print(i)
+
+      parms.simu <- matrix(nom,nrow=1)
+      parms.simu[id.active] <- as.matrix(UQ.sample[i,])
+      res.simu <- simu.model(parms.simu)
+      resout[i,1] <- res.simu$time.maxU
+      resout[i,2] <- res.simu$value.maxU
+      resout[i,3] <- res.simu$cumul.DR
+    }
+    return(resout)
```

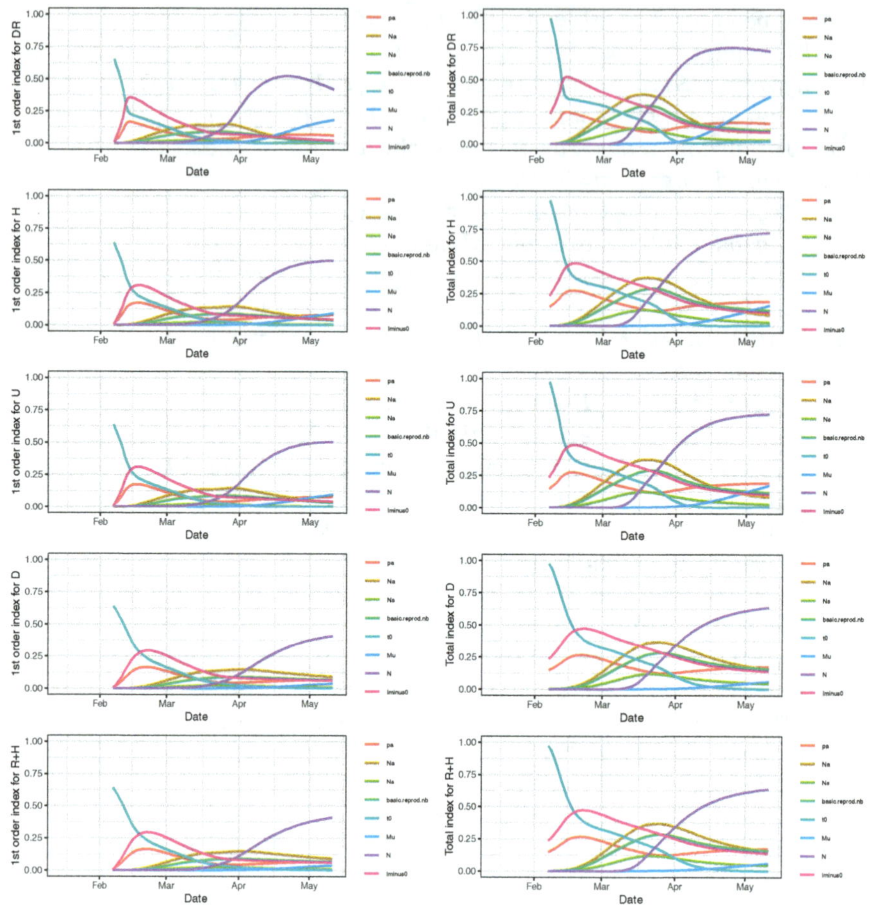


**Figure 7.6.** *Ubiquitous first-order and total Sobol' indices for (from top to bottom) daily reported, H, U, D, and* $R^{+H}$.

```
+ }
> X1 <- generate.sample.UQ(generate.basic.sample.MC(nGSA,id.reduced),
+                          lb.all,ub.all,nom.all,id.reduced)
> X2 <- generate.sample.UQ(generate.basic.sample.MC(nGSA,id.reduced),
+                          lb.all,ub.all,nom.all,id.reduced)
> # First-order sensitivity indices
> x <- sobolEff(model = NULL, X1, X2,order=1)
> y <- GSA.scalars(x$X,nom.all,id.reduced)
> s.time.maxU <- tell(x,y[,1,drop=F])
> s.value.maxU <- tell(x,y[,2,drop=F])
> s.cumul.DR <- tell(x,y[,3,drop=F])
> rownames(s.time.maxU$S) <- rownames(s.value.maxU$S) <-
+   rownames(s.cumul.DR$S) <- parms.name.reduced
> gstimeU <-ggplot(s.time.maxU) +
+   ylab("Date Max. U") +
+   theme(axis.text.x = element_text(angle = 30, hjust = 1, size = 6))
> gsmaxU <- ggplot(s.value.maxU) +
```

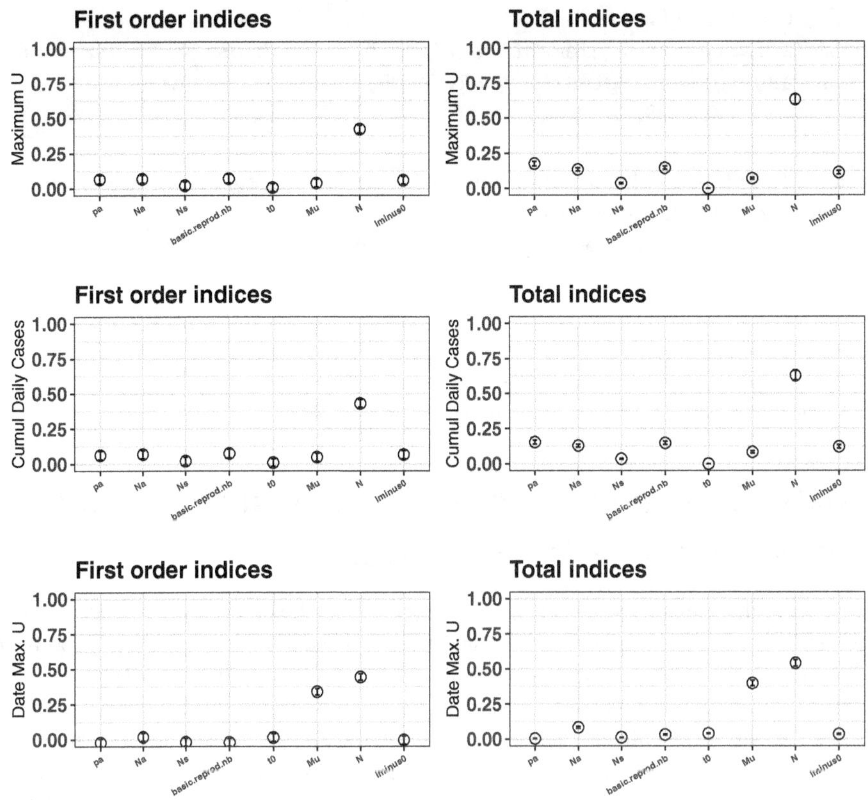


**Figure 7.7.** *First-order and total Sobol' indices for maximum value of U, total reported cases, and time where maximum value of U is reached.*

```
+   ylab("Maximum U") +
+   theme(axis.text.x = element_text(angle = 30, hjust = 1, size = 6))
> gsDR <- ggplot(s.cumul.DR) +
+   ylab("Cumul Daily Cases") +
+   theme(axis.text.x = element_text(angle = 30, hjust = 1, size = 6))
> # Total sensitivity indices
> x <- sobolEff(model = NULL, X1, X2,order=0)
> y <- GSA.scalars(x$X,nom.all,id.reduced)
> t.time.maxU <- tell(x,y[,1,drop=F])
> t.value.maxU <- tell(x,y[,2,drop=F])
> t.cumul.DR <- tell(x,y[,3,drop=F])
> rownames(t.time.maxU$S) <- rownames(t.value.maxU$S) <-
+   rownames(t.cumul.DR$S) <- parms.name.reduced
> gttimeU <- ggplot(t.time.maxU) +
+   ylab("Date Max. U") +
+   theme(axis.text.x = element_text(angle = 30, hjust = 1, size = 6))
> gtmaxU <- ggplot(t.value.maxU) +
+   ylab("Maximum U")+
+   theme(axis.text.x = element_text(angle = 30, hjust = 1, size = 6))
> gtDR <- ggplot(t.cumul.DR) +
```

```
+   ylab("Cumul Daily Cases") +
+   theme(axis.text.x = element_text(angle = 30, hjust = 1, size = 6))
> grid.arrange(gsmaxU, gtmaxU, gsDR, gtDR, gstimeU, gttimeU,
+              nrow=3, ncol=2)
```

It is clear from Fig. 7.7 that the most influential parameters are $\mu$ and $N$ related to the transmission rate. Since all the parameters have a maximum value of total Sobol' index not equal to 0 for at least one compartment, we do not fix any parameter at this stage and for the rest of the study:

```
> id.posterior <- id.reduced
> parms.name.posterior <- parms.name.reduced
> nparms.posterior <- length(parms.name.posterior)
```

## 7.3 ▪ Data calibration and posterior sensitivity analysis

### 7.3.1 ▪ Collected data and deterministic calibration

We have seen in the previous section that the uncertainty on the compartment dynamics is large. However, in March 2020, French agencies started to publish daily data on the total reported cases as well as data coming from hospitals (admissions, ICUs, deaths, and recovered). It is thus possible to reduce the uncertainty on the model parameters by constraining the compartmental model to reproduce as best as possible all available data (see, e.g., [438]). For illustration purposes, we place ourselves as of April 30 and collect data up to this date in order to produce forecasts for the last two weeks before the end of the lockdown. We start by loading the number of daily reported cases:

```
> # Cumulated reported cases
> # from https://www.eficiens.com/coronavirus-statistiques/
> data <- read.table('data.csv',header=T,sep=",",
+                    stringsAsFactors = F)
> data.date <- as.Date(paste0(data$date,"/20"),format="%d/%m/%y")
> data.Rep <- data.frame(date=data.date,nb=data$nbre)
> data.DailyRep <- data.frame(date=data.date[2:length(data.date)],
+                             nb=diff(data$nbre))
> init.calib.DailyRep <-  as.Date("25/02/20",format="%d/%m/%y")
> end.calib.DailyRep <-  as.Date("30/04/20",format="%d/%m/%y")
> ind.calib.data.DailyRep <-
+   which(data.DailyRep$date>=init.calib.DailyRep &
+           data.DailyRep$date<=end.calib.DailyRep)
```

Notice that daily cases are available starting February 25. We then gather data from hospitals, starting March 18:

```
> library(dplyr)
> # Data from hospitals from
> # https://www.data.gouv.fr/fr/organizations/sante-publique-france/
> data.hosp <-
```

```
+   read.table('donnees-hospitalieres-covid19-2020-05-15-19h00.csv',
+               header=T, sep=";", stringsAsFactors = F)
> data.hosp <- data.hosp[data.hosp$sexe==0,]
> data.H.temp <- data.hosp %>% group_by(jour) %>% tally(hosp)
>               # Cases in hospitals
> data.date.H <- as.Date(data.H.temp$jour,format="%Y-%m-%d")
> data.H <- data.frame(date=data.date.H,nb=data.H.temp$n)
> data.U.temp <- data.hosp %>% group_by(jour) %>% tally(rea)
>               # Cases in ICU
> data.U <- data.frame(date=data.date.H,nb=data.U.temp$n)
> data.D.temp <- data.hosp %>% group_by(jour) %>% tally(dc)
>               # Cumulated deaths in hospitals
> data.D <- data.frame(date=data.date.H,nb=data.D.temp$n)
> data.R.temp <- data.hosp %>% group_by(jour) %>% tally(rad)
>               # Cumulated cured in hospitals
> data.R <- data.frame(date=data.date.H,nb=data.R.temp$n)
> init.calib.hosp <-  as.Date("18/03/20",format="%d/%m/%y")
> end.calib.hosp <-  as.Date("30/04/20",format="%d/%m/%y")
> ind.calib.data.hosp <-
+   which(data.H$date>=init.calib.hosp & data.H$date<=end.calib.hosp)
```

We first illustrate in Fig. 7.8 the data that will be calibrated:

```
> colors <- c("DR" = rgb(0,100,0,maxColorValue = 255),
+             "H" = rgb(147,112,219,maxColorValue = 255),
+             "U" = rgb(205,133,63,maxColorValue = 255),
+             "D" = rgb(220,20,60,maxColorValue = 255),
+             "R+H"=rgb(70,130,180,maxColorValue = 255))
> ggplot() +
+   geom_point(data=data.DailyRep,aes(x=date,y=nb,color="DR")) +
+   geom_point(data=data.H,aes(x=date,y=nb,color="H")) +
+   geom_point(data=data.U,aes(x=date,y=nb,color="U")) +
+   geom_point(data=data.D,aes(x=date,y=nb,color="D")) +
+   geom_point(data=data.R,aes(x=date,y=nb,color="R+H")) +
+   scale_colour_manual(values = colors)  +
+   ylab("Number of Cases") + xlab("Date") + theme_bw() +
+   theme(legend.position = "top",legend.title = element_blank())
```

Remember that even if data collection continues in May, we do not use this information for calibration and only use data until April 30. We will first define a minimization problem with an objective function $OF$ given as the sum of squared differences between model outputs and collected data, with a specific normalization,

$$OF(x) = \sum_{i=1}^{n_{DR}} f_{DR}(x,s_i)^2 + \sum_{i=1}^{n_H} f_H(x,t_i)^2 + \sum_{i=1}^{n_H} f_U(x,t_i)^2 + \sum_{i=1}^{n_H} f_D(x,t_i)^2 + \sum_{i=1}^{n_H} f_R(x,t_i)^2,$$

where $s_i$ ($i = 1,\dots,n_{DR}$) are the dates where the daily reported cases are collected ($s_1 = 02/25$, $s_{n_{DR}} = 04/30$), $t_i$ ($i = 1,\dots,n_H$) are the dates where data from hospitals are collected

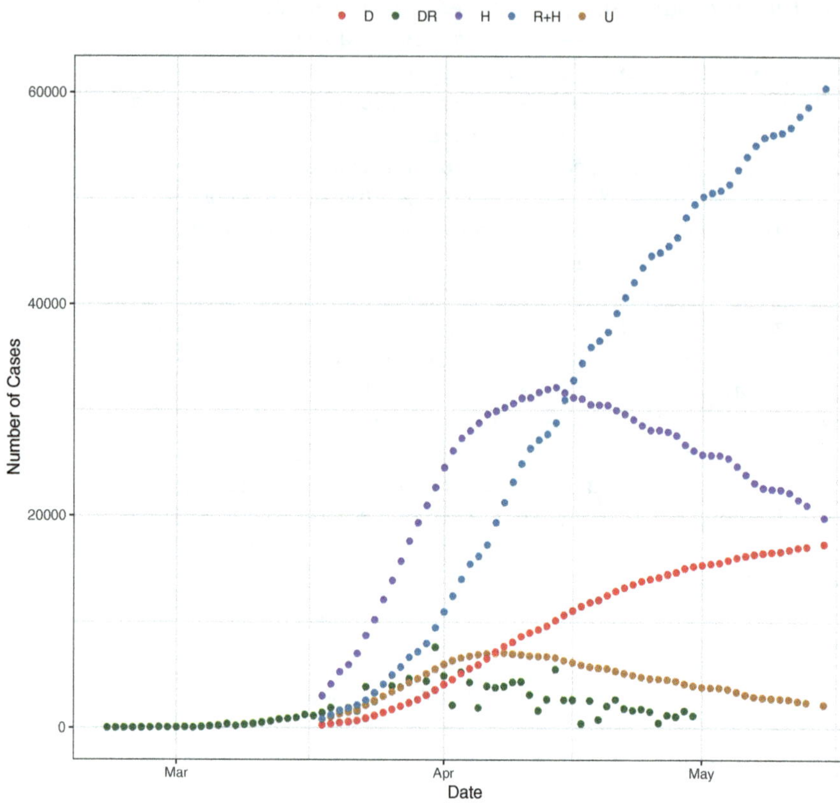


**Figure 7.8.** *Collected data for daily reported, H, U, D, and $R^{+H}$ to be calibrated.*

($t_1 = 03/18, t_{n_H} = 04/30$), and the residuals $f$ are defined as

$$\begin{aligned}
f_{DR}(x, s_i) &= (\mathsf{simu.model}_{DR}(x, s_i) - \mathrm{data}_{DR}(s_i)) / \sqrt{\mathrm{data}_{DR}(s_i)}, \\
f_{H}(x, t_i) &= (\mathsf{simu.model}_{H}(x, t_i) - \mathrm{data}_{H}(t_i)) / \sqrt{\mathrm{data}_{H}(t_i)}, \\
f_{U}(x, t_i) &= (\mathsf{simu.model}_{U}(x, t_i) - \mathrm{data}_{U}(t_i)) / \sqrt{\mathrm{data}_{U}(t_i)}, \\
f_{D}(x, t_i) &= (\mathsf{simu.model}_{D}(x, t_i) - \mathrm{data}_{D}(t_i)) / \sqrt{\mathrm{data}_{D}(t_i)}, \\
f_{R}(x, t_i) &= (\mathsf{simu.model}_{R^{+H}}(x, t_i) - \mathrm{data}_{R}(t_i)) / \sqrt{\mathrm{data}_{R}(t_i)},
\end{aligned}$$

with $\mathsf{simu.model}_C(x, u)$ the number of people in compartment $C$ at time $u$ computed by the compartmental model for the parameters equal to $x$. These residuals $f$ inside the objective function are defined in the following command lines:

```
> obj.func.all.res <- function(x,nom,id.active){
+   parms.simu <- matrix(nom,nrow=1)
+   parms.simu[id.active] <- x
+   res.simu <- simu.model(parms.simu)
+   DR.simu <- res.simu$DR; H.simu <- res.simu$H
+   U.simu <- res.simu$U; D.simu <- res.simu$D
+   R.simu <- res.simu$RplusHU
```

```
+   # Standard deviation of error is proportional to
+   # square root of the measure
+   res <- NULL
+   # Residuals DailyRep (DR)
+   ind.calib.times.DailyRep <- which(times.date>=init.calib.DailyRep &
+                                 times.date<=end.calib.DailyRep)
+   res <- c(res,(DR.simu[ind.calib.times.DailyRep] -
+                 data.DailyRep[ind.calib.data.DailyRep,2])/
+             sqrt(data.DailyRep[ind.calib.data.DailyRep,2]))
+   # Residuals Hospitals (H, U, D, R+H)
+   ind.calib.times.hosp <-
+     which(times.date>=init.calib.hosp & times.date<=end.calib.hosp)
+   res <- c(res,(H.simu[ind.calib.times.hosp] -
+                 data.H$nb[ind.calib.data.hosp]) /
+                 sqrt(data.H$nb[ind.calib.data.hosp]))
+   res <- c(res,(U.simu[ind.calib.times.hosp] -
+                 data.U$nb[ind.calib.data.hosp]) /
+                 sqrt(data.U$nb[ind.calib.data.hosp]))
+   res <- c(res,(D.simu[ind.calib.times.hosp] -
+                 data.D$nb[ind.calib.data.hosp]) /
+                 sqrt(data.D$nb[ind.calib.data.hosp]))
+   res <- c(res,(R.simu[ind.calib.times.hosp] -
+                 data.R$nb[ind.calib.data.hosp]) /
+                 sqrt(data.R$nb[ind.calib.data.hosp]))
+   return(res)
+ }
```

In order to solve this minimization problem, we use the Levenberg–Marquardt algorithm from the minpack.lm package, which only necessitates the knowledge of the residuals; see [326]. Since it is a local algorithm, we adopt a multistart approach where the initial values are obtained via an optimized design of experiments (see Appendix A):

```
> library(DiceDesign)
> library(minpack.lm)
> # Multistart optimization with Levenberg-Marquardt
> lb.optim <- lb.all[id.posterior]; ub.optim <- ub.all[id.posterior]
> nparms.optim <- length(id.posterior)
> n.multistart.all <- nparms.optim*10
> # Multistart values are obtained with an optimized DOE
> DOE.all <- lhsDesign(n.multistart.all, nparms.optim)$design
> DOE.all <- discrepSA_LHS(DOE.all)$design
> params.all.init <- matrix(NA,n.multistart.all,nparms.optim)
> for (i in 1:nparms.optim){
+   params.all.init[,i] <- lb.optim[i] +
+     (ub.optim[i]-lb.optim[i]) * DOE.all[,i]
+ }
> params.all.opt <- matrix(NA,n.multistart.all,nparms.optim)
> obj.all.opt <- matrix(NA,n.multistart.all,1)
> # Launch multistart
> for (i in 1:n.multistart.all){
```

```
+     oo <- nls.lm(params.all.init[i,], fn=obj.func.all.res,
+                  lower=lb.optim, upper=ub.optim,
+                  control=nls.lm.control(maxiter = 1000),
+                  nom=nom.all, id.active=id.posterior)
+     params.all.opt[i,] <- oo$par
+     obj.all.opt[i] <- sum(oo$fvec^2)
+ }
> # Retrieve best fit
> id.best <- which.min(obj.all.opt)
> params.all.best <- params.all.opt[id.best,]
> res.best <- obj.func.all.res(params.all.best,nom.all,id.posterior)
```

After having identified the parameter values which best reproduce the collected data, we can run the compartmental model to check the fit, which seems rather satisfactory (see Fig. 7.9):

```
> # Check calibration
> parms.simu <- matrix(nom.all,nrow=1)
> parms.simu[id.posterior] <- params.all.best
> res.simu <- simu.model(parms.simu)
> H.calib <- res.simu$H; U.calib <- res.simu$U
> D.calib <- res.simu$D; DR.calib <- res.simu$DR
> R.calib <- res.simu$RplusHU
> dfH.calib <- data.frame(date=times.date,H=c(H.calib))
> dfU.calib <- data.frame(date=times.date,U=c(U.calib))
> dfD.calib <- data.frame(date=times.date,D=c(D.calib))
> dfDR.calib <- data.frame(date=times.date,DR=c(DR.calib))
> dfR.calib <- data.frame(date=times.date,R=c(R.calib))
> ggplot() +
+     geom_line(data=dfH.calib,aes(x=date,y=H,color="H"),size=1.5) +
+     geom_line(data=dfU.calib,aes(x=date,y=U,color="U"),size=1.5) +
+     geom_line(data=dfD.calib,aes(x=date,y=D,color="D"),size=1.5)+
+     geom_line(data=dfDR.calib,aes(x=date,y=DR,color="DR"),size=1.5) +
+     geom_line(data=dfR.calib,aes(x=date,y=R,color="R+H"),size=1.5) +
+     geom_point(data=data.H[ind.calib.data.hosp,],
+                aes(x=date,y=nb,color="H")) +
+     geom_point(data=data.U[ind.calib.data.hosp,],
+                aes(x=date,y=nb,color="U")) +
+     geom_point(data=data.D[ind.calib.data.hosp,],
+                aes(x=date,y=nb,color="D")) +
+     geom_point(data=data.DailyRep[ind.calib.data.DailyRep,],
+                aes(x=date,y=nb,color="DR"))+
+     geom_point(data=data.R[ind.calib.data.hosp,],
+                aes(x=date,y=nb,color="R+H"))+
+     scale_colour_manual(values = colors)  +
+     ylab("Number of Cases") + xlab("Date") + theme_bw() +
+     theme(legend.position = "top",legend.title = element_blank())
```

One can remark that there seems to be outliers for the number of daily reported individuals: One potential explanation is that we observed delays for reported cases in France on weekends,

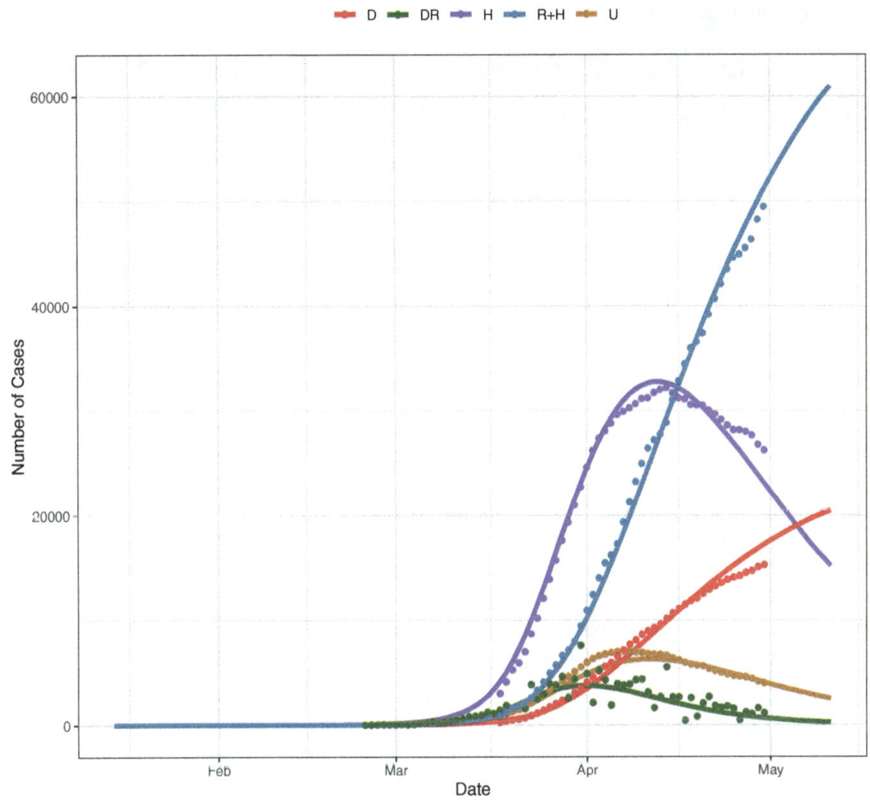


**Figure 7.9.** *Collected data and best model fit for daily reported, H, U, D, and $R^{+H}$ after calibration.*

which leads to virtually smaller numbers on Saturdays and Sundays and larger ones on Mondays. Such a seasonality could be incorporated in the model to improve the fit.

## 7.3.2 ▪ Probabilistic calibration

To go further and update the prior uncertainty on the parameters with collected data, we can adopt a Bayesian point of view for the calibration (see, e.g., [438]). The posterior uncertainty on the parameters $x$ is written as

$$p(x|\text{data}) \propto p(\text{data}|x)\, p_0(x),$$

where $p(\text{data}|x)$ is the likelihood of the data given $x$, $p_0(x)$ is the prior distribution on the parameters, and $p(x|\text{data})$ is the posterior distribution of the parameters given the data. The variance of the observation errors being unknown, we include it inside the problem and end up with the joint posterior

$$\begin{aligned} p(x, \sigma_{DR}^2, \sigma_H^2, \sigma_U^2, \sigma_D^2, \sigma_R^2|\text{data}) \propto\ & p(\text{data}|x, \sigma_{DR}^2, \sigma_H^2, \sigma_U^2, \sigma_D^2, \sigma_R^2) \\ & p_0(x)p_0(\sigma_{DR}^2)p_0(\sigma_H^2)p_0(\sigma_U^2)p_0(\sigma_D^2)p_0(\sigma_R^2), \end{aligned}$$

where $\sigma_C^2$ is the observation error for compartment $C$ and $p_0(\sigma_C^2)$ is the prior distribution on the variance of the observation error for compartment $C$. Concerning the data likelihood, we assume the following statistical model:

$$\begin{aligned}
\text{data}_{DR}(s_i) &= \mathsf{simu.model}_{DR}(x^*, s_i) + \sqrt{\text{data}_{DR}(s_i)}\,\epsilon_i^{DR},\\
\text{data}_{H}(t_i) &= \mathsf{simu.model}_{H}(x^*, t_i) + \sqrt{\text{data}_{H}(t_i)}\,\epsilon_i^{H},\\
\text{data}_{U}(t_i) &= \mathsf{simu.model}_{U}(x^*, t_i) + \sqrt{\text{data}_{U}(t_i)}\,\epsilon_i^{U},\\
\text{data}_{D}(t_i) &= \mathsf{simu.model}_{D}(x^*, t_i) + \sqrt{\text{data}_{D}(t_i)}\,\epsilon_i^{D},\\
\text{data}_{R}(t_i) &= \mathsf{simu.model}_{R}(x^*, t_i) + \sqrt{\text{data}_{R}(t_i)}\,\epsilon_i^{R},
\end{aligned}$$

where $\epsilon_i^C \sim \mathcal{N}(0, \sigma_C^2)$ are independent Gaussian errors for compartment $C$ with $\epsilon_i^C$ and $\epsilon_j^{C'}$ independent for all $i, j, C, C'$. In this case, the log-likelihood

$$\log \mathcal{L} = \log p(\text{data}|x, \sigma_{DR}^2, \sigma_H^2, \sigma_U^2, \sigma_D^2, \sigma_R^2)$$

is written as

$$\log \mathcal{L} = \sum_{i=1}^{n_{DR}} \frac{f_{DR}(x, s_i)^2}{\sigma_{DR}^2} + \sum_{i=1}^{n_H} \frac{f_H(x, t_i)^2}{\sigma_H^2} + \sum_{i=1}^{n_H} \frac{f_U(x, t_i)^2}{\sigma_U^2} + \sum_{i=1}^{n_H} \frac{f_D(x, t_i)^2}{\sigma_D^2} + \sum_{i=1}^{n_H} \frac{f_R(x, t_i)^2}{\sigma_R^2},$$

where we use the $f$ residuals defined above.

Since the fit from the previous section can be used to provide an estimate of the Gaussian error variances (by computing the variance of the residuals at the optimum), we assign them an informative prior. More precisely, denoting $s_{0,C}^2$ the estimate of the variance for compartment $C$ and $n_{0,C}$ the number of observations used to build this estimate, the prior distribution for $\sigma_C^2$ is a scaled inverse chi-squared distribution defined as

$$p_0(\sigma_C^2) \propto \frac{1}{\sigma_C^{n_{0,C}+2}} \exp\left[-\frac{n_{0,C}s_{0,C}^2}{2\sigma_C^2}\right].$$

This choice of prior is arbitrary but is made here because it is often used as a conjugate prior (for the variance parameter of a normal distribution). The prior distribution on the parameters $x$ being given by the uniform distributions defined before, the log-posterior distribution is finally written in R as follows:

```
> # Estimation of noise variance with residuals
> ndata.DR <- length(ind.calib.data.DailyRep)
> ndata.H <- length(ind.calib.data.hosp)
> sig2.DR <- var(res.best[1:ndata.DR])
> sig2.H <- var(res.best[(ndata.DR+1):(ndata.DR+ndata.H)])
> sig2.U <- var(res.best[(ndata.DR+ndata.H+1):(ndata.DR+2*ndata.H)])
> sig2.D <- var(res.best[(ndata.DR+2*ndata.H+1):(ndata.DR+3*ndata.H)])
> sig2.R <- var(res.best[(ndata.DR+3*ndata.H+1):(ndata.DR+4*ndata.H)])
> # MCMC on both model parameters and
> # standard deviations (with a scaled inverse chi-squared distribution)
> logpost.all.with.stdev <- function(x){
+   if (all(x[1:nparms.posterior]>=lb.optim) &
+       all(x[1:nparms.posterior]<=ub.optim)){
+     sq.residuals <- obj.func.all.res(x[1:nparms.posterior],
+                                       nom.all,id.posterior)^2/
+       c(rep(sig2.DR,ndata.DR),
+         rep(c(sig2.H,sig2.U,sig2.D,sig2.R),each=ndata.H))
```

```
+       res <- - sum(sq.residuals)/2 -
+         0.5*(ndata.DR+2)*log(x[nparms.posterior+1]) -
+         0.5*ndata.DR*sig2.DR/x[nparms.posterior+1] -
+         0.5*(ndata.H+2)*log(x[nparms.posterior+2]) -
+         0.5*ndata.H*sig2.H/x[nparms.posterior+2] -
+         0.5*(ndata.H+2)*log(x[nparms.posterior+3]) -
+         0.5*ndata.H*sig2.U/x[nparms.posterior+3] -
+         0.5*(ndata.H+2)*log(x[nparms.posterior+4]) -
+         0.5*ndata.H*sig2.D/x[nparms.posterior+4] -
+         0.5*(ndata.H+2)*log(x[nparms.posterior+5]) -
+         0.5*ndata.H*sig2.R/x[nparms.posterior+5]
+     }else{
+       res <- - Inf
+     }
+     return(res)
+ }
```

The posterior distribution does not have an explicit expression since it involves calls to the compartmental model in the likelihood. Consequently, we have to resort to approximate sampling algorithms. Here we simply use a Metropolis–Hastings sampling scheme with the mcmc package; see, e.g., [480]. After an initial burn-in period and a large thinning to limit autocorrelation between the samples, we can generate a sample of the posterior distribution of the parameters by marginalizing out the variances:

```
> library(mcmc)
> # Burnin
> mcmc.sample.burnin <- metrop(logpost.all.with.stdev,
+                       initial=c(params.all.init[1,],
/                                 sig2.DR,sig2.H,sig2.U,sig2.D,sig2.R),
+                       scale=c((ub.optim-lb.optim)/150,
+                               sig2.DR/100,sig2.H/100,sig2.U/100,
+                               sig2.D/100,sig2.R/100),
+                       nbatch=5e4,blen=1)
> # Posterior sample
> nsample.posterior <- 5000
> mcmc.sample <- metrop(mcmc.sample.burnin,
+                       scale=c((ub.optim-lb.optim)/150,
+                               sig2.DR/100,sig2.H/100,sig2.U/100,
+                               sig2.D/100,sig2.R/100),
+                       nbatch-nsample.posterior,blen-200)
> # We discard standard deviation samples
> UQ.sample.posterior <- mcmc.sample$batch[,1:nparms.posterior]
```

Most of the parameters have been well constrained by the data, except $N_s$ (number of illness days of unhospitalized symptomatic people). The uncertainties on $t_0$ (epidemic starting date) and on the parameters related to the transmission rate ($\mu$ and $N$) have been largely reduced.

We can then compare the prior and posterior marginal distributions in Fig. 7.10:

```
> UQ.sample.gg <- as.data.frame(UQ.sample)
> UQ.sample.posterior.gg <- as.data.frame(UQ.sample.posterior)
> colnames(UQ.sample.gg) <- parms.name
```

```
> colnames(UQ.sample.posterior.gg) <- parms.name.posterior
> pl <- lapply(1:8,function(i){
+   ggplot() +
+   geom_histogram(data=UQ.sample.gg[,id.posterior[i],drop=F],
+                  aes_string(x=parms.name.posterior[i]),
+                  bins=30,alpha=0.2,fill="blue") +
+   geom_histogram(data=UQ.sample.posterior.gg[,i,drop=F],
+                  aes_string(x=parms.name.posterior[i]),
+                  bins=30,alpha=0.2,fill="red") +
+   ylab("") + theme_bw()
+ })
> grid.arrange(grobs=pl,nrow=4,ncol=2)
```

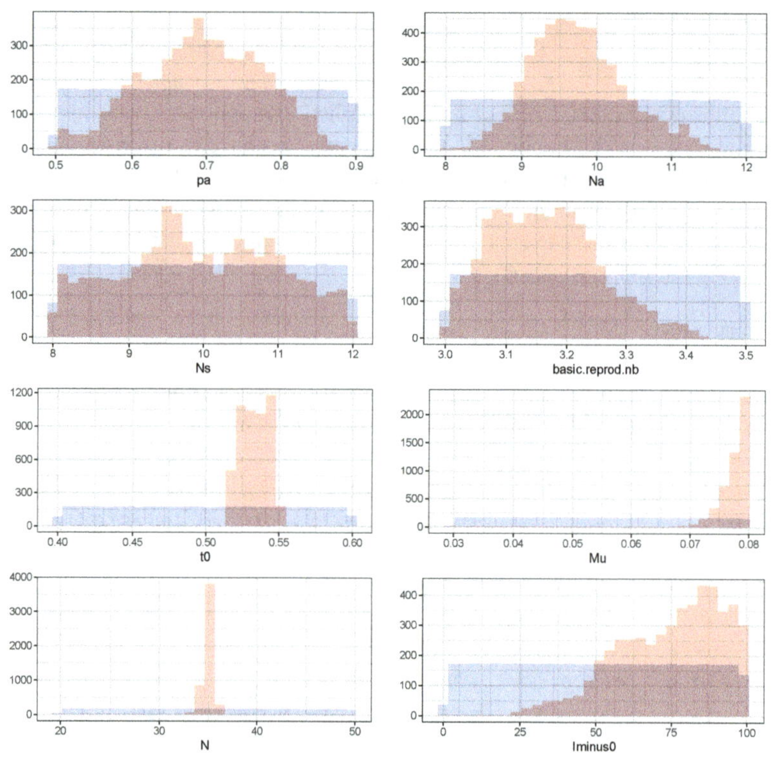


**Figure 7.10.** *Histograms of prior (blue) and posterior (red) probability distributions of the calibrated parameters.*

We can also investigate the posterior parameter correlations:

```
> library(corrplot)
> dfcor <- as.data.frame(UQ.sample.posterior)
> colnames(dfcor) <- parms.name.posterior
> cortest <- cor.mtest(dfcor, conf.level = .99)
> corrplot(cor(dfcor), p.mat = cortest$p, sig.level = .01,
+          method = "ellipse", order = "AOE",tl.srt = 45,insig = "blank")
```

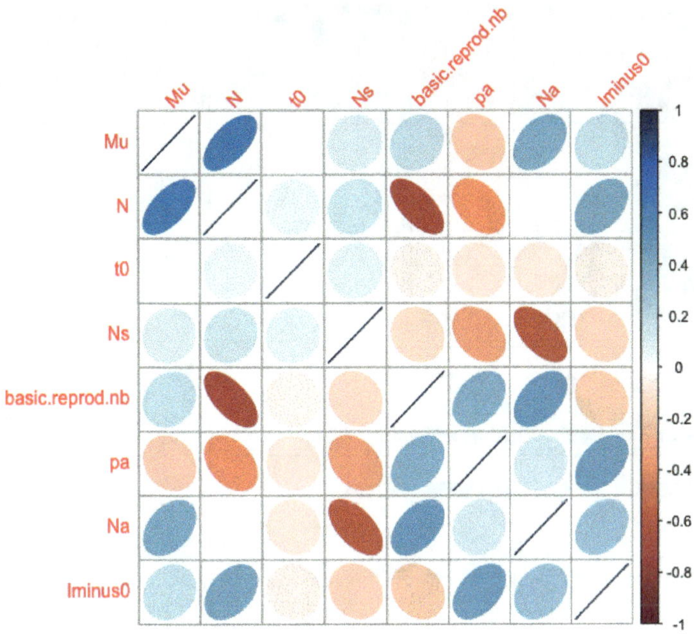


**Figure 7.11.** *Linear correlation coefficients (in color levels) of the calibrated parameters on the posterior sample.*

Figure 7.11 clearly shows that some parameters are highly correlated after calibration, as $N$ and $\mu$, and $N$ and $R_0$ (i.e., basic.reprod.nb).

A summary of the parameter posterior probability distribution is given in Fig. 7.12, obtained by

```
> library(viridis)
> library(GGally)
> # Lower visualization
> # 2D pdf estimate
> lowerfun <- function(data,mapping){
+   df <- subset(data,sample=="Posterior",select=-sample)
+   ggplot(data = df, mapping = mapping)+
+     stat_density_2d(aes(fill = ..level..), geom = "polygon") +
+     scale_fill_viridis() + theme_bw()
+ }
> # Diagonal visualization
> # Prior/posterior marginal histograms
> diagfun <- function(data,mapping){
+   x <- GGally::eval_data_col(data, mapping$x)
+   df <- data.frame(x1=x[1:nsample],
+                    x2=x[(nsample+1):(nsample+nsample.posterior)])
+   ggplot(df, mapping = mapping) +
+     geom_histogram(aes(x=x1),bins=30,alpha=0.2,fill="blue") +
+     geom_histogram(aes(x=x2),bins=30,alpha=0.2,fill="red") + theme_bw()
+ }
```

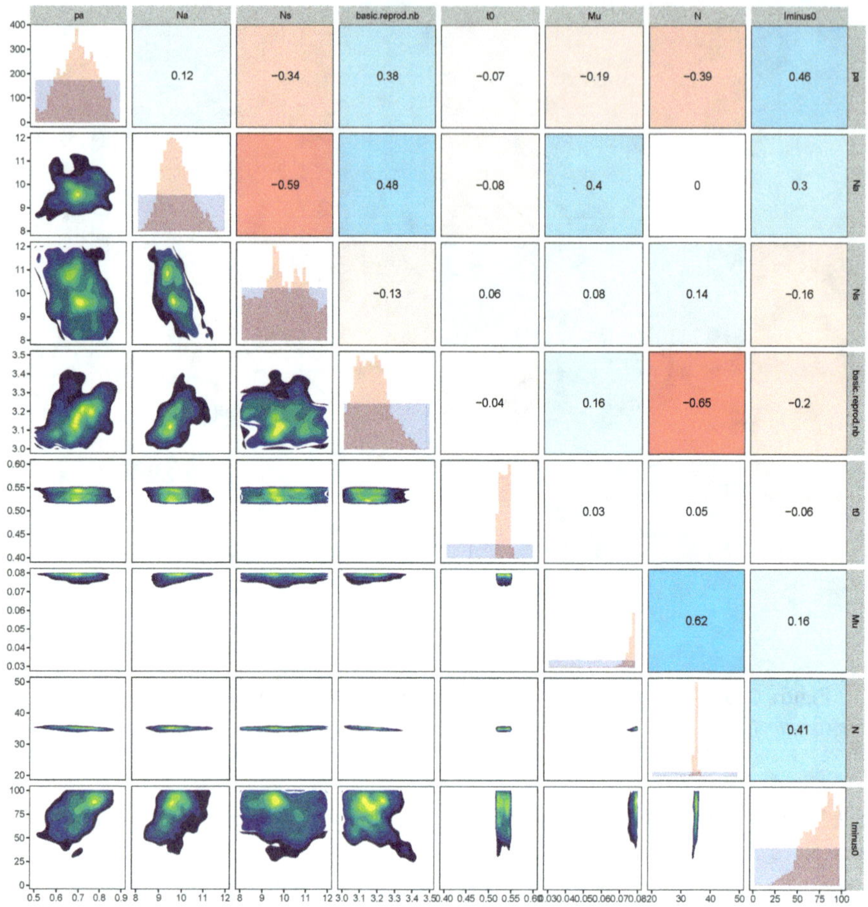


**Figure 7.12.** *Summary of parameter posterior distribution. Upper panel: pairwise correlation. Diagonal: prior (blue) and posterior (red) marginal histograms. Lower panel: pairwise joint pdf estimation given as a heatmap (blue: low density value; yellow: high density value).*

```
> # Upper visualization
> # Pairwise correlation
> # Source: https://stackoverflow.com/questions/57450913/
> # removing-background-color-for-column-labels-keeping-plot-background-color
> upperfun <- function(data, mapping, method, ...){
+   x <-
+     eval_data_col(data, mapping$x)[(nsample+1):(nsample+nsample.posterior)]
+   y <-
+     eval_data_col(data, mapping$y)[(nsample+1):(nsample+nsample.posterior)]
+   corr <- cor(x, y)
+   colFn <- colorRampPalette(c("brown1", "white", "dodgerblue"),
+                             interpolate ='spline')
+   fill <- colFn(100)[findInterval(corr, seq(-1, 1, length = 100))]
+   ggally_text(
+     label = as.character(round(corr, 2)),
```

```
+      mapping = aes(),
+      xP = 0.5, yP = 0.5,
+      color = 'black',
+      xrange = range(x, na.rm = T),
+      yrange = range(y, na.rm = T),
+      ...) +
+      theme_void() +
+      theme(panel.background = element_rect(fill = fill))
+ }
> UQ1 <- UQ.sample.gg[,id.posterior]; UQ1$sample <- "Prior"
> UQ2 <- UQ.sample.posterior.gg; UQ2$sample <- "Posterior"
> p <- ggpairs(rbind(UQ1,UQ2),columns = 1:nparms.posterior,
+               lower = list(continuous = wrap(lowerfun)),
+               upper=list(continuous=wrap(upperfun)),
+               diag=list(continuous = wrap(diagfun)))
> # Change axis limits
> for(i in 2:p$nrow) {
+   for(j in 1:(i-1)) {
+     p[i,j] <- p[i,j] +
+       scale_x_continuous(limits = c(lb.optim[j], ub.optim[j])) +
+       scale_y_continuous(limits = c(lb.optim[i], ub.optim[i]))
+   }
+ }
> for (i in 1:p$nrow){
+   p[i,i] <- p[i,i] +
+     scale_x_continuous(limits = c(lb.optim[i], ub.optim[i]))
+ }
> # Remove grid
> p + theme(panel.grid.major = element_blank(),
+             panel.grid.minor = element_blank())
```

At present, the obtained posterior sample can be propagated through the compartmental model to update the uncertainty on the outputs:

```
> Iminus.sample.posterior <- Iplus.sample.posterior <-
+   Rminus.sample.posterior <- RplusI.sample.posterior <-
+   RplusHU.sample.posterior <- H.sample.posterior <-
+   U.sample.posterior <- D.sample.posterior <-
+   DR.sample.posterior <- R.sample.posterior <-
+   R.sample.posterior <- matrix(0,nsample.posterior,ntimes.date)
> time.maxU.sample.posterior <- value.maxU.sample.posterior <-
+   cumul.DR.sample.posterior <- matrix(0,nsample.posterior,1)
> for (i in 1:nsample.posterior){
+   if (!as.logical(mod(i,100)))
+     print(paste0("Sample ",i,"/",nsample.posterior))
+
+   parms.simu <- matrix(nom.all,nrow=1)
+   parms.simu[id.posterior] <- as.matrix(UQ.sample.posterior[i,])
+   res.simu <- simu.model(parms.simu)
+
+   Iminus.sample.posterior[i,] <- res.simu$Iminus
```

```
+   Iplus.sample.posterior[i,] <- res.simu$Iplus
+   Rminus.sample.posterior[i,] <- res.simu$Rminus
+   RplusI.sample.posterior[i,] <- res.simu$RplusI
+   RplusHU.sample.posterior[i,] <- res.simu$RplusHU
+   H.sample.posterior[i,] <- res.simu$H
+   U.sample.posterior[i,] <- res.simu$U
+   D.sample.posterior[i,] <- res.simu$D
+   DR.sample.posterior[i,] <- res.simu$DR
+   time.maxU.sample.posterior[i,] <- res.simu$time.maxU
+   value.maxU.sample.posterior[i,] <- res.simu$value.maxU
+   cumul.DR.sample.posterior[i,] <- res.simu$cumul.DR
+ }
> p.low <- 0.25
> p.up <- 0.75
> Iminus.med.posterior <-
+   apply(Iminus.sample.posterior,2,median,na.rm=T)
> Iplus.med.posterior <-
+   apply(Iplus.sample.posterior,2,median,na.rm=T)
> Rminus.med.posterior <-
+   apply(Rminus.sample.posterior,2,median,na.rm=T)
> RplusI.med.posterior <-
+   apply(RplusI.sample.posterior,2,median,na.rm=T)
> RplusHU.med.posterior <-
+   apply(RplusHU.sample.posterior,2,median,na.rm=T)
> H.med.posterior <-
+   apply(H.sample.posterior,2,median,na.rm=T)
> U.med.posterior <-
+   apply(U.sample.posterior,2,median,na.rm=T)
> D.med.posterior <-
+   apply(D.sample.posterior,2,median,na.rm=T)
> DR.med.posterior <-
+   apply(DR.sample.posterior,2,median,na.rm=T)
> Iminus.qlow.posterior <-
+   apply(Iminus.sample.posterior,2,quantile,probs=p.low,na.rm=T)
> Iplus.qlow.posterior <-
+   apply(Iplus.sample.posterior,2,quantile,probs=p.low,na.rm=T)
> Rminus.qlow.posterior <-
+   apply(Rminus.sample.posterior,2,quantile,probs=p.low,na.rm=T)
> RplusI.qlow.posterior <-
+   apply(RplusI.sample.posterior,2,quantile,probs=p.low,na.rm=T)
> RplusHU.qlow.posterior <-
+   apply(RplusHU.sample.posterior,2,quantile,probs=p.low,na.rm=T)
> H.qlow.posterior <-
+   apply(H.sample.posterior,2,quantile,probs=p.low,na.rm=T)
> U.qlow.posterior <-
+   apply(U.sample.posterior,2,quantile,probs=p.low,na.rm=T)
> D.qlow.posterior <-
+   apply(D.sample.posterior,2,quantile,probs=p.low,na.rm=T)
> DR.qlow.posterior <-
+   apply(DR.sample.posterior,2,quantile,probs=p.low,na.rm=T)
```

```
> Iminus.qup.posterior <-
+   apply(Iminus.sample.posterior,2,quantile,probs=p.up,na.rm=T)
> Iplus.qup.posterior <-
+   apply(Iplus.sample.posterior,2,quantile,probs=p.up,na.rm=T)
> Rminus.qup.posterior <-
+   apply(Rminus.sample.posterior,2,quantile,probs=p.up,na.rm=T)
> RplusI.qup.posterior <-
+   apply(RplusI.sample.posterior,2,quantile,probs=p.up,na.rm=T)
> RplusHU.qup.posterior <-
+   apply(RplusHU.sample.posterior,2,quantile,probs=p.up,na.rm=T)
> H.qup.posterior <-
+   apply(H.sample.posterior,2,quantile,probs=p.up,na.rm=T)
> U.qup.posterior <-
+   apply(U.sample.posterior,2,quantile,probs=p.up,na.rm=T)
> D.qup.posterior <-
+   apply(D.sample.posterior,2,quantile,probs=p.up,na.rm=T)
> DR.qup.posterior <-
+   apply(DR.sample.posterior,2,quantile,probs=p.up,na.rm=T)
> dfUQ.post <- data.frame(Min=c(H.qlow.posterior,U.qlow.posterior,
+                               D.qlow.posterior,RplusHU.qlow.posterior,
+                               DR.qlow.posterior),
+                         Med=c(H.med.posterior,U.med.posterior,
+                               D.med.posterior,RplusHU.med.posterior,
+                               DR.med.posterior),
+                         Max=c(H.qup.posterior,U.qup.posterior,
+                               D.qup.posterior,RplusHU.qup.posterior,
+                               DR.qup.posterior),
+                         Compartment=rep(c("H","U","D","R+H","DR"),
+                                         each=ntimes.date),
+                         Date=rep(times.date,5))
```

The posterior uncertainty on the compartments which are calibrated can then be visualized:

```
> ggplot() +
+   geom_line(data=dfUQ.post,
+             aes(x=Date,y=Med,colour=Compartment),size=1) +
+   geom_ribbon(data=dfUQ.post,
+               aes(x=Date,ymin=Min,ymax=Max,fill=Compartment),
+               alpha=0.2,size=0) +
+   scale_color_manual(values =
+                       c("H"=rgb(147,112,219,maxColorValue = 255),
+                         "D"=rgb(220,20,60,maxColorValue = 255),
+                         "U"=rgb(205,133,63,maxColorValue = 255),
+                         "DR"=rgb(0,100,0,maxColorValue = 255),
+                         "R+H"=rgb(70,130,180,maxColorValue = 255))) +
+   scale_fill_manual(values =
+                      c("H"=rgb(147,112,219,maxColorValue = 255),
+                        "D"=rgb(220,20,60,maxColorValue = 255),
+                        "U"=rgb(205,133,63,maxColorValue = 255),
+                        "DR"=rgb(0,100,0,maxColorValue = 255),
+                        "R+H"=rgb(70,130,180,maxColorValue = 255))) +
```

```
+    geom_point(data=data.H[ind.calib.data.hosp,],
+                aes(x=date,y=nb,color="H")) +
+    geom_point(data=data.U[ind.calib.data.hosp,],
+                aes(x=date,y=nb,color="U")) +
+    geom_point(data=data.D[ind.calib.data.hosp,],
+                aes(x=date,y=nb,color="D")) +
+    geom_point(data=data.DailyRep[ind.calib.data.DailyRep,],
+                aes(x=date,y=nb,color="DR"))+
+    geom_point(data=data.R[ind.calib.data.hosp,],
+                aes(x=date,y=nb,color="R+H"))+
+    ylab("Number of cases")  + theme_bw() + theme(legend.position="top")
```

The residual uncertainty on the calibrated compartments is very small, as shown in Fig. 7.13. Indeed, confidence intervals are so small that they are mingled with the median curves. This leads to the same result as in Fig. 7.9.

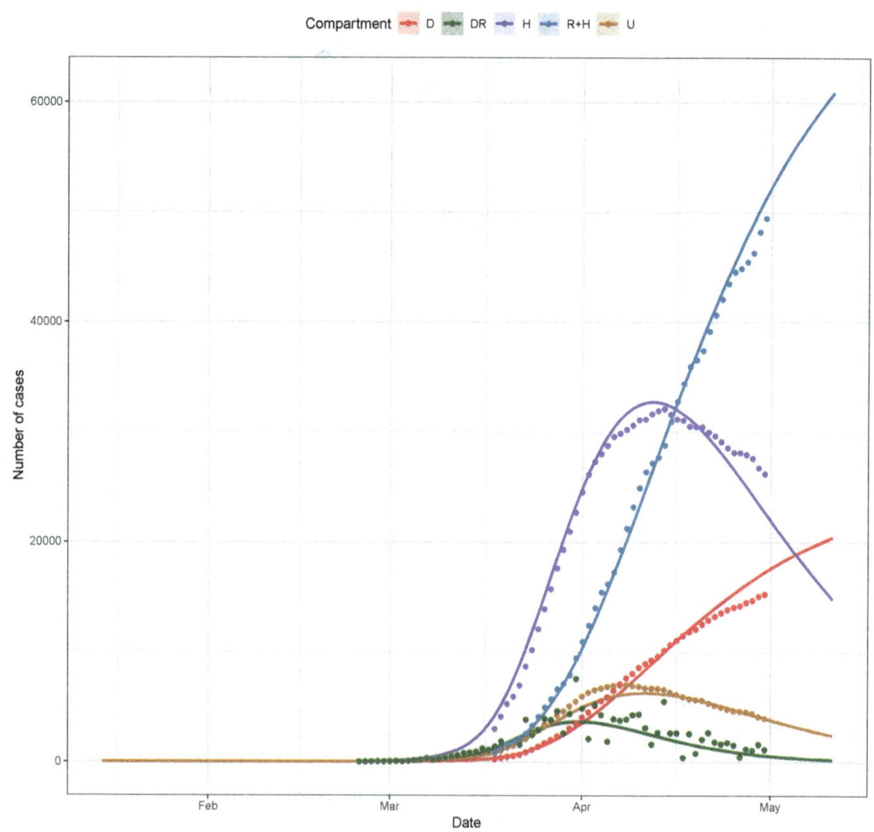


**Figure 7.13.** *Posterior uncertainty quantification for compartments daily reported, H, U, and D.*

We can also compare the posterior uncertainty on the unobserved compartments, in particular the ones with undetected people (infected and recovered):

```
> dfUQ.post.compare <-
+   data.frame(Min=c(Iminus.qlow.posterior,Rminus.qlow.posterior,
```

```
+                    Iminus.qlow,Rminus.qlow),
+              Med=c(Iminus.med.posterior,Rminus.med.posterior,
+                    Iminus.med,Rminus.med),
+               Max=c(Iminus.qup.posterior,Rminus.qup.posterior,
+                     Iminus.qup,Rminus.qup),
+               Compartment=rep(c("Post. Iminus","Post. Rminus",
+                                 "Prior Iminus","Prior Rminus"),
+                               each=ntimes.date),Date=rep(times.date,4))
> ggplot() +
+   geom_line(data=dfUQ.post.compare,
+             aes(x=Date,y=Med,colour=Compartment,
+                 linetype=Compartment),size=1) +
+   scale_linetype_manual(values=c("Post. Iminus"="solid",
+                                  "Post. Rminus"="solid",
+                                  "Prior Iminus"="dotted",
+                                  "Prior Rminus"="dotted")) +
+   geom_ribbon(data=dfUQ.post.compare,
+               aes(x=Date,ymin=Min,ymax=Max,fill=Compartment),
+               alpha=0.2,size=0) +
+   scale_color_manual(
+     values = c("Post. Iminus"=rgb(147,112,219,maxColorValue = 255),
+                "Post. Rminus"=rgb(205,133,63,maxColorValue = 255),
+                "Prior Iminus"=rgb(147,112,219,maxColorValue = 255),
+                "Prior Rminus"=rgb(205,133,63,maxColorValue = 255))) +
+   scale_fill_manual(
+     values = c("Post. Iminus"=rgb(147,112,219,maxColorValue = 255),
+                "Post. Rminus"=rgb(205,133,63,maxColorValue = 255),
+                "Prior Iminus"=rgb(147,112,219,maxColorValue = 255),
+                "Prior Rminus"=rgb(205,133,63,maxColorValue = 255))) +
+   ylab("Number of cases")  + theme_bw() + theme(legend.position="top")
```

Figure 7.14 shows that while the prior uncertainty on these compartments was large, it is largely reduced after calibration.

### 7.3.3 ▪ Posterior sensitivity analysis with Shapley effects

Finally, we investigate which parameters have an impact on the outputs after calibration since their influence may be different from what was observed when using their prior probability distribution in Section 7.2.3. In particular, the uncertainty has been greatly reduced for some parameters ($t_0$, $\mu$, $N$; see Fig. 7.10). On the other hand, we have observed that after calibration, some of them exhibit strong correlation: It is thus recommended to use sensitivity analysis tools which handle dependent inputs. Then, to perform this *factor prioritization* setting, we focus on Shapley effects introduced in Section 5.5. As a sample is available from the probabilistic calibration step, we use the given-data estimation algorithm of Broto et al. [56] with the sobolshap_knn() (which uses a fast approximate nearest-neighbor algorithm) function of the sensitivity package:

```
> library(RANN)
> x <- sobolshap_knn(model = NULL, X = UQ.sample.posterior, U = NULL,
+                    method = "knn", n.knn = 3,
+                    return.shap = T, n.limit = 2000)
> tell(x,cbind(H.sample.posterior,U.sample.posterior,
```

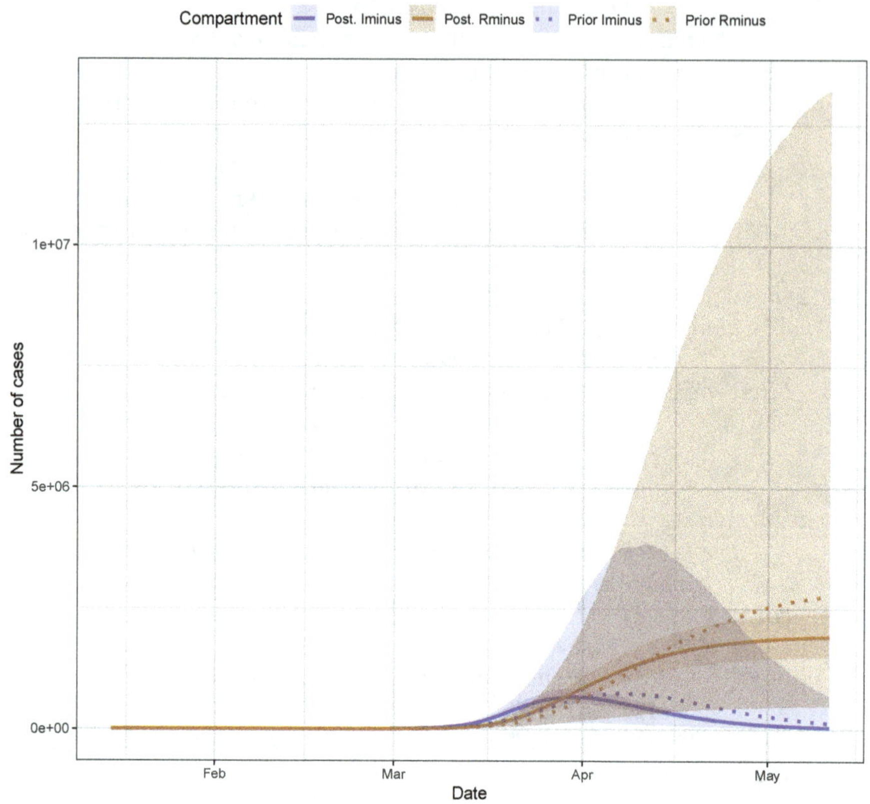


**Figure 7.14.** *Comparison of prior and posterior uncertainty quantification for undetected infected and recovered individuals. Dashed line is prior median, and solid line is posterior median.*

```
+                D.sample.posterior,DR.sample.posterior))
> Shapley.H <- as.data.frame(x$Shap[1:ntimes.date,])
> Shapley.U <- as.data.frame(x$Shap[(ntimes.date+1):(2*ntimes.date),])
> Shapley.D <- as.data.frame(x$Shap[(2*ntimes.date+1):(3*ntimes.date),])
> Shapley.DR <- as.data.frame(x$Shap[(3*ntimes.date+1):(4*ntimes.date),])
> colnames(Shapley.H) <- colnames(Shapley.U) <-
+   colnames(Shapley.D) <- colnames(Shapley.DR) <- parms.name.posterior
> dfShapley.H <- cbind(melt(Shapley.H,measure.vars=1:nparms.posterior),
+                       data.frame(times=rep(times.date,nparms.posterior)))
> dfShapley.U <- cbind(melt(Shapley.U,measure.vars=1:nparms.posterior),
+                       data.frame(times=rep(times.date,nparms.posterior)))
> dfShapley.D <- cbind(melt(Shapley.D,measure.vars=1:nparms.posterior),
+                       data.frame(times=rep(times.date,nparms.posterior)))
> dfShapley.DR <- cbind(melt(Shapley.DR,measure.vars=1:nparms.posterior),
+                        data.frame(times=rep(times.date,nparms.posterior)))
> gH <- ggplot(dfShapley.H,aes(x=times,y=value,colour=variable)) +
+   geom_line(size=1) +
+   coord_cartesian(ylim=c(0,0.5)) +
+   ylab("Shapley for H") + xlab("Date") + theme_bw() +
```

```
+   theme(legend.position="right",
+         legend.text = element_text(size=6),
+         legend.title = element_blank())
> gU <- ggplot(dfShapley.U,aes(x=times,y=value,colour=variable)) +
+   geom_line(size=1) +
+   coord_cartesian(ylim=c(0,0.5)) +
+   ylab("Shapley for U") + xlab("Date") + theme_bw() +
+   theme(legend.position="right",
+         legend.text = element_text(size=6),
+         legend.title = element_blank())
> gD <- ggplot(dfShapley.D,aes(x=times,y=value,colour=variable)) +
+   geom_line(size=1) +
+   coord_cartesian(ylim=c(0,0.5)) +
+   ylab("Shapley for D") + xlab("Date") + theme_bw() +
+   theme(legend.position="right",
+         legend.text = element_text(size=6),
+         legend.title = element_blank())
> gDR <- ggplot(dfShapley.DR,aes(x=times,y=value,colour=variable)) +
+   geom_line(size=1) +
+   coord_cartesian(ylim=c(0,0.5)) +
+   ylab("Shapley for DR") + xlab("Date") + theme_bw() +
+   theme(legend.position="right",
+         legend.text = element_text(size=6),
+         legend.title = element_blank())
> grid.arrange(gH, gU,gD, gDR, nrow=2, ncol=2)
```

Figure 7.15 shows that some parameters which had large impact in the prior SA such as $t_0$ and $I_0^-$ now have a much smaller influence after calibration.

We can also compute the Shapley effects for the scalar outputs of interest (see Fig. 7.16):

```
> x <- sobolshap_knn(model = NULL, X = UQ.sample.posterior, U = NULL,
+                    method = "knn", n.knn = 3, return.shap = T,
+                    n.limit = 2000)
> shap.time.maxU <- tell(x,time.maxU.sample.posterior)
> x <- sobolshap_knn(model = NULL, X = UQ.sample.posterior, U = NULL,
+                    method = "knn", n.knn = 3, return.shap = T,
+                    n.limit = 2000)
> shap.value.maxU <- tell(x,value.maxU.sample.posterior)
> x <- sobolshap_knn(model = NULL, X = UQ.sample.posterior, U = NULL,
+                    method = "knn", n.knn = 3, return.shap = T,
+                    n.limit = 2000)
> shap.cumul.DR <- tell(x,cumul.DR.sample.posterior)
> colnames(shap.time.maxU$Shap) <- colnames(shap.value.maxU$Shap) <-
+   colnames(shap.cumul.DR$Shap) <- parms.name.posterior
> gshaptimeU <- ggplot(shap.time.maxU, ylim = c(0, 0.5)) +
+   ylab("Date Max. U") +
+   theme(axis.text.x = element_text(angle = 30, hjust = 1, size = 8))
> gshapmaxU <- ggplot(shap.value.maxU, ylim = c(0, 0.5)) +
+   ylab("Maximum U")+
+   theme(axis.text.x = element_text(angle = 30, hjust = 1, size = 8))
> gshapDR <- ggplot(shap.cumul.DR, ylim = c(0, 0.5)) +
```

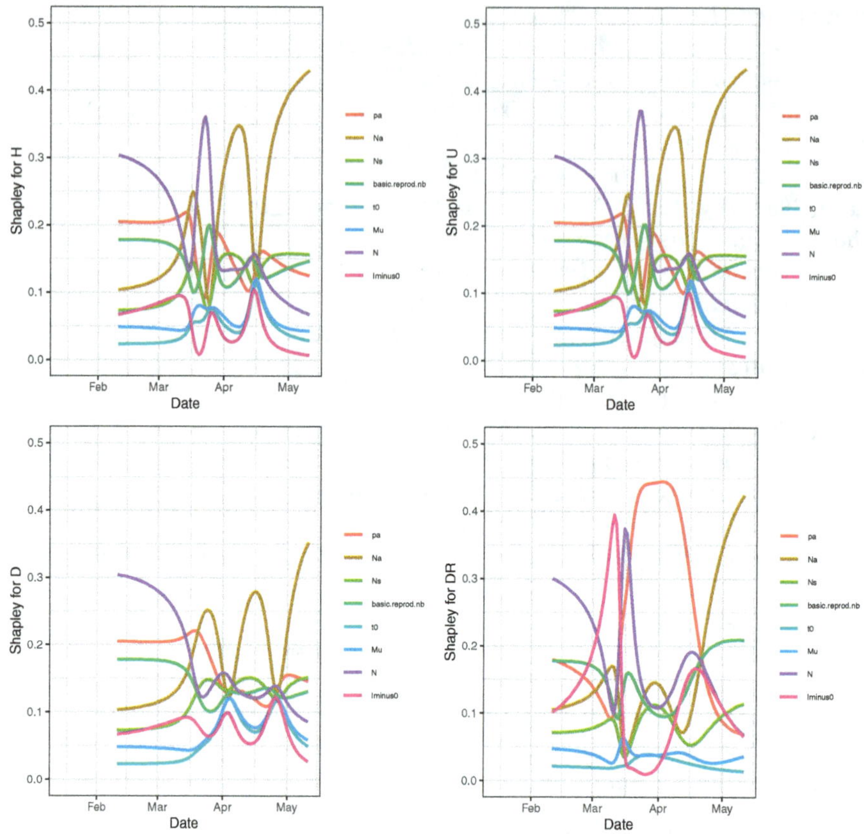


**Figure 7.15.** *Shapley effects for calibrated outputs with the posterior parameter distribution.*

```
+   ylab("Cumul Daily Cases") +
+   theme(axis.text.x = element_text(angle = 30, hjust = 1, size = 8))
> grid.arrange(gshapmaxU, gshapDR, gshaptimeU, nrow=3, ncol=1)
```

As a conclusion, one can state that the parameters $N$ (date of effect for social distancing), $N_a$ (number of illness days of symptomatic people), and $p_a$ (probability of having no symptoms in infected people) are the most influential on our model. These are the parameters to be characterized first in order to reduce the uncertainties of the model outputs. As an example, $N_a$ and $p_a$, which are related to people with low symptoms, could be estimated more precisely with extensive survey data.

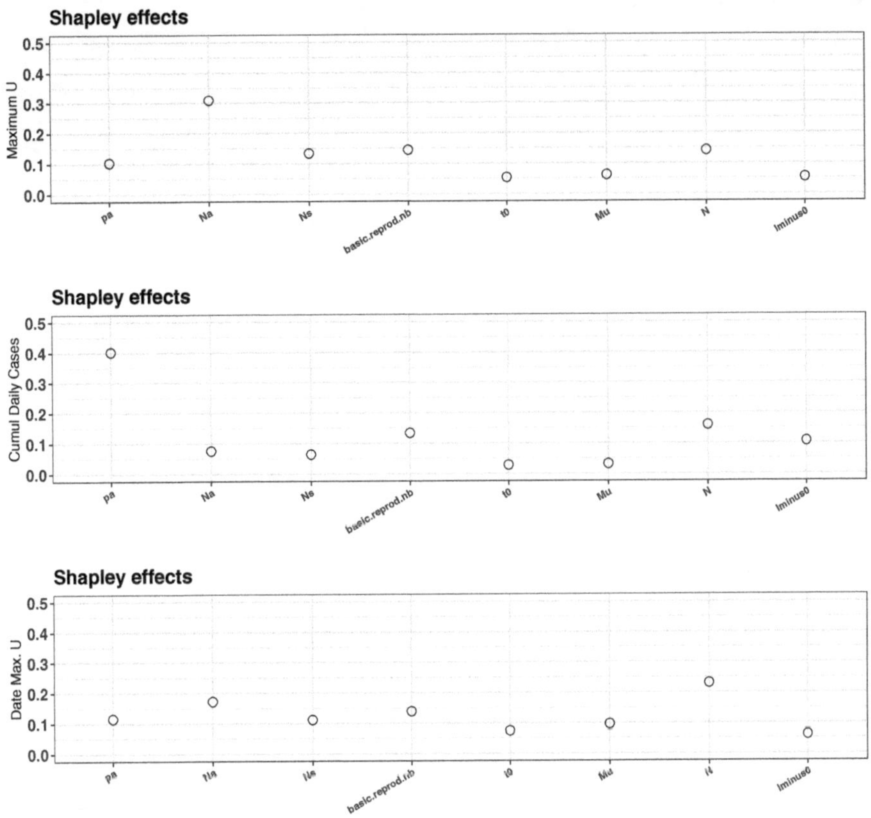


**Figure 7.16.** *Shapley effects for the maximum value of U, the total reported cases, and the time where the maximum value of U is reached.*

## Appendix A

# Sampling designs for model exploration in R

## A.1 ▪ Introduction

In SA, simulations are based on building samples in the input space, as it can be done by the Monte Carlo method. This is first briefly reviewed in this section. However, it is common to deal with problems with a small sample size (on the order of hundreds) and a large number of inputs (several dozen). Using efficient sampling methods, alternatives to the Monte Carlo method, is then crucial not only for accelerating the convergence of estimators but also for ensuring that the supports of the inputs are well sampled [333]. In this way, it is expected to maximize the chance to discover particular values and behavior of the model output.

Engineers often need to explore a physical system or a phenomenon by inferring knowledge from well-chosen experimental trials. An exploratory design is a design which tries to get across information on response variables (called "outputs") to the system/phenomenon in the support of some input variables (often called "inputs") by way of experiments (which can be physical or numerical). It often starts from a total ignorance of the output behavior and then exploring the input domain in the most efficient way. One of its first goals is to identify the noninfluential inputs with a small number of experiments while making realistic hypotheses on the phenomenon complexity.

The classical theory of design of experiments is devoted to this task and was developed a century ago for physical experiments [325]. It includes several techniques that answer screening requirements (such as the well-known fractional factorial designs), whose main goal is to obtain qualitative information of the inputs effects on the outputs using a minimal number of experiments [111]. For this purpose, methods are most often based on a discretization of the inputs in a small number of levels (e.g., two or three), and fitting linear models is one of the main objectives.

Another point of view which comes from the field of computer experiments is to consider the so-called space-filling criteria [245], whose aim is to perform an initial exploration of a physical numerical model depending on $d$ inputs using $N$ calculations. Indeed, the user has often no knowledge about who the influential (and thus noninfluential) inputs of the numerical model are. So it might be good to be able to choose the $N$ calculation points in a way that covers well the $d$-dimensional space: No area should be undersampled or oversampled (so as not to waste calculation points). Designs that allow one to fulfill this mission are called space-filling designs (SFD) [245, 127]. It has also been recognized that space-filling criteria are useful when considering physical experiments (for example, for spatially located measurements [336]).

Among the required properties of a design of experiments, two are essential:

1. Space-filling properties in the $d$-dimensional space. Two main types of criteria are classical: those based on the uniformity of the point distributions and those based on geometrical distances between points.

2. Space-filling properties of all the subprojection spaces. This is the space-filling robustness for the projections over lower-dimensional subspaces (note that there are $2^d - 2$ of such subspaces). Indeed, to capture precisely an interaction effect between several inputs, a good coverage onto the subspace spanned by these inputs is required. Moreover, a preliminary screening step may reveal some useless (i.e., noninfluential) input variables that can be neglected in the subsequent analysis. The new sample, solely including the retained input variables, has to keep good space-filling properties; see, e.g., [67, 380].

Two other properties can be considered depending on the context of the study. First, the orthogonality between columns of the design matrix makes diagonal the information matrix and allows us to provide good estimates of the regression coefficients of a linear model. Second, sometimes the sequential construction of the design could be useful. Indeed, if the runs are performed sequentially, it would be interesting to ensure good properties of a subdesign (just containing a first part of the design matrix).

We present here an introduction to SFD (see [200, 377, 333, 489, 152, 375] for more complete overviews) on the $[0,1]^d$ ($d$ is the number of variables) experimental domain. Deterministic or probabilistic transformation of samples supported on $[0,1]^d$ allow us to obtain samples following any other distribution. Section A.2 introduces the basic sampling methods. Then Section A.3 describes the expected properties of an exploratory design. Various ways to build space-filling designs, in order to offer a good coverage of the input domain by the experimental points, are presented in Section A.4. Finally, Section A.5 introduces a tool for design classification.

In this chapter, the DiceDesign package [119] is used for explaining some designs and algorithms. After launching the R software and installing the DiceDesign package, loading the package requires us to write the following command:

```
> library(DiceDesign)
```

To know the features of the package, type

```
> help(package="DiceDesign")
```

## A.2 ▪ Basic sampling techniques

### A.2.1 ▪ Monte Carlo sampling

#### Main principles

The Monte Carlo method depends on our ability to generate samples of random variables that follow specific probability distributions. The first ingredient consists in simulating uniformly distributed "random" numbers in the interval $[0,1]$. In practice, computers can only generate deterministic sequences of numbers that can be exactly reproduced knowing the used algorithm and its initial values. We therefore talk of pseudorandom number generators. It allows us to simulate samples that approximately share the statistical properties of the target distribution [244, 273].

From simulations of the uniform distribution, generation of variables following any distribution $p$ can be performed using various methods. Here, we discuss the most classical one (in the case of mutually independent random variables) involving the cumulative distribution function.

It involves two steps:

1. Generate independent and identically distributed (i.i.d.) pseudorandom variables that "imitate" the uniform distribution in the interval $[0, 1]$.

2. Apply a quantile transform (inverse function of the cumulative distribution function associated with probability distribution $p$) to these i.i.d. uniform random variables. The output variables from this transform follow the distribution $p$.

Note that the method can take into account some dependencies between variables, though the samples drawn remain independent of each other.

Consider now the design of a numerical experiment with $N$ simulations and $d$ inputs. The $N$ $d$-uples form an experiment matrix $\mathbf{X}^N = (\mathbf{X}^{(i)})_{i=1,\dots,N}$ of size $N \times d$, whose $i$th row $\mathbf{X}^{(i)}$ contains the set of values of the inputs. Monte Carlo sampling consists in the random generation of the experiment matrix $\mathbf{X}^N$. This is done by generating independently each column (corresponding to a sample of size $N$ for each input variable). In a Monte Carlo sample, each pseudorandom draw is supposed to be independent of the others. We therefore simulate a sample of size $N$ for $d$ variables, themselves mutually independent.

### Example in R

The two following examples with $N = 9$ and $d = 2$ in R considers uniform $\mathcal{U}(0, 1)$ and Gaussian $\mathcal{N}(0, 1)$ distributions. It allows us to draw Fig. A.1:

```
> set.seed(12345) # To reproduce the same results
> d <- 2 ; N <- 9
> XU <- matrix(runif(d * N), nrow = N)
> XN <- matrix(rnorm(d * N), nrow = N)
> colnames(XU) <- colnames(XN) <- c("X1", "X2")

> par(mfrow = c(1,2))
> plot(XU, xlim = c(0,1), ylim = c(0,1))
> plot(XN, xlim = c(-2,2), ylim = c(-2,2))
```

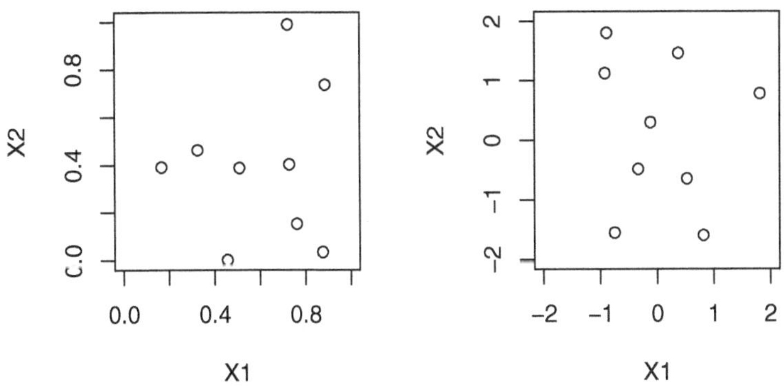


**Figure A.1.** *2D Monte Carlo samples with uniform (left) and Gaussian (right) distributions.*

### Application to integral estimation

The Monte Carlo method is very useful for the numerical approximation of nonexplicit computations. For example, it can be used to solve partial differential equations, linear systems, and optimization problems. The most simple and classical illustration of the method is the numerical computation of the integral $I$ of a function $h$ on the interval $[0, 1]$:

$$I = \int_0^1 h(x)dx.$$

The Monte Carlo method consists first in writing this quantity as

$$I = \mathbb{E}h(X),$$

where $X$ is a random variable with uniform distribution on $[0, 1]$. The numerical estimation of $I$ then follows from the law of large numbers. This fundamental result of probability theory states that if $(X^{(i)})_{i\in\mathbb{N}}$ is a sequence of independent random variables from the uniform distribution on $[0, 1]$, then

$$\frac{1}{n}\sum_{i=1}^{n} h(X^{(i)}) \xrightarrow[n\to+\infty]{\text{a.s.}} \mathbb{E}h(X).$$

In other words, if $x^{(1)}, x^{(2)}, \ldots, x^{(n)}$ are randomly drawn from $[0, 1]$, then for $n$ large enough, $I$ is well approximated by

$$\bar{I} = \frac{1}{n}\sum_{i=1}^{n} h(x^{(i)}).$$

The rate of convergence of a Monte Carlo estimation is given by the central limit theorem, which stipulates that if $Y$ is a random value with mean $\mu$ and finite variance $\sigma^2$, then the distribution of the mean of $n$ i.i.d. random draws $Y^{(i)}$ distributed as this random variable converges in distribution toward a standard Gaussian (when $n \to \infty$). If $\bar{Y} = \frac{1}{n}\sum_{i=1}^{n} Y^{(i)}$, we have

$$\frac{\bar{Y} - \mu}{\sigma/\sqrt{n}} \xrightarrow[n\to+\infty]{\mathcal{D}} \mathcal{N}(0, 1).$$

This theorem makes no assumptions on the form of the distribution of $Y$ except that it has finite variance. It can therefore be directly applied on the output of a numerical model $Y = G(\mathbf{X})$.

The rate of convergence of the Monte Carlo estimator is thus $\frac{\sigma}{\sqrt{n}}$, which can be quite slow. For instance, to improve the precision of the estimation by a factor of 10, the number of calculations has to be multiplied by 100. However, there are two important advantages of the Monte Carlo method that result from this rate of convergence:

- It does not depend on the dimension $d$ of $\mathbf{X}$ (though increasing the number of random variables may increase the standard deviation $\sigma$ of the output).
- It does not depend on the regularity of the function $G$ being integrated, as long as it is square integrable.

Using more efficient sampling methods, alternatives to a Monte Carlo sample, is crucial in many situations. For instance, in uncertainty quantification, the Monte Carlo convergence of estimators needs to be accelerated. In a domain exploration context, a good sampling in the input supports has to be ensured [333]. Indeed, it is expected to maximize the chance to discover particular values and behavior of the system/phenomenon outputs.

## A.2.2 ▪ Latin hypercube sampling

### Main principles

The Latin hypercube sampling (LHS) method [320, 454, 282] consists in placing the sample points in a much more uniform way, across the whole domain of values for each input, than does the crude Monte Carlo. For this, the range interval for each input $X_j$ is subdivided into $N$ segments of equal probability $1/N$; then a value is randomly drawn in each segment. This means that only one random point is found in each segment. The $N$ values thus obtained for $X_1$ are then randomly added to the $N$ found for $X_2$, forming a matrix of size $N \times 2$, and so on for $X_3$, etc., up to the $d$th variable. This algorithm implicitly supposes independence between input variables, though extensions have been developed for performing LHS with dependent variables [203, 366].

We should distinguish LHS from two very similar notions defined below: Latin hypercube designs and centered uniform Latin hypercube samples.

**Definition A.1.** *A Latin hypercube design (LHD) for $N$ simulations and $d$ variables is an $N \times d$ matrix for which each column is a random permutation of the sequence $[1 : N] = \{1, \dots, N\}$. We denote it by* $\mathrm{LHD}(N, d)$.

**Definition A.2.** *A centered (uniform) Latin hypercube sample (cLHS) for $N$ simulations and $d$ variables is an $N \times d$ matrix for which each column (except perhaps the first) is a random permutation of the sequence $\{(1-0.5)/N, \dots, (N-0.5)/N\}$. We denote it by* $\mathrm{cLHS}(N, d)$.

To apply the LHS method, we first generate a design $D = \mathrm{LHD}(N, d)$ with $d$ independent random permutations of $[1 : N]$. Then we do the following:

- A $\mathrm{cLHS}(N, d)$ is obtained by calculating $D' = (D - 0.5J)/N$, where $J$ is an $N \times d$ matrix with all entries equal to 1. This is called *centering*.
- An $\mathrm{LHS}(N, d)$ is obtained by subtracting from each element of $D$ the result of an independent uniform random draw on $[0, 1/N]$. This is called a *randomized* LHS. In the general case of mutually independent random variables $X_1, \dots, X_d$ with invertible cumulative distribution functions $F_{X_1}, \dots, F_{X_d}$, the $i$th sample of the $j$th variable can be constructed by
$$x_j^{(i)} = F_{X_j}^{-1}\left(\frac{\pi_j^{(i)} - \xi_j^{(i)}}{N}\right),$$
where $(\pi_j)_{j=1,\dots,d}$ are independent random permutations of integers $[1 : N]$ and $(\xi_j^{(i)})_{i=1,\dots,N}$ are independent draws from $\mathcal{U}[0,1]$ that are also independent of the $\pi_j$.

### Example in R

The two following examples with $N = 9$ and $d = 2$ considers $\mathcal{U}(0,1)$-distributed cLHS and LHS. It allows us to draw Fig. A.2 (the seed argument in lhsDesign() allows us to reproduce the results):

```
> library(DiceDesign)
> d <- 2 ; N <- 9
> XcLHS <- lhsDesign(N, d, randomized = FALSE, seed = 12345)$design
> XLHS <- lhsDesign(N, d, randomized = TRUE, seed = 12345)$design
> colnames(XcLHS) <- colnames(XLHS) <- c("X1", "X2")

> par(mfrow = c(1,2))
> plot(XcLHS, xlim=c(0,1), ylim = c(0,1), pch = 20)
> abline(h=c(0,seq(N)/N), lty=2) ; abline(v=c(0,seq(N)/N), lty=2)
> plot(XLHS, xlim=c(0,1), ylim = c(0,1), pch = 20)
> abline(h=c(0,seq(N)/N), lty=2) ; abline(v=c(0,seq(N)/N), lty=2)
```

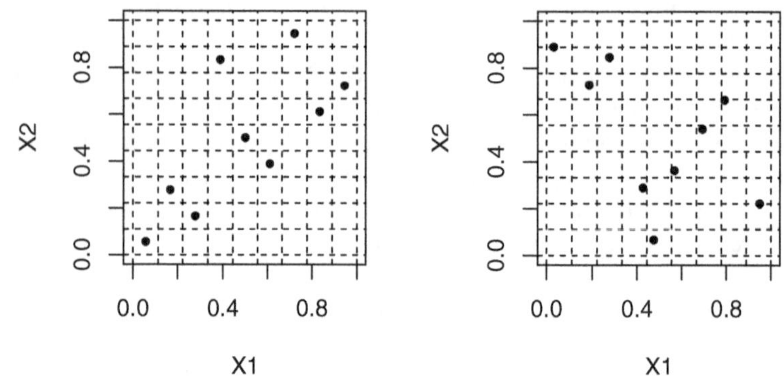


**Figure A.2.** *2D centered (left) and randomized (right) Latin hypercube samples with uniform distributions.*

For the LHS, we can recover the permutations of $[1 : N]$ that have been generated in all the dimensions by

```
> t(( XcLHS * N + 0.5))

    [,1] [,2] [,3] [,4] [,5] [,6] [,7] [,8] [,9]
X1     3    8    2    4    7    1    9    5    6
X2     2    6    3    8    9    1    7    5    4
```

The LHS methods guarantee a uniform coverage across the domain of values for each input so that it avoids an undersampling of some segments and oversampling of others. However, the standard LHS method in no way controls the quality of the joint distribution of samples coming from two or more input variables.

### A.2.3 ▪ Orthogonal arrays

An orthogonal array guarantees a good distribution of observations or simulations when considering jointly several input variables [273, 282]. Its formal definition is given as follows.

**Definition A.3.** *An orthogonal array in dimension d, with q levels, strength $t \leq d$, and index $\lambda$, is a matrix with $n = \lambda q^t$ rows and d columns such that in every $n \times t$ submatrix, each of the $q^t$ possible rows (i.e., the distinct t-uples $(l_1, \ldots, l_t)$, where the $l_i$'s take their values in the set of the q levels), occurs exactly the same number $\lambda$ of times.*

Note that it is possible to make a hybrid construction by combining orthogonal arrays associated with Latin hypercubes (i.e., uniformity of marginals) with those of orthogonal arrays (i.e., uniformity across more than one dimension) [474]. However, this kind of technique imposes some constraints on the sample size $N$ with respect to the dimension $d$ and is therefore more difficult to apply than basic LHS-based techniques in most engineering problems.

**Remark 41.** *Orthogonal arrays can be built in* R *using the* DoE.base, planor, *and* DiceDesign *packages. Orthogonal array LHS can also be obtained via the* lhs *and* DiceDesign *packages.*

## A.3 ▪ Space-filling criteria

We consider a sequence of $N$ points $\mathbf{x}^N = (\mathbf{x}^{(i)})_{i=1,\dots,N}$, which can also be called the input sample or the design of experiments. In this section, for the sake of simplicity, $\mathbf{x}^N \in [0,1]^d$. Each component $x_j$ ($j = 1, \dots, d$) of $\mathbf{x}$ is bounded by minimal and maximal real values, and a simple scaling process is applied to transform the variables in $[0,1]$ if necessary.

### A.3.1 ▪ Discrepancy criteria

Let us first emphasize that uniformity is not ensured via regular grids: Except in dimension 1, a regular discretization of the domain does not create a uniform sample. The discrepancy measure, introduced in the following definition, is a quantitative information quantifying uniformity.

**Definition A.4.** *The discrepancy at the origin of a sequence of points $\mathbf{x}^N$, denoted by $D^*(\mathbf{x}^N)$, is defined as*

$$D^*(\mathbf{x}^N) = \sup_{\mathbf{b}\in[0,1[^d} \left| \frac{1}{N} \sum_{k=1}^{N} \mathbf{1}_{\{\mathbf{x}^{(k)}\in[\mathbf{0},\mathbf{b}]\}} - \mathit{Volume}([\mathbf{0},\mathbf{b}]) \right| , \tag{A.1}$$

*where $[\mathbf{0},\mathbf{b}] = [0,b_1] \times \cdots \times [0,b_d]$ and $\mathbf{0}$ is a vector of zeros of dimension d.*

The geometric interpretation of this definition is obvious. It acts as a comparison between the volumes of all subintervals (attached to the origin) of the domain of the $\mathbf{x}$ (Volume($[\mathbf{0},\mathbf{b}]$)) and the proportion of points contained in these intervals ($\frac{1}{N}\sum_{k=1}^{N} \mathbf{1}_{\{\mathbf{x}^{(k)}\in[\mathbf{0},\mathbf{b}]\}}$). The "sup" norm in the definition allows us to quantify the maximum deviation for a given sequence. The discrepancy is therefore the uniform deviation of the distribution of the sample points from the uniform one. In other words, it measures the irregularity of the sampling distribution.

The discrepancy at the origin is unfortunately not an easily computable criterion due to the $\mathbb{L}^\infty$ norm (with the sup) taken in Eq. (A.1). In order to compute discrepancies, it is therefore necessary to change the norm, and in practice the $\mathbb{L}^2$ norm is chosen [273]. For example, the $\mathbb{L}^2$-discrepancy at the origin is written as

$$D_2^*(\mathbf{x}^N) = \left\{ \int_{[0,1[^d} \left[ \frac{1}{N} \sum_{k=1}^{N} \mathbf{1}_{\{\mathbf{x}^{(k)}\in[\mathbf{0},\mathbf{b}]\}} - \text{Volume}([\mathbf{0},\mathbf{b}]) \right]^2 d\mathbf{b} \right\}^{\frac{1}{2}} . \tag{A.2}$$

Different settings of intervals (based on the origin and corners of the intervals we consider) allow us to simplify Eq. (A.2) [194]. Various $\mathbb{L}^2$-discrepancy criteria are then obtained with easy-to-calculate analytic formulas [127, 128]. We note that the centered $\mathbb{L}^2$-discrepancy (which considers intervals attached to the origin) and the wraparound $\mathbb{L}^2$-discrepancy (which connects each boundary to the opposite one so that intervals cut at the boundary can be continued on the other side) are particularly useful. Indeed, they appear to have robustness properties when considering projections of the sample on lower-dimensional spaces [127, 103, 128].

#### Example in R

With $d = 2$ and $N = 9$, the Monte Carlo and LHS designs (simulated in Sections A.2.1 and A.2.2) are compared in terms of their discrepancy criteria, adding a comparison with those of a full factorial design, created by the following R commands:

```
> library(DiceDesign)
> Xfact <- factDesign(d, 3)$design
> colnames(Xfact) <- c("X1", "X2")
```

We then compute, by using the discrepancyCriteria() function of the DiceDesign package, the centered ("C2") and wraparound ("W2") $\mathbb{L}^2$-discrepancy criteria of the three designs shown in Fig. A.3:

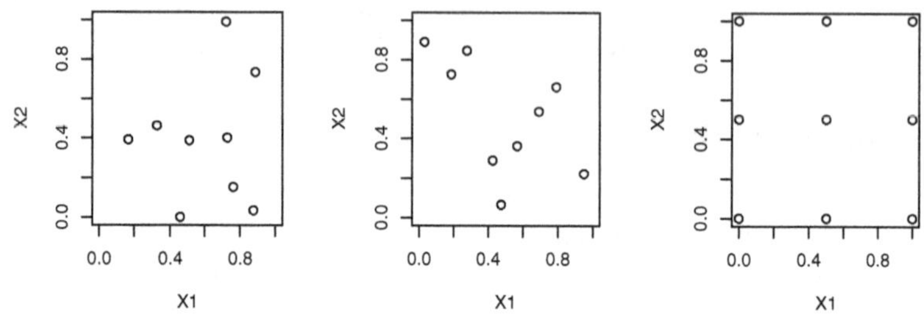


**Figure A.3.** *Designs with $N = 9$ and $d = 2$: Monte Carlo sample (left), Latin hypercube sample (middle), and full factorial design (right).*

```
> discrepancyCriteria(XU, type = c("C2", "W2"))

$DisC2
[1] 0.2082509

$DisW2
[1] 0.2147351

> discrepancyCriteria(XLHS, type = c("C2", "W2"))

$DisC2
[1] 0.0984695

$DisW2
[1] 0.1177514

> discrepancyCriteria(Xfact, type = c("C2", "W2"))

$DisC2
[1] 0.2468943

$DisW2
[1] 0.3888889
```

Results clearly show the nonuniform disposition of the points for both the full factorial design and the Monte Carlo one. Indeed, their discrepancy is larger than that of the LHS.

**Remark 42.** *The* sensitivity *package offers a C++ implementation of the* discrepancyCriteria *function via the* discrepancyCriteria_cplus() *function. It strongly accelerates the computation of discrepancies in cases of large number of points and dimension.*

## A.3.2 ▪ Geometric criteria: Maximin and minimax

The geometric criteria [226] are based on distances between points to measure the quality of an experimental design $(\mathbf{x}^{(i)})_{i=1,\dots,N}$. A particularly interesting criteria, named *minimax*, is based on the maximal distance between any point in the domain and any point in the design. The minimax criteria is denoted $\phi_{\rm mM}(\cdot)$ and given by

$$\phi_{\rm mM}(\mathbf{x}^N) = \max_{\mathbf{x}\in[0,1]^d} \min_{\mathbf{x}^{(i)}} \|\mathbf{x} - \mathbf{x}^{(i)}\|.$$

The euclidean norm is usually used. A low value of $\phi_{\rm mM}$ for a given experimental design means that no point in the domain is too far from a point from the design. For this reason, $\phi_{\rm mM}$ would seem to be particularly appropriate regarding the question of space-filling.

However, minimizing the $\phi_{mM}$ criteria implies being able to compute the distances between any point in the domain and any point in the design. In practice, discretization of the domain allows us to approximate $\phi_{\rm mM}$. But this approximation becomes expensive to compute in more than three dimensions. Therefore, this criterion is not used when building numerical experimental design. Nevertheless, this situation will evolve as recent research finds alternative strategies to compute numerically $\phi_{mM}$ (see [377, 375, 299] for recent results and algorithms).

**Remark 43.** *Several* R *packages propose ways to build minimax designs as the* minimaxdesign *package and the* cover.design() *function of the* fields *package.*

In order to build a space-filling design, a more practical approach is to maximize the minimal distance separating pairs of points in the design. We denote this distance by $d_{ij} = \|\mathbf{x}^{(i)} - \mathbf{x}^{(j)}\|$ and usually take the euclidean norm. The minimal distance criteria, named *mindist* and written $\phi_{\rm Mm}(\cdot)$, is then given by

$$\phi_{\rm Mm}(\mathbf{x}^N) = \min_{i,j=1,\dots,N, i\neq j} d_{ij}.$$

A design $\mathbf{x}^N$ is *maximin* whenever it maximizes $\phi_{\rm Mm}$ and minimizes the number of pairs of points separated by exactly the minimal distance (it is generally not unique). The aim is thus to obtain a design as close as possible to the maximin one. Roughly speaking, a high value of $\phi_{\rm Mm}$ enforces locating the design points as far away from each other as possible (meaning better coverage of the space by the design points).

Figure A.4 (left) shows a maximin design in dimension 2. Note that optimization of the $\phi_{\rm Mm}$ criteria enforces the points to leave the center and move to the boundaries. This occurs especially in higher dimensions. Figure A.4 (right) shows a minimax design in dimension 2. Compared with the maximin design, we see that the points, closer to each other, minimize the volume of empty areas and are farther from the edges.

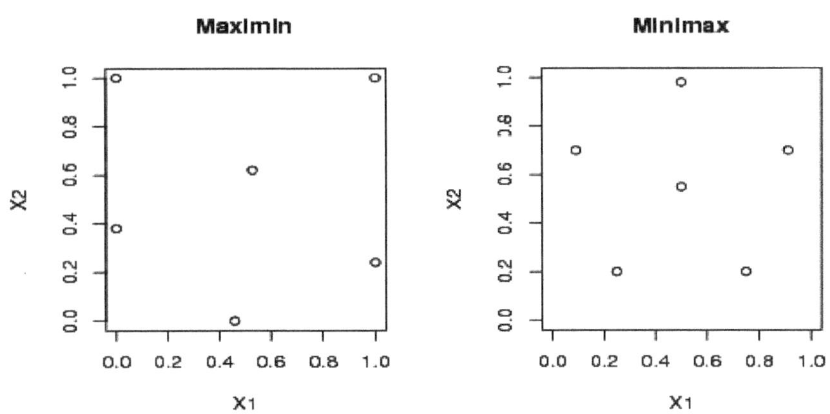


**Figure A.4.** *Maximin (left) and minimax (right) designs with $N = 6$ and $d = 2$.*

#### Example in R

The mindist criterion can be computed via the mindist() function of the DiceDesign package on the three designs shown in Fig. A.3:

```
> print(c(mindist(XU),mindist(XLHS),mindist(Xfact)))

[1] 0.1645076 0.1498919 0.5000000
```

In this case, the mindist criterion is maximal for the full factorial design because of its perfect regular structure in terms of point location.

**Remark 44.** *The sensitivity package offers a C++ implementation of the mindist function via the maximin_cplus() function. It accelerates the mindist computation in cases of large number of points and dimension.*

### A.3.3 ▪ Criteria based on distances between probability distributions

The discrepancy criterion introduced above aims at evaluating how far a sequence of points is from uniformity using volume comparisons. A simple yet powerful generalization consists in performing the same kind of evaluation by looking at the distance between the sequence of points (seen as an empirical probability distribution) and uniformity (that is, the uniform distribution on the hypercube $[0,1]^d$). For the sequence of points $\mathbf{x}^N$, such a criterion takes the general form

$$D_d(\mathbf{x}^N) = d^2\left(\frac{1}{N}\sum_{i=1}^{N}\delta_{\mathbf{x}^{(i)}}, \mathcal{U}[0,1]^d\right), \tag{A.3}$$

where $d(P,Q)$ is any distance between probability distributions $P$ and $Q$. Note that the criterion has been indexed by $d$ to emphasize that it highly depends on the distance choice. The difficult issue is to estimate this distance in high dimensions (curse of dimensionality). Thanks to the kernel trick, the maximal mean discrepancy (MMD)-based criteria are ways to tackle this problem.

#### MMD-based criteria

We focus on the kernel-based distance MMD between two probability measures $P$ and $Q$ [439]:

$$\begin{aligned}\mathrm{MMD}^2(\mathrm{P},\mathrm{Q};\mathcal{H}) = {} & \mathbb{E}_{\xi,\xi'\sim\mathrm{P}}k(\xi,\xi') + \mathbb{E}_{\zeta,\zeta'\sim\mathrm{Q}}k(\zeta,\zeta') \\ & -2\mathbb{E}_{\xi\sim\mathrm{P},\zeta\sim\mathrm{Q}}k(\xi,\zeta).\end{aligned} \tag{A.4}$$

Here, $k$ is the kernel of the RKHS $\mathcal{H}$ (see [34], for example), and all the considered random variables $(\xi,\xi',\zeta,\zeta')$ are independent. Via the kernel trick, the MMD allows us to compare an infinity of features of a P-distributed random variable and a Q-distributed random variable. The space-filling criterion is then

$$D_{\mathcal{H}}(\mathbf{x}^N) = \mathrm{MMD}^2\left(\frac{1}{N}\sum_{i=1}^{N}\delta_{\mathbf{x}^{(i)}}, \mathcal{U}[0,1]^d; \mathcal{H}\right).$$

Interestingly, for some specific kernels $k$, $D_{\mathcal{H}}$ boils down to some discrepancy criteria detailed above (see [193]).

In addition, by choosing $k(x,x') = \frac{1}{2}(\|x\| + \|x'\| - \|x - x'\|)$, the MMD distance (A.4) is equivalent to the energy distance developed by Székely and Rizzo [471, 472] (see [429] for the equivalence theorem) and given by

$$d_E^2(\mathrm{P},\mathrm{Q}) = 2\mathbb{E}\|\xi - \zeta\| - \mathbb{E}\|\xi - \xi'\| - \mathbb{E}\|\zeta - \zeta'\|$$

with $\xi, \xi' \sim \mathrm{P}$, $\zeta, \zeta' \sim \mathrm{Q}$ (all these random variables are assumed to be independent) and by using the euclidean norm. Its interpretation allows us to understand that we compute the mean distance between two (P- and Q-distributed) random variables and subtract from the mean distance between two P-distributed random variables plus the mean distance between two Q-distributed random variables.

It is worth noting that the energy distance has been used in a more general setting to propose the concept of *support points* [300]. Indeed, the energy distance is a kernel discrepancy for nonuniform distributions. Instead of focusing on space-filling designs, which by construction involve the uniform distribution in Eq. (A.3), Mak and Joseph [300] consider any probability distribution to be approximated by a discrete sequence $\mathbf{x}^N$. In particular, their framework readily accommodates empirical distributions as well. In this context, the problem is equivalent to finding a small sample $\mathbf{x}^N$ representative of a much larger sample. This is closely related to the problem of quantization or kernel herding; see also [80].

**Remark 45.** *The MMD criterion is available in* R *via the* kmmd() *function of the* kernlab *package. The energy distance is available via the* edist() *function of the* energy *package and the* e_dist() *function of the* support *package.*

#### Entropy-based criteria

Some authors, such as Jourdan and Franco [228, 229], have proposed using the Kullback–Leibler (KL) divergence to quantify the difference between the density function of the sample $f$ and the uniform density function. This leads to the criterion

$$I_{\mathrm{KL}}(f) = \int f(\mathbf{x}) \ln(f(\mathbf{x}))\, d\mathbf{x},$$

which has to be minimized (it corresponds to maximizing the Shannon entropy) to build a so-called minimal KL information design. However, in this approach, difficulties arise with the classical issues of the density approximation in high dimensions.

## A.4 ▪ Building space-filling designs

### A.4.1 ▪ Low-discrepancy sequences

The so-called quasi-Monte Carlo samples have been proposed to improve the convergence rate of Monte Carlo samples when estimating integrals, replacing the probabilistic error bounds with deterministic ones [273]. The idea is to find deterministic sequences $(\mathbf{x}^{(i)})_{i\geq 1}$ that allow us to approximate integrals using formulas of the form

$$\int_{[0,1]^d} h(\mathbf{x}) d\mathbf{x} \simeq \lim_{N\to+\infty} \frac{1}{N}[h(\mathbf{x}^{(1)}) + \cdots + h(\mathbf{x}^{(N)})].$$

Here, $h$ is a function on the interval $[0,1]$. If $h$ has a certain level of regularity, it is possible to find such sequences (called low-discrepancy sequences) which allow us to provide approximations with a rate of convergence on the order of $\log(N)^d/N$, better than that of the crude Monte Carlo method ($1/\sqrt{N}$). The discrepancy criterion (to be minimized) is then the key element to finding these sequences, which belong in the class of the space-filling designs.

A sequence $(\mathbf{x}^{(i)})_{i\geq 1}$ is said to be uniformly distributed on $[0,1]^d$ if $D^*(\mathbf{x})$ (Eq. (A.1)) tends to zero when $N$ tends to infinity. A sequence is said to be low-discrepancy if its discrepancy is asymptotically better than that of a random sequence. This kind of sequence has the property of uniformly and regularly filling the unit cube. Another feature of low-discrepancy sequences is theoretical in that the error bound for numerical integration of a function with finite variation $V(h)$ in the Hardy and Krause sense [338] is proportional to the discrepancy of the integration design. Unlike random sequences, which lead to confidence intervals for given probabilities, this deterministic upper bound is difficult to compute and often far from the true error.

The construction of some low-discrepancy sequences, based on number theory concepts, is briefly presented in this section. More details can be found, for example, in Lemieux [273]. To begin with, we recall the definition of the $p$-adic decomposition of a given integer.

**Definition A.5.** *For an integer $p > 1$, the p-adic decomposition of the integer $L > 0$ is the sequence $(a_k)_{k=1,\dots,r}$ such that*

$$L = \sum_{k=0}^{r} a_k p^k \text{ , with } 0 \leq a_k < p \;\; \forall 0 \leq k < r \text{ and } 0 < a_r < p.$$

The essence of this decomposition is to write a number $L$ in base $p$ and then to convert this expansion into a number in $[0,1]$. This leads to Van Der Corput sequences.

### Van Der Corput sequences

If we note

$$\xi_p(L) = \sum_{k=0}^{r} \frac{a_k}{p^{k+1}},$$

then $(\xi_p(i))_{i=1,\dots,N}$ is a Van Der Corput sequence of length $N$ in base $p$. This is a 1D sequence, and in practice the most common choice is $p = 2$.

### Halton sequences

Halton sequences [176] are multidimensional generalizations of Van Der Corput sequences. These consist of taking a different base for each dimension. Let $p_1, \dots, p_d$ be the first $d$ prime numbers not less than 2. A Halton sequence of size $N$ is written $(\mathbf{x}^{(i)})_{i=1,\dots,N}$, with

$$\mathbf{x}^{(i)} = (\xi_{p_1}(i), \dots, \xi_{p_d}(i)) \;\; \forall i = 1, \dots, N,$$

where $\left(\xi_{p_j}(i)\right)_{i=1,\dots,N}$ is the Van Der Corput sequence in base $p_j$. Taking these bases within the prime numbers minimizes the discrepancy and leads to uniform sequences [338].

### Example in R

The halton() function of the randtoolbox package in R can be used, for example, to generate a sequence of $N = 100$ points in dimension $d = 10$:

```
> library(randtoolbox) ; library(DiceDesign)
> d <- 10 ; N <- 100
> Xhal <- halton(n = N, dim = d)
> discrepancyCriteria(Xhal, type = "C2")
```

```
$DisC2
[1] 0.1989843

> set.seed(12345) # To reproduce the same results
> X <- matrix(runif(d * N), nrow = N) # Compare with random sample
> discrepancyCriteria(X, type = "C2")

$DisC2
[1] 0.2763681
```

As expected, the Halton sequence discrepancy is lower than that of a Monte Carlo sample. Despite the uniformity of the distribution in the unit cube for smaller-valued bases, the Halton sequences exhibit pathological behavior in higher dimensions. Figure A.5 illustrates this by projecting the previous Halton sequences in coordinates $(1,2)$ (Fig. A.5, left) and $(8,10)$ (Fig. A.5, right). While the $(1,2)$ projection shows a uniform distribution of points, strong alignments of points (along one particular direction) occur in the projection $(8,10)$. This leads to poorly covered areas.

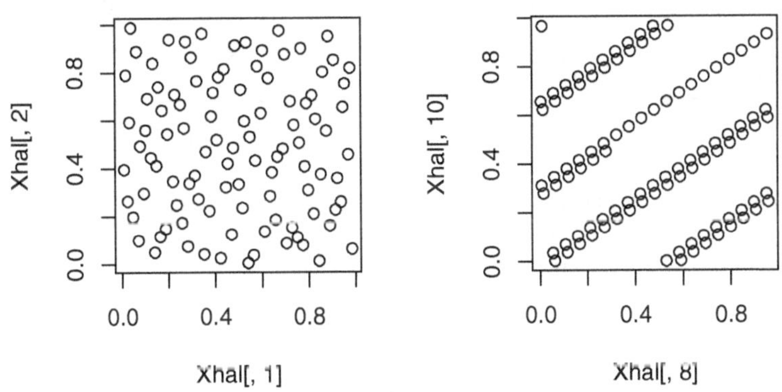


**Figure A.5.** *Two bidimensional projections of the Halton sequence of* 100 *points in dimension* 10*: coordinates* $(1,2)$ *(left) and* $(8,10)$ *(right).*

### Other sequences

The Hammersley sequences [177] are a slight variation of the Halton ones and have the same pathologies. To overcome the pathological problems on subprojections, more sophisticated sequences have been proposed. For example, the Faure sequences [129] (available in the R package DiceDesign) are based on the same principles as the Halton ones but use only one prime number for basis with a permutation rule for each dimension.

Sobol' sequences [440] are considered to be among the best low-discrepancy sequences (along with Niederreiter sequences [338], which have better asymptotic properties) due to their solid theoretical properties and their good performance in numerical trials when the dimension increases. Sobol' sequences are constructed using linear recurrence in modulo 2 arithmetic on the finite field $\mathbb{Z}_2 = \{0,1\}$ and primitive polynomials.

### Example in R

A Sobol' sequence can be generated in R using the sobol() function of the randtoolbox package ($N = 100$, $d = 10$ in the following example):

```
> Xsob <- sobol(n = N, dim = d)
> discrepancyCriteria(Xsob, type= "C2")

$DisC2
[1] 0.1530425
```

The Sobol' sequence discrepancy is lower than the one of the Halton sequence. Figure A.6 shows the same projections as those with the Halton sequence (Fig. A.5): Alignments of points have disappeared. However, Sobol' sequences are also known to present subprojection pathologies in high dimensions. In order to avoid this behavior, scrambling techniques can also be applied [347, 273].

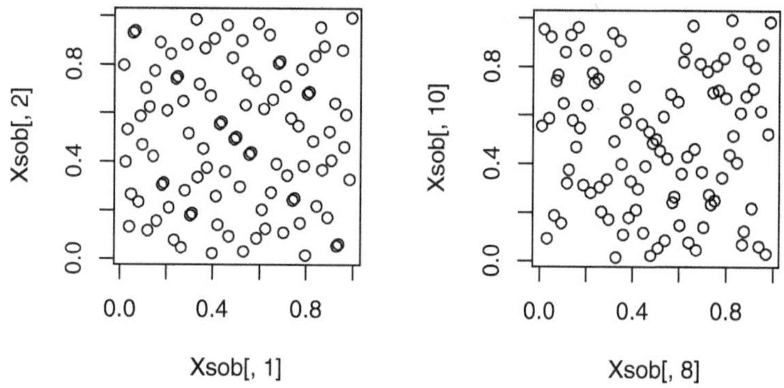


**Figure A.6.** *Two bidimensional projections of the Halton sequence of* 100 *points in dimension* 10*: coordinates* (1, 2) *(left) and* (8, 10) *(right).*

## A.4.2 ▪ Optimized Latin hypercube sampling

Another way to obtain space-filling samples is to develop design optimization strategies with respect to a well-tailored space-filling criterion (for instance, discrepancy or maximin; see Section A.3). In most of the algorithms, the optimization is performed in the set of all LHS (see [200, 489], for example). Indeed, it allows us to ensure uniform projection on each dimension (a desired property in design of numerical experiments), while a constraint-free optimization process would break this property.

Columnwise-pairwise exchange, simulated annealing, and genetic algorithms have been proposed in the literature. The application of a simulated annealing algorithm to optimize LHS has been proposed in Morris and Mitchell [330]. It is deeply studied in Damblin et al. [103]. It is known to be efficient and rather simple to use (in terms of tuning parameters choice). It allows us to obtain either

1. a maximin LHS, by maximizing the mindist criteria [330] (in fact by minimizing an equivalent criterion much more easy to optimize); or

2. a low-discrepancy LHS by minimizing an $\mathbb{L}^2$-discrepancy [223].

Of course, other criteria can be considered, but some practical studies suggest using Latin hypercubes with wraparound $\mathbb{L}^2$-discrepancy or low centered $\mathbb{L}^2$ discrepancy (for their robustness on subprojections of dimensions greater than or equal to 2) [127, 103].

### Example in R

The following R commands apply the LHS simulated annealing algorithm by minimizing the wraparound $\mathbb{L}^2$-discrepancy. The tuning parameters are the initial temperature T0, the temperature decrease rate c, and the total number of iterations it:

```
> library(DiceDesign)
> d <- 2 ; N <- 9
> XLHS <- lhsDesign(N, d, seed = 12345)$design
> Design <- discrepSA_LHS(XLHS,T0=1,c=0.99,it=1000,criterion="W2")
> XLHSopt <- Design$design
> discrepancyCriteria(XLHSopt, type = c("C2", "W2"))

$DisC2
[1] 0.07943895

$DisW2
[1] 0.1113406

> par(mfrow = c(1,2))
> plot(XLHSopt, xlim = c(0,1), ylim = c(0,1))
> plot(Design$critValues, type = "l", ylab = "Discrepancy")
```

As expected, the discrepancy criteria of the optimized LHS are smaller than those of the random LHS. Figure A.7 (left) gives the design point arrangement of the optimized LHS (to be compared to the ones in Fig. A.3). Figure A.7 (right) illustrates the convergence of the optimized criterion.

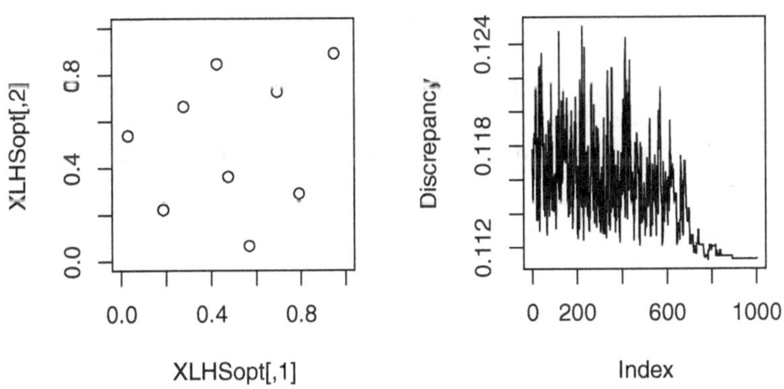


**Figure A.7.** *Left: optimized centered $\mathbb{L}^2$-discrepancy Latin hypercube sample ($N = 9$ and $d = 2$). Right: evolution of the optimized criterion along the simulated annealing algorithm iterations.*

We now compare designs in dimension 10 and size 100:

```
> d <- 10 ; N <- 100
> XLHS <- lhsDesign(N, d, seed = 12345)$design
> Design <- discrepSA_LHS(XLHS,T0=1,c=0.99,it=5000,criterion="C2")
> Xopt <- Design$design
> discrepancyCriteria(Xopt, type = "C2")

$DisC2
[1] 0.1348863
```

The discrepancy of the optimized LHS is lower than those of the Sobol' sequence obtained in the previous section.

**Remark 46.** *The* lhs *package of* R *proposes some algorithms to build optimized LHS as the* maximinLHS() *function to build maximin LHS. However, this function is based on a sequential point addition algorithm [25] that results in poor optimized designs compared to the ones obtained by nonsequential optimization algorithms. It is illustrated below by taking the default values for the tuning parameters of each algorithm and then comparing the* mindist criterion *of each design:*

```
> d <- 10 ; N <- 100
> library(DiceDesign)
> XLHS <- lhsDesign(N, d, seed = 12345)$design
> Design <- maximinSA_LHS(XLHS)
> XmaxSA <- Design$design
> library(lhs)
> Xbuild <- maximinLHS(N, d, method = "build")
> Xiter <- maximinLHS(N, d, method = "iterative")
> print(c(mindist(XLHS),mindist(XmaxSA),
+          mindist(Xbuild),mindist(Xiter)))

[1] 0.3893240 0.7465982 0.5031989 0.5788820
```

## A.4.3 ▪ Maximum projection designs

Low wraparound and centered $\mathbb{L}^2$-discrepancy LHS are known to be robust when projected on low-dimensional subspaces [127, 103]. However, the maximin LHS does not ensure such properties, as numerically shown by Damblin et al. [103]. Joseph et al. [227] have therefore proposed a geometric criterion that takes into account the distances between points in all the subprojection spaces. The maximum projection (denoted *maxpro)* criterion (that has to be minimized) is

$$\phi_{\text{maxpro}}(\mathbf{x}^N) = \left[ \binom{N}{2}^{-1} \sum_{i=1}^{N-1} \sum_{j=i+1}^{N} \frac{1}{\prod_{l=1}^{d} (x_l^{(i)} - x_l^{(j)})} \right]^{1/d}.$$

Joseph et al. [227] have then proposed some optimization algorithms (such as the simulated annealing one) to build *maxpro*-optimized designs and maxpro LHS (LHS optimizing the *maxpro* criterion).

### Example in R

The MaxPro package of R is devoted to maxpro-type criteria and designs:

```
> set.seed(12345) # To reproduce the same results
> library(MaxPro)
> d <- 10 ; N <- 100
> XMP <- MaxProLHD(N, d, itermax = 100, total_iter = 1e4)
> MaxProMeasure(XMP$Design)

[1] 44.10265
```

```
> print(c(MaxProMeasure(Xopt),MaxProMeasure(Xsob),
+         MaxProMeasure(XLHS)))
```

```
[1]  61.50675  73.99304 109.18538
```

As expected, the *maxpro* criterion of the maxpro LHS is the best among the ones of the other designs. The centered $\mathbb{L}^2$-discrepancy optimized LHS has also a better maxpro criterion than those of the Sobol' sequence and the random LHS.

### A.4.4 ▪ Sequential space-filling designs

Sequential construction of a design can be useful in order to add points in the design when the results reveal something lacking or when an additional computational budget is available after a first study [432, 430]. Other practical situations can benefit from a sequential design, e.g., for adding test points [205].

Low-discrepancy deterministic sequences introduced in Section A.4.1 are sequential, but their space-filling pathologies are often annoying. Iterative constructions of Sobol' sequences have also been discussed in Gilquin et al. [162].

To build sequential space-filling designs, different empirical strategies have been proposed in Iooss et al. [205], Crombecq et al. [91], and Li et al. [280]. Shang and Apley [430] have deeply improved the CADEX algorithm of Kennard and Stone [234]. Also called the "coffee-house design" [336], CADEX is a sequential selection algorithm of points (inside a database) which are as far away from previously selected points as possible. The three algorithms proposed by Shang and Apley [430] are applied in the package FSSF of R.

Developed for different purposes, nested SFD [381, 383] and sliced SFD [384] can also be used for building sequential designs. For example, Sheikholeslami and Razavi [432] have developed the progressive LHS from the concept of sliced LHS [382, 17].

Finally, more theoretical insights on building sequential space-filling samples can be found in Abtini et al. [1] and Pronzato and Zhigljavsky [378].

## A.5 ▪ Design classification using minimum spanning tree

The minimum spanning tree (MST) [121] has been introduced for studying SFD in Franco et al. [141] in order to analyze a design according to the distances between its points. Regarding design points as vertices, an MST is a tree that connects all the vertices together and whose sum of edge lengths is minimal. Once the MST of a design is built, the mean and standard deviation of the edge lengths can be calculated. As it focuses on global characteristics of the design arrangement, the MST is more robust (to particular point locations) than other geometric criteria such as the mindist one. However, it cannot be used for design building due to its much larger evaluation cost.

#### Example in R

To provide an illustration, the MST is applied by using the DiceDesign package of R on two designs with $N = 9$ and $d = 2$: the optimized wraparound $\mathbb{L}^2$-discrepancy LHS (Section A.4.2) and the Monte Carlo sample (Section A.2.1):

```
> library(DiceDesign)
> mstLHSopt <- mstCriteria(XLHSopt)
> mstLHSopt$stats
```

```
[1] 0.30102486 0.03447905
```

```
> mstU <- mstCriteria(XU)
> mstU$stats
```

```
[1] 0.25254151 0.07718965
```

The results of the argument `stats` are the mean and standard deviation of the MST edge lengths. The optimized LHS shows a larger mean and a smaller standard deviation than the random design, which proves the more regular distribution of its points. This is confirmed by Fig. A.8, which displays the MST of each design.

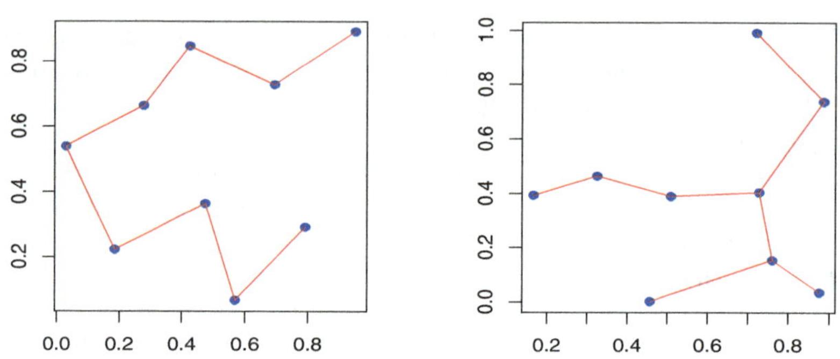


**Figure A.8.** *Minimum spanning tree of designs with* $N = 9$ *and* $d = 2$. *Optimized Latin hypercube sample (left) and random sample (right).*

The MST has been mainly used in computer experiments for design classification [141, 419, 103], as shown by Fig. A.9. Designs described as quasi-periodic are characterized by large mean and small standard deviation compared with random designs or nonoptimized LHS. Such quasi-periodic designs, which correspond to SFD, fill the space efficiently from the point-distance perspective: Large mean is related to large interpoint distance, and small standard deviation means that the minimal interpoint distances are similar.

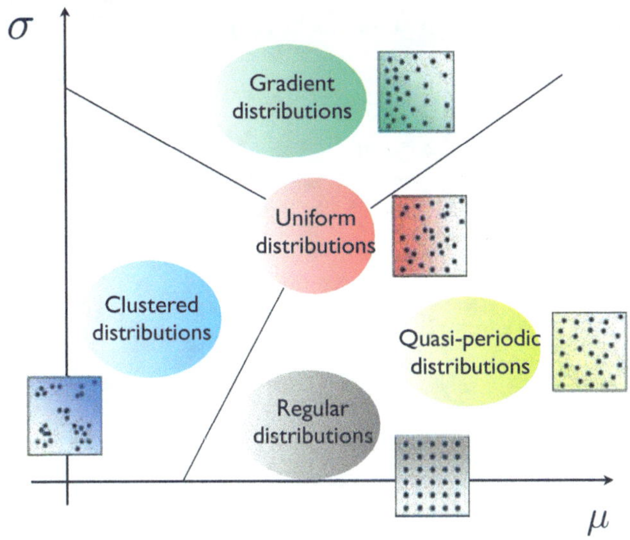


**Figure A.9.** *Classification of the different design classes as a function of the minimum spanning tree edge lengths mean $\mu$ and standard deviation $\sigma$. Uniform (resp., quasic-periodic) distributions correspond to uniform Monte Carlo (resp., space-filling) designs. Courtesy of Laurent Gilquin (IFPEN) and inspired from Franco et al. [141].*

# Appendix B
# Reproducing kernel Hilbert space

Let $\mathcal{X}$ be an arbitrary set and $\mathcal{H}$ a Hilbert space of real-valued functions $f : \mathcal{X} \to \mathbb{R}$ on $\mathcal{X}$ with inner product $\langle\cdot,\cdot\rangle_{\mathcal{H}}$ (denoted $\langle\cdot,\cdot\rangle$ in what follows). For every $x \in \mathcal{X}$, we define the *evaluation functional* $L_x : \mathcal{H} \to \mathbb{R}$ as $f \mapsto L_x(f) = f(x)$.

**Definition B.1.** *A Hilbert space $\mathcal{H}$ is a reproducing kernel Hilbert space (RKHS) if the evaluation functionals are continuous.*

More details on RKHS can be found in the monograph of Berlinet and Thomas-Agnan [34].

**Definition B.2.** *A function $k : \mathcal{X} \times \mathcal{X} \to \mathbb{R}$ is a kernel if*

1. *$k$ is symmetric, i.e., for all $x, x' \in \mathcal{X}$, $k(x, x') = k(x', x)$;*
2. *$k$ is positive definite, i.e., for any $n \in \mathbb{N}$, $x_1, \dots, x_n \in \mathcal{X}$ and $\alpha_1, \dots, \alpha_n \in \mathbb{R}$,*

$$\sum_{i,j=1}^{n} \alpha_i \alpha_j k(x_i, x_j) \geq 0.$$

**Definition B.3.** *The reproducing kernel of an RKHS $\mathcal{H}$ is a function $k : \mathcal{X} \times \mathcal{X} \to \mathbb{R}$ such that*

1. *for all $x \in \mathcal{X}$, $k(x, \cdot) \in \mathcal{H}$;*
2. *for all $f \in \mathcal{H}$ and for all $x \in \mathcal{X}$, $\langle f, k(x, \cdot)\rangle = f(x)$ (reproducing property).*

*In particular, the reproducing kernel is a kernel; i.e., it is symmetric and positive definite.*

**Theorem B.4.** *Every symmetric positive definite kernel defines a unique RHKS, of which it is the reproducing kernel.*

**Theorem B.5 (Mercer; see [10]).** *Suppose $k$ is a symmetric positive definite kernel on a compact set $\mathcal{X}$, and consider the integral operator $T_k : \mathbb{L}^2(\mathcal{X}) \to \mathbb{L}^2(\mathcal{X})$ defined by*

$$(T_k f)(x) = \int_{\mathcal{X}} k(x, u) f(u) du.$$

*Then there is an orthonormal basis $\{e_i\}$ of $\mathbb{L}^2(\mathcal{X})$ consisting of eigenfunctions of $T_k$ such that the corresponding sequence of eigenvalues $\{\lambda_i\}$ is nonnegative. The eigenfunctions corresponding* *to nonzero eigenvalues are continuous on $\mathcal{X}$, and $k$ has the representation*

$$k(x,x') = \sum_{i=1}^{\infty} \lambda_i e_i(x) e_i(x'),$$

*where the convergence is absolute and uniform.*

**Theorem B.6 (representer's theorem; see [435]).** *Let $\mathcal{H}$ be an RKHS with kernel $k$, and consider the minimization problem*

$$\min_{f\in\mathcal{H}} \sum_{i=1}^{n} l(f(x_i), y_i) + \lambda \|f\|^2,$$

*where $l(\cdot,\cdot)$ is any loss function and $\lambda > 0$. Then if $l(\cdot,\cdot)$ is convex with respect to its second argument, the solution of the minimization problem is unique and written as*

$$f^*(x) = \sum_{i=1}^{n} a_i k(x_i, x).$$

**Definition B.7.** *Let $\xi$ denote a random variable on $\mathcal{X}$ with probability distribution $\mathrm{P}_\xi$ and $\mathcal{H}$ an RKHS with kernel $k$. The kernel embedding of the distribution $\mathrm{P}_\xi$ is the function in $\mathcal{H}$ given by the kernel mean*

$$\mu_{\mathrm{P}_\xi} = \mathbb{E}_\xi\, k(\xi, \cdot).$$

**Definition B.8.** *A kernel is characteristic if the kernel embedding $\mu : \mathcal{M}_1^+ \to \mathcal{H}$ is injective.*

## Appendix C

# Toy functions

## C.1 ▪ Arctangent temporal function

The arctangent temporal function [11] gives a 1D functional output toy function. It is defined by the equation

$$G(t, \mathbf{X}) = \arctan(X_1)\cos(t) + \arctan(X_2)\sin(t) ,$$

where $(X_1, X_2) \in [-7, 7]^2$ and $t$ represents the time which is regularly sampled in the interval $[0, 2\pi]$.

## C.2 ▪ Bratley model

This analytical function, introduced in Bratley et al. [52], is defined as

$$G(\mathbf{x}) = G(x_1, \ldots, x_d) = \sum_{i=1}^{d}(-1)^i \prod_{j=1}^{i} x_j ,$$

with $x_i \in [0, 1]$ for $i = 1, \ldots, d$.

## C.3 ▪ HeterDisc function

This analytical function is defined as

$$G(\mathbf{x}) = \begin{cases} \sin\left(\pi\frac{X_1}{5}\right) + \frac{2}{10}\cos\left(4\pi\frac{X_1}{5}\right) + \frac{X_3}{100}(X_2 - 10) \text{ if } X_1 < 10 , \\ \frac{X_3}{100}(X_2 - 10) + \frac{1}{10}(X_4 - 20) \text{ if } X_1 \geq 10 , \end{cases}$$

with $x_i \in [0, 20]$ for $i = 1, \ldots, 4$.

## C.4 ▪ Ishigami function

We consider the Ishigami function currently used in sensitivity studies [414],

$$G(\mathbf{X}) = \sin(X_1) + 7\sin^2(X_2) + 0.1X_3^4\sin(X_1) ,$$

with $x_i \in [-\pi, \pi]$ for $i = 1, \ldots, 3$.

## C.5 ▪ Linkletter decreasing function

The Linkletter decreasing test function [285] is defined by the equation

$$G(\mathbf{X}) = \sum_{i=1}^{8} \frac{0.2}{2^{i-1}} X_i \,,$$

with $x_i \in [0,1]$ for $i = 1, \ldots, d$. Of course, $X_9$ and $X_{10}$ are dummy variables in the Linkletter decreasing function.

## C.6 ▪ Morris function

The Morris test function [329] is defined by the equation

$$G(\mathbf{X}) = \beta_0 + \sum_{i=1}^{20} \beta_i w_i + \sum_{1=i<j}^{20} \beta_{i,j} w_i w_j + \sum_{1=i<j<l}^{20} \beta_{i,j,l} w_i w_j w_l + \sum_{1=i<j<l<s}^{20} \beta_{i,j,l,s} w_i w_j w_l w_s \,,$$

where $w_i = 2(x_i - \frac{1}{2})$ except for $i = 3, 5, 7$, and where $w_i = 2\big(1.1\frac{x_i}{x_i+0.1} - \frac{1}{2}\big)$ and $x_i \in [0,1]$ for $i = 1, \ldots, 20$. The coefficient values are $\beta_i = 20$ for $i = 1, 2, \ldots, 10$, $\beta_{i,j} = -15$ for $i, j = 1, 2, \ldots, 6$, $i < j$, $\beta_{i,j,l} = -10$ for $i, j, l = 1, 2, \ldots, 5$, $i < j < l$, and $\beta_{1,2,3,4} = 5$. Each of the remaining zero-order ($\beta_0$), first-order ($\beta_i$), and second-order ($\beta_{i,j}$) coefficients is obtained from a simulation of an $\mathcal{N}(0,1)$-variable. Each of the remaining third-order ($\beta_{i,j,l}$) and fourth-order ($\beta_{i,j,l,s}$) coefficients is worth zero.

## C.7 ▪ Sobol' $g$-function

We consider the analytical multiplicative function introduced in Sobol' [442], the Sobol' $g$-function, depending on $d$ variables $(x_1, \ldots, x_d)$,

$$G(\mathbf{x}) = \prod_{i=1}^{d} g_i(x_i) \,,$$

with $x_i \in [0,1]$, $g_i(x_i) = \frac{|4x_i - 2| + a_i}{1 + a_i}$, and $a_i > 0$ for any $i = 1, \ldots, d$. In this book, the numerical tests with the Sobol' $g$-function are applied with $a_1 = 0$, $a_2 = 1$, $a_3 = 4.5$, $a_4 = 9$, $a_i = 99$ for any $i = 5, \ldots, 8$ and $g_i(x_i) = 0$ for $i > 8$ (in case $d > 8$).

# Appendix D
# Software

Devoted to parametric/exploratory studies, design/optimization issues, risk assessment, or UQ of computer models, numerous software has been developed by several institutes and researchers (e.g., DAKOTA,[8] UQLab,[9] Uranie,[10] Lagun[11]). Most of these platforms contain an SA module, while several software packages are specialized on SA (e.g., SALib,[12] SAFE[13]). In the R environment, many packages are devoted to some SA issues. This appendix first lists some of them, focusing on the one developed by the authors of this book (the sensitivity package[14]). Then the platform OpenTURNS[15] for UQ and robust optimization studies, codeveloped by the company of one author of this book, is presented.

## D.1 ▪ Packages for sensitivity analysis in the R environment

### D.1.1 ▪ The sensitivity package

For global SA, the sensitivity package [206] includes a collection of functions for factor screening, sensitivity indices estimation, and reliability sensitivity analysis of model outputs. It implements (in version 1.24.0)

- a few screening techniques such as the sequential bifurcations (see Section 2.1.3) and the Morris method (see Section 2.1.4);
- the main sampling-based procedures such as linear regression coefficients, partial correlation coefficients, semipartial correlation coefficients, rank transform–based coefficients, and general dominance analysis measures LMG and PMVD (see Section 2.2.2);
- the distributed evaluation of local sensitivity analysis (see Section 2.3.2),
- the variance-based sensitivity indices (Sobol' indices), by various schemes of the so-called pick-freeze method (see Section 3.2) and replicated orthogonal array–based Latin hyper-

[8]See https://dakota.sandia.gov/.
[9]https://www.uqlab.com/
[10]https://sourceforge.net/projects/uranie/
[11]https://gitlab.com/drti/lagun
[12]See https://salib.readthedocs.io/en/latest/.
[13]See https://www.safetoolbox.info/.
[14]See https://cran.r-project.org/web/packages/sensitivity/index.html.
[15]See https://openturns.github.io/www/.

cube sample approach (see Section 3.2), and the extended-FAST method (see Section 4.1), etc.;

- the estimation of the Sobol' indices with a Gaussian process metamodel coming from the R package DiceKriging (see Section 4.2.3);
- the Poincaré constants (see Section 3.1.4) and generalized coefficients for the derivative-based global sensitivity measures;
- the Shapley effects by using the direct estimation algorithms (see Section 5.5), the nearest-neighbor or ranking technique, and efficient implementations in the linear Gaussian framework;
- Sobol' indices for multidimensional outputs: aggregated Sobol' indices (see Section 3.3.2) and functional (1D) Sobol' indices (see Section 3.3.1);
- the sensitivity indices based on Csiszar $f$-divergence (see Section 6.1.2) and Hilbert–Schmidt independence criterion (see Section 6.2.1), also relevant for multidimensional outputs;
- the first-order quantile-oriented sensitivity indices (see Section 6.3.3);
- several weight transformation functions of the output sample in order to perform target sensitivity analysis (see Section 6.3.1);
- the robustness analysis by the perturbed law–based indices (see Section 6.4.1);
- the support index functions of Fruth et al. [145].

### D.1.2 ▪ Other packages

Other packages, focusing on a few particular methods, have been developed on SA in R:

- The planor package allows one to build in fractional factorial design (see Section 2.1.2).
- The FME package contains basic SA and local SA methods, while the spartan package contains basic methods for exploring stochastic numerical models.
- The ATmet, pse, and ODEsensitivity packages interface several sensitivity package functions for, respectively, metrology applications, parameter space exploration, and sensitivity analysis in ODE models.
- The multisensi package allows one to apply sensitivity analyses to numerical models with multivariate output, while the package safi implements SA methods of models with functional inputs.
- The fast package is fully devoted to the FAST method, while the sensobol package provides rapid computations of first, total, and, if desired, second- and third-order Sobol' indices.
- The GPC package computes Sobol' indices using the polynomial chaos metamodel (see Section 4.2.3), while the tgp package uses the treed Gaussian process metamodel.
- The mistral package implements the FORM importance factors for sensitivity studies of failure probability computation (see Section 6.3.2).

As a global and generic platform allowing one to include all the methods of these different R packages, the mtk package [496] is also available. It is an object-oriented framework which aims at dealing with external simulation platforms and managing all the different tasks of uncertainty and sensitivity analyses.

## D.2 ▪ A generic platform for uncertainty treatment: OpenTURNS

### D.2.1 ▪ Introduction

The needs to assess robust performances for complex systems and to answer tighter regulatory processes (security, safety, environmental control, health impacts, etc.) have led to the emergence of a new industrial simulation challenge: to take uncertainties into account when dealing with complex numerical simulation frameworks [438]. Many attempts at treating uncertainty in industrial applications have involved different mathematical approaches and standards (sometimes domain specific): metrology, structural reliability, variational approaches, design of experiments, global sensitivity analysis, machine learning approaches, etc. However, facing the questioning of their certification authorities in an increasing number of different domains, a generic methodology has emerged from the joint effort of several industrial companies and academic institutions [360]. The specific organizational challenges attached are transparency (with respect to safety authorities), genericity (multiapplicative domain issues), modularity (easy integration from the open-source community), multiaccessibility (different levels of use and users), and industrial computing capabilities.

As no software was fully answering the mathematical techniques and organizational challenges mentioned above, EDF R&D,[16] Airbus Group,[17] and Phimeca Engineering[18] started a collaboration at the beginning of 2005, joined by IMACS[19] in 2014 and ONERA[20] in 2019, for the development of an open-source software platform dedicated to uncertainty propagation by probabilistic methods, named OpenTURNS, for open-source treatment of uncertainty, Risk 'N Statistics [24].

### D.2.2 ▪ Algorithms

Based on a probabilistic representation of the model input variables, OpenTURNS aims to integrate all the useful algorithms for

- the probabilistic modeling, including the dependence structure via the copula theory;
- the statistical sampling tools, including various numerical designs of experiments;
- the uncertainty propagation and quantification step;
- the sensitivity analysis step;
- the metamodeling solution;
- the parameter calibration and model validation techniques;
- the robust optimization issues.

[16]Research and Development Division of the company Electricité de France, https://www.edf.fr/.

[17]European company in aerospace, https://www.airbus.com/.

[18]Research and engineering company in optimization and reliability of products and structures, https://www.phimeca.com/.

[19]Research and technology company on scientific computing and mathematical engineering, https://imacs.polytechnique.fr/.

[20]French research institute in aerospace, https://www.onera.fr/en.

All these algorithms (see [438] for a comprehensive review) can be considered in a broad class of techniques called "simulation analytics" [337], which refers to mathematical and statistical tools for analyzing simulation ensembles and extracting meaningful information from simulation results.

The quantification of uncertainties on scalar inputs can use parametric (e.g., maximum likelihood or method of moments) or nonparametric methods (e.g., kernel smoothing). The quantification of uncertainties on a scalar output may be performed by evaluating the mean, variance, or a probability of exceeding a threshold. The library provides iterative algorithms to estimate the quantity of interest based on a predefined accuracy. When the input or output is a stochastic process, the library provides dimension reduction features (e.g., Karhunen–Loève) and a comprehensive collection of covariance models.

Most common SA algorithms are provided in OpenTURNS, including those based on linear regression, e.g., Pearson/Spearman correlations and SRC/SRRC, or based on Taylor expansion. Sobol' indices can be estimated with four estimators including Saltelli, Jansen, Mauntz–Kucherenko, and Martinez. Moreover, the distribution of the estimators is computed by either bootstrap or with the asymptotic distribution. The indices can be estimated with an incremental algorithm which is based on an accuracy defined by the user. The library can compute the FAST and ANCOVA indices. For reliability problems, several sensitivity indices are provided, including importance factors and Hasover indices. Screening can be performed based on the Morris method. In order to experiment with new algorithms on 30 benchmark problems with reference solutions, the otbenchmark module can help.

The library provides several types of designs of experiments, starting from a simple random sample from any random vector (including dependencies). The library provides quasi-Monte Carlo sequences, LHS, and optimized LHS using two different algorithms (Monte Carlo and simulated annealing) and discrepancy and geometric criteria. Stratified designs of experiments and Gaussian quadratures designs are provided, with arbitrary marginal distributions. Graphical features allow one to plot these multidimensional designs.

Within the OpenTURNS library, future improvements concern the nonlinear optimization features, a building block for several algorithms, including, for example, computer calibration. New developments are also planned in order to provide reliability algorithms based on kriging and various sampling techniques (Monte Carlo, importance sampling, and subset sampling). Regarding SA, future algorithms will propose the Shapley effects, the $F$-divergence-based indices, the PLI, and the HSIC.

### D.2.3 ▪ Integration and community

OpenTURNS is supported by its core team and its user community via `www.openturns.org`. With hundreds of classes, approximately 200,000 lines of code, and hundreds of users, the project needed a new communication system. Starting in 2020, a new forum, a chat, and a Stackoverflow tag were created for users to share their knowledge.

At EDF, OpenTURNS is the reference software on UQ issues and for methodological dissemination in the business units and engineering operational departments. Numerous efforts have been made for its integration into the various computing environments. It is thus integrated into the SALOME platform [408], allowing users to couple many field-physics models. Since 2017, OpenTURNS has facilitated the coupling of any system model thanks to the use of the Functional Mock-Up Interface (FMI) standard, an API widely used by most modeling languages, such as Modelica.

The OpenTURNS Python module, built on top of the C++ library, is the interface that most users commonly know in practice. However, there are situations where a UQ study has to be realized without using a programming language. This is why Persalys,[21] the graphical user interface (GUI) of OpenTURNS, has been developed [23]. Its main goal is to increase the use of OpenTURNS and, more generally, of the UQ methodology in industrial practice. Based on the Paraview framework, some advanced graphical features for sensitivity analysis are also available in the GUI:

- the interactive plot matrix view;
- the parallel coordinate plot (also known as the cobweb plot);
- the functional boxplot, which allows a joint interactive statistical analysis of functional outputs (as thermal-hydraulic transient outputs, for example; see [396]).

Finally, the interface can be used within the HPC context, with only a limited input from the user. The integration of the graphical interface within SALOME will also improve, e.g., in order to ease the management of distributed computations on clusters, even in cases of failures of the computer code or the hardware.

[21] See `https://persalys.fr`.

# Bibliography


[1] M. Abtini, C. Helbert, F. Musy, L. Pronzato, and M.-J. Rendas. Sequential design for prediction with Gaussian process models. *Preprint*, 2020, https://hal.archives-ouvertes.fr/hal-02471110/. (Cited on p. 245)

[2] J. Aitchison and C.G.G. Aitken. Multivariate binary discrimination by the kernel method. *Biometrika*, 63(3):413–420, 1976. (Cited on p. 43)

[3] A. Alexanderian, P. A. Gremaud, and R. C. Smith. Variance-based sensitivity analysis for time-dependent processes. *Reliability Engineering & System Safety*, 196:106722, 2020. (Cited on pp. 87, 188)

[4] H. Andersson and T. Britton. *Stochastic Epidemic Models and Their Statistical Analysis*, Springer Science & Business Media, 2012. (Cited on p. 187)

[5] B. Ankenman, R.C.H. Cheng, and S.M. Lewis. Screening for dispersion effects by sequential bifurcation. *ACM Transactions on Modeling and Computer Simulation*, 25:1, 2014. (Cited on p. 23)

[6] B. Ankenman, B.L. Nelson, and J. Staum. Stochastic kriging for simulation metamodeling. *Operations Research*, 58:371–382, 2010. (Cited on p. 91)

[7] A. Antoniadis. Analysis of variance on function spaces. *Math. Operationsforsch. u. Statist., Ser. Statist.*, 15:59–71, 1984. (Cited on pp. 49, 90)

[8] A. Antoniadis, S. Lambert-Lacroix, and J-M. Poggi. Random forests for global sensitivity analysis: A selective review. *Reliability Engineering & System Safety*, 206:107312, 2021. (Cited on p. 3)

[9] S.K. Au. Reliability-based design sensitivity by efficient simulation. *Computers & Structures*, 83:1048–1061, 2005. (Cited on p. 171)

[10] J.P. Aubin. *Applied Functional Analysis*, 2nd ed. Wiley-Interscience, New York, 2000. (Cited on p. 249)

[11] B. Auder. *Classification et modélisation de sorties fonctionnelles de codes de calcul.* Ph.D. thesis, University Paris VI, France, 2011. (Cited on p. 251)

[12] B. Auder, A. de Crécy, B. Iooss, and M. Marquès. Screening and metamodeling of computer experiments with functional outputs. Application to thermal-hydraulic computations. *Reliability Engineering & System Safety*, 107:122–131, 2012. (Cited on pp. 10, 18, 78)

[13] B. Auder and B. Iooss. Global sensitivity analysis based on entropy. In S. Martorell, C. Guedes Soares, and J. Barnett, editors, *Safety, Reliability and Risk Analysis—Proceedings of the ESREL 2008 Conference*, pages 2107–2115, Valencia, Spain, September 2009. CRC Press, 2009. (Cited on pp. 149, 152)

[14] M. Azadkia and S. Chatterjee. A simple measure of conditional dependence. *Preprint*, 2019, arXiv:1910.12327. (Cited on p. 140)

[15] S. Azzi, Y. Huang, B. Sudret, and J. Wiart. Surrogate modeling of stochastic functions: Application to computational electromagnetic dosimetry. *International Journal for Uncertainty Quantification*, 9:351–363, 2019. (Cited on p. 91)

[16] S. Azzi, B. Sudret, and J. Wiart. Sensitivity analysis for stochastic simulators using differential entropy. *International Journal for Uncertainty Quantification*, 10:25–33, 2020. (Cited on p. 91)

[17] S. Ba, W.R. Myers, and W.A. Brenneman. Optimal sliced Latin hypercube designs. *Technometrics*, 57:479–487, 2015. (Cited on p. 245)

[18] E. Baker, P. Barbillon, A. Fadikar, R.B. Gramacy, R. Herbei, D. Higdon, J. Huang, L.R. Johnson, P. Ma, A. Mondal, B. Pires, J. Sacks, and V. Sokolov. Analyzing stochastic computer models: An overview with opportunities. *Preprint*, 2020, https://arxiv.org/pdf/2002.01321.pdf. (Cited on p. 91)

[19] D. Bakry, F. Barthe, P. Cattiaux, and A. Guillin. A simple proof of the Poincaré inequality for a large class of probability measures including the log-concave case. *Electronic Communications Probability*, 13:60–66, 2008. (Cited on p. 56)

[20] D. Bakry, I. Gentil, and M. Ledoux. *Analysis and Geometry of Markov Diffusion Operators*, volume 348 of *Grundlehren der Mathematischen Wissenschaften [Fundamental Principles of Mathematical Sciences]*. Springer, Cham, 2014. (Cited on p. 57)

[21] M. Baucells and E. Borgonovo. Invariant probabilistic sensitivity analysis. *Management Science*, 59(11):2536–2549, 2013. (Cited on p. 150)

[22] M. Baudin, K. Boumhaout, T. Delage, B. Iooss, and J.-M. Martinez. Numerical stability of Sobol' indices estimation formula. In *Proceedings of the 8th International Conference on Sensitivity Analysis of Model Output (SAMO 2016)*, Le Tampon, Réunion Island, France, December 2016. (Cited on p. 74)

[23] M. Baudin, T. Delage, A. Dutfoy, A. Geay, O. Mircescu, A. Ladier, J. Schueller, and T. Yalamas. OpenTURNS and its graphical interface. In *Eccomas Procedia UNCECOMP (2019)*, pages 467–489, Crete, Greece, June 2019. (Cited on p. 257)

[24] M. Baudin, A. Dutfoy, B. Iooss, and A.-L. Popelin. Open TURNS: An industrial software for uncertainty quantification in simulation. In R. Ghanem, D. Higdon, and H. Owhadi, editors, *Springer Handbook on Uncertainty Quantification*, pages 2001–2038. Springer, 2017. (Cited on pp. 1, 255)

[25] B.K. Beachkofski and R.V. Grandhi. Improved distributed hypercube sampling. In *Proceedings of 43rd AIAA/ASME/ASCE/AHS/ASC Structures, Structural Dynamics, and Materials Conference, AIAA 2002-1274*, Denver, CO, April 2002. (Cited on p. 244)

[26] W. Becker and A. Saltelli. Design for sensitivity analysis. In A. Dean, M. Morris, J. Stufken, and D. Bingham, editors, *Handbook of Design and Analysis of Experiments*, Handbooks of Modern Statistical Methods. Chapman & Hall/CRC, 2015. (Cited on p. 3)

[27] J. Bect, D. Ginsbourger, L. Li, V. Picheny, and E. Vazquez. Sequential design of computer experiments for the estimation of a probability of failure. *Statistics and Computing*, 22:773–793, 2012. (Cited on p. 170)

[28] J. Bect, L. Li, and E. Vazquez. Bayesian subset simulation. *SIAM/ASA Journal on Uncertainty Quantification*, 5:762–786, 2017. (Cited on p. 170)

[29] C. Bénard, S. Da Veiga, and E. Scornet. MDA for random forests: Inconsistency, and a practical solution via the Sobol-MDA. *Preprint*, 2021, arXiv:2102.13347. (Cited on p. 3)

[30] N. Benoumechiara. *Treatment of dependency in sensitivity analysis for industrial reliability*. Ph.D. thesis, Sorbonne University, France, 2019. (Cited on pp. 3, 109)

[31] N. Benoumechiara and K. Elie-Dit-Cosaque. Shapley effects for sensitivity analysis with dependent inputs: Bootstrap and kriging-based algorithms. *ESAIM: Proceedings and Surveys*, 65:266–293, 2019. (Cited on pp. 118, 129, 132, 133, 137, 138)

[32] N. Benoumechiara, B. Michel, P. Saint-Pierre, and N. Bousquet. Detecting and modeling worst-case dependence structures between random inputs of computational reliability models. *Dependence Modeling*, 8:263–297, 2020. (Cited on p. 185)

[33] J.O. Berger. An overview of robust Bayesian analysis (with discussion). *Test*, 3:5–124, 1994. (Cited on p. 13)

[34] A. Berlinet and C. Thomas-Agnan. *Reproducing Kernel Hilbert Spaces in Probability and Statistics*. Springer, 2004. (Cited on pp. 126, 238, 249)

[35] B. Bettonvil and J.P.C. Kleijnen. Searching for important factors in simulation models with many factors: Sequential bifurcation. *European Journal of Operational Research*, 96:180–194, 1997. (Cited on pp. 20, 23)

[36] R.N. Bhattacharya and R.R. Rao. *Normal Approximation and Asymptotic Expansions*, volume 64 in Classics in Applied Mathematics. SIAM, Philadelphia, 1986. (Cited on p. 70)

[37] J. Bien, J. Taylor, and R. Tibshirani. A lasso for hierarchical interactions. *Annals of Statistics*, 41(3):1111, 2013. (Cited on p. 34)

[38] P. Billingsley. *Probability and Measure*. Wiley Series in Probability and Statistics. Wiley, 2012. (Cited on p. 47)

[39] M. Binois, R.B. Gramacy, and M. Ludkovski. Practical heteroscedastic Gaussian process modeling for large simulation experiments. *Journal of Computational and Graphical Statistics*, 27:808–821, 2018. (Cited on p. 91)

[40] P. Bjerager and S. Krenk. Parametric sensitivity in first order reliability theory. *Journal of Engineering Mechanics*, 115:1577–1582, 1989. (Cited on p. 171)

[41] G. Blatman and B. Sudret. Adaptive sparse polynomial chaos expansion based on least angle regression. *Journal of Computational Physics*, 230(6):2345–2367, 2011. (Cited on p. 34)

[42] E. Borgonovo. A new uncertainty importance measure. *Reliability Engineering & System Safety*, 92:771–784, 2007. (Cited on pp. 118, 149, 152, 185)

[43] E. Borgonovo. *Sensitivity Analysis: An Introduction for the Management Scientist*. Springer, 2017. (Cited on pp. 3, 4, 13, 37)

[44] E. Borgonovo, W. Castaings, and S. Tarantola. Moment independent importance measures: New results and analytical test cases. *Risk Analysis*, 31:404–428, 2011. (Cited on pp. 118, 149)

[45] E. Borgonovo and B. Iooss. Moment-independent and reliability-based importance measures. In R. Ghanem, D. Higdon, and H. Owhadi, editors, *Springer Handbook on Uncertainty Quantification*, pages 1265–1287. Springer, 2017. (Cited on p. 149)

[46] E. Borgonovo, X. Lu, E. Plischke, O. Rakovec, and M.C. Hill. Making the most out of a hydrological model data set: Sensitivity analyses to open the model black-box. *Water Resources Research*, 53:7933–7950, 2017. (Cited on pp. 3, 29, 39)

[47] E. Borgonovo, M.D. Morris, and E. Plischke. Functional ANOVA with multiple distributions: Implications for the sensitivity analysis of computer experiments. *SIAM/ASA Journal of Uncertainty Quantification*, 6:397–427, 2018. (Cited on p. 184)

[48] E. Borgonovo and E. Plischke. Sensitivity analysis: A review of recent advances. *European Journal of Operational Research*, 248:869–887, 2016. (Cited on pp. 3, 18, 37)

[49] R.C. Bose and K.A. Bush. Orthogonal arrays of strength two and three. *Annals of Mathematical Statistics*, pages 508–524, 1952. (Cited on p. 66)

[50] S. Boucheron, G. Lugosi, and P. Massart. *Concentration Inequalities: A Nonasymptotic Theory of Independence*. Oxford University Press, 2013. (Cited on pp. 69, 70)

[51] G.E. Box and N.R. Draper. *Empirical Model Building and Response Surfaces*. Wiley Series in Probability and Mathematical Statistics. Wiley, 1987. (Cited on p. 2)

[52] P. Bratley, B.L. Fox, and H. Niederreiter. Implementation and tests of low-discrepancy sequences. *ACM Transactions on Modeling and Computer Simulation*, 2(3):195–213, 1992. (Cited on p. 251)

[53] F. Brauer. Mathematical epidemiology: Past, present, and future. *Infectious Disease Modelling*, 2(2):113–127, 2017. (Cited on p. 187)

[54] T. Britton, E. Pardoux, F. Ball, C. Laredo, D. Sirl, and V.C. Tran. *Stochastic Epidemic Models with Inference*. Springer, 2019. (Cited on p. 187)

[55] B. Broto, F. Bachoc, L. Clouvel, and J.-M. Martinez. Block-diagonal covariance estimation and application to the Shapley effects in sensitivity analysis. *Preprint*, 2020, https://arxiv.org/abs/1907.12780. (Cited on p. 143)

[56] B. Broto, F. Bachoc, and M. Depecker. Variance reduction for estimation of Shapley effects and adaptation to unknown input distribution. *SIAM/ASA Journal on Uncertainty Quantification*, 8:693–716, 2020. (Cited on pp. 90, 116, 118, 137, 138, 139, 140, 144, 223)

[57] B. Broto, F. Bachoc, M. Depecker, and J.-M. Martinez. Sensitivity indices for independent groups of variables. *Mathematics and Computers in Simulation*, 163:19–31, 2019. (Cited on pp. 118, 139, 143)

[58] T. Browne, J.C. Fort, B. Iooss, and L. Le Gratiet. Estimate of quantile-oriented sensitivity indices. *Working paper*, 2017, https://hal.archives-ouvertes.fr/hal-01450891. (Cited on p. 172)

[59] T. Browne, B. Iooss, L. Le Gratiet, J. Lonchampt, and E. Remy. Stochastic simulators based optimization by Gaussian process metamodels: Application to maintenance investments planning issues. *Quality and Reliability Engineering International Journal*, 32:2067–2080, 2016. (Cited on pp. 7, 91)

[60] D.G. Cacuci. Sensitivity theory for nonlinear systems. I. Nonlinear functional analysis approach. *Journal of Mathematical Physics*, 22:2794, 1981. (Cited on pp. 37, 39)

[61] D.G. Cacuci. *Sensitivity and Uncertainty Analysis: Theory*. Chapman & Hall/CRC, 2003. (Cited on pp. 37, 76)

[62] R. Caflisch, W. Morokoff, and A. Owen. Valuation of mortgage backed securities using Brownian bridges to reduce effective dimension. *Journal of Computational Finance*, 1:27–46, 1997. (Cited on pp. 54, 55, 103)

[63] P. Calmon. Trends and stakes of NDT simulation. *Journal of Non Destructive Evaluation*, 31:339–341, 2012. (Cited on p. 8)

[64] F. Campolongo and R. Braddock. The use of graph theory in the sensitivity analysis of model output: A second order screening method. *Reliability Engineering & System Safety*, 64:1–12, 1999. (Cited on p. 27)

[65] F. Campolongo, J. Cariboni, and A. Saltelli. An effective screening design for sensitivity analysis of large models. *Environmental Modelling & Software*, 22:1509–1518, 2007. (Cited on pp. 25, 27)

[66] Y. Caniou and B. Sudret. Covariance-based sensitivity indices based on polynomial chaos functional decomposition. In *Proceedings of the 7th International Conference on Sensitivity Analysis of Model Output (SAMO 2013)*, Nice, France, July 2013. (Cited on p. 125)

[67] C. Cannamela, J. Garnier, and B. Iooss. Controlled stratification for quantile estimation. *Annals of Applied Statistics*, 2:1554–1580, 2008. (Cited on p. 230)

[68] J. Castro, D. Gómez, and J. Tejada. Polynomial calculation of the Shapley value based on sampling. *Computers & Operations Research*, 36(5):1726–1730, 2009. (Cited on pp. 133, 135, 136)

[69] V. Chabridon. *Reliability-oriented sensitivity analysis under probabilistic model uncertainty: Application to aerospace systems*. Ph.D. thesis, University Clermont Auvergne, France, 2018. (Cited on p. 171)

[70] V. Chabridon, M. Balesdent, J.-M. Bourinet, J. Morio, and N. Gayton. Reliability-based sensitivity estimators of rare event probability in the presence of distribution parameter uncertainty. *Reliability Engineering & System Safety*, 178:164–178, 2018. (Cited on p. 185)

[71] V. Chabridon, M. Balesdent, G. Perrin, J. Morio, J.-M. Bourinet, and N. Gayton. Global reliability-oriented sensitivity analysis under distribution parameter uncertainty. In C. Gogu, editor, *Mechanical Engineering under Uncertainties: From Classical Approaches to Some Recent Developments*, pages 237–277. Wiley, 2021. (Cited on pp. 170, 185)

[72] V. Chabridon, A. Marrel, and B. Iooss. Tools for global and target sensitivity analyses in the context of high-dimensional thermal-hydraulic numerical experiments. In *ANS Best Estimate Plus Uncertainty International Conference (BEPU 2020)*, Giardini Naxos, Italy, June 2021, https://hal.archives-ouvertes.fr/hal-02877385/. (Cited on p. 161)

[73] M. Champion, G. Chastaing, S. Gadat, and C. Prieur. $L_2$-boosting for sensitivity analysis with dependent inputs. *Statistica Sinica*, 25:1477–1502, 2015. (Cited on p. 125)

[74] A. Charpentier, R. Elie, M. Laurière, and V.C. Tran. COVID-19 pandemic control: Balancing detection policy and lockdown intervention under ICU sustainability. *Mathematical Modelling of Natural Phenomena*, 15(57), 2020. (Cited on pp. 188, 189, 190, 193, 195)

[75] G. Chastaing, F. Gamboa, and C. Prieur. Generalized Hoeffding-Sobol decomposition for dependent variables: Application to sensitivity analysis. *Electronic Journal of Statistics*, 6:2420–2448, 2012. (Cited on pp. 118, 123, 124, 125, 126)

[76] G. Chastaing, F. Gamboa, and C. Prieur. Generalized Sobol sensitivity indices for dependent variables: Numerical methods. *Journal of Statistical Computation and Simulation*, 85(7):1306–1333, 2015. (Cited on p. 125)

[77] G. Chastaing and L. Le Gratiet. ANOVA decomposition of conditional Gaussian processes for sensitivity analysis with dependent inputs. *Journal of Statistical Computation and Simulation*, 85:2164–2186, 2015. (Cited on p. 126)

[78] S. Chatterjee. A new coefficient of correlation. *Journal of the American Statistical Association, In press*, 2021. (Cited on p. 140)

[79] W. Chen, R. Jin, and A. Sudjianto. Analytical metamodel-based global sensitivity analysis and uncertainty propagation for robust design. *Journal of Mechanical Design*, 127:875–886, 2005. (Cited on p. 109)

[80] Y. Chen, M. Welling, and A. Smola. Super-samples from kernel herding. In *Proceedings of the Twenty-Sixth Conference on Uncertainty in Artificial Intelligence*, pages 109–116. AUAI Press, 2010. (Cited on p. 239)

[81] J.-P. Chilès and P. Delfiner. *Geostatistics: Modeling Spatial Uncertainty*. Wiley, New-York, 1999. (Cited on p. 111)

[82] R. Christensen. *Linear Models for Multivariate, Time Series and Spatial Data*. Springer-Verlag, 1990. (Cited on p. 32)

[83] H.L. Cleaves, A. Alexanderian, H. Guy, R.C. Smith, and M. Yu. Derivative-based global sensitivity analysis for models with high-dimensional inputs and functional outputs. *SIAM Journal on Scientific Computing*, 41:A3524–A3551, 2019. (Cited on p. 88)

[84] W.S. Cleveland. LOWESS: A program for smoothing scatterplots by robust locally weighted regression. *The American Statistician*, 25:54, 1981. (Cited on p. 28)

[85] L. Clouvel, P. Mosca, J.M. Martinez, and G. Delipei. Shapley and Johnson values for sensitivity analysis of PWR power distribution in fast flux calculation. In *M&C 2019*, Portland, OR, August 2019. (Cited on p. 33)

[86] Y. Collette, N. Hansen, G. Pujol, D.S. Aponte, and R. Le Riche. Object-oriented programming of optimizers: Examples in Scilab. In P. Breitkopf and R.F. Coelho, editors, *Multidisciplinary Design Optimization in Computational Mechanics*, pages 499–538. Wiley, 2013. (Cited on p. 14)

[87] P.G. Constantine and P. Diaz. Global sensitivity metrics from active subspaces. *Reliability Engineering & System Safety*, 162:1–13, 2017. (Cited on p. 43)

[88] R. Cont, R. Deguest, and G. Scandolo. Robustness and sensitivity analysis of risk measurement procedures. *Quantitative Finance*, 10:593–606, 2010. (Cited on p. 185)

[89] I. Covert, S. Lundberg, and S.-I. Lee. Understanding global feature contributions with additive importance measures. *Advances in Neural Information Processing Systems*, 33, 2020. (Cited on p. 142)

[90] T. Crestaux, O. Le Maître, and J.-M. Martinez. Polynomial chaos expansions for uncertainties quantification and sensitivity analysis. *Reliability Engineering & System Safety*, 94:1161–1172, 2009. (Cited on p. 107)

[91] K. Crombecq, E. Laermans, and T. Dhaene. Efficient space-filling and non-collapsing sequential design strategies for simulation-based modelling. *European Journal of Operational Research*, 214:683–696, 2011. (Cited on p. 245)

[92] R.A. Cropp and R. Braddock. The new Morris method: An efficient second-order screening method. *Reliability Engineering & System Safety*, 78:77–83, 2002. (Cited on p. 27)

[93] I. Csiszár. Information-type measures of difference of probability distributions and indirect observations. *Studia Scientiarum Mathematicarum Hungarica*, 2:299–318, 1967. (Cited on p. 150)

[94] L.J. Cui, Z.Z. Lu, and X.P. Zhao. Moment-independent importance measure of basic random variable and its probability density evolution solution. *Science China Technological Sciences*, 53:1138–1145, 2010. (Cited on p. 168)

[95] H. Cukier, C.M. Fortuin, K. Shuler, A.G. Petschek, and J.H. Schaibly. Study of the sensitivity of coupled reaction systems to uncertainties in rate coefficients: Theory. *Journal of Chemical Physics*, 59:3873–3878, 1973. (Cited on pp. 53, 90, 95, 96, 100)

[96] H. Cukier, R.I. Levine, and K. Shuler. Nonlinear sensitivity analysis of multiparameter model systems. *Journal of Computational Physics*, 26:1–42, 1978. (Cited on pp. 90, 95, 96, 100)

[97] H. Cukier, J.H. Schaibly, and K. Shuler. Study of the sensitivity of coupled reaction systems to uncertainties in rate coefficients: Analysis of the approximations. *Journal of Chemical Physics*, 63:1140–1149, 1975. (Cited on pp. 90, 96, 100)

[98] M. Cuturi, J.-P. Vert, O. Birkenes, and T. Matsui. A kernel for time series based on global alignments. In *2007 IEEE International Conference on Acoustics, Speech and Signal Processing—ICASSP'07*, volume 2, pages II-413–II-416. IEEE, 2007. (Cited on p. 167)

[99] S. Da Veiga. Global sensitivity analysis with dependence measures. *Journal of Statistical Computation and Simulation*, 85:1283–1305, 2015. (Cited on pp. 13, 35, 36, 150, 161, 201)

[100] S. Da Veiga. Kernel-based ANOVA decomposition and Shapley effects—application to global sensitivity analysis. *Preprint*, 2021, arXiv:2101.05487. (Cited on pp. 158, 159, 161, 162, 164, 165)

[101] S. Da Veiga and F. Gamboa. Efficient estimation of sensitivity indices. *Journal of Nonparametric Statistics*, 25(3):573–595, 2013. (Cited on p. 67)

[102] S. Da Veiga, F. Wahl, and F. Gamboa. Local polynomial estimation for sensitivity analysis on models with correlated inputs. *Technometrics*, 51(4):452–463, 2009. (Cited on pp. 28, 106, 117)

[103] G. Damblin, M. Couplet, and B. Iooss. Numerical studies of space filling designs: Optimization of Latin hypercube samples and subprojection properties. *Journal of Simulation*, 7:276–289, 2013. (Cited on pp. 235, 242, 244, 246)

[104] G. Damblin and A. Ghione. Adaptive use of replicated Latin hypercube designs for computing Sobol' sensitivity indices. *Reliability Engineering & System Safety*, 212:107507, 2021. (Cited on p. 90)

[105] M. De Lozzo and A. Marrel. Estimation of the derivative-based global sensitivity measures using a Gaussian process metamodel. *SIAM/ASA Journal on Uncertainty Quantification*, 4:708–738, 2016. (Cited on p. 105)

[106] M. De Lozzo and A. Marrel. New improvements in the use of dependence measures for sensitivity analysis and screening. *Journal of Statistical Computation and Simulation*, 86:3038–3058, 2016. (Cited on p. 161)

[107] M. De Lozzo and A. Marrel. Sensitivity analysis with dependence and variance-based measures for spatio-temporal numerical simulators. *Stochastic Environmental Research and Risk Assessment*, 31:1437–1453, 2017. (Cited on pp. 13, 78)

[108] E. de Rocquigny. La maîtrise des incertitudes dans un contexte industriel. 1ère partie: Une approche méthodologique globale basée sur des exemples. *Journal de la Société Française de Statistique*, 147(3):33–71, 2006. (Cited on p. 6)

[109] E. de Rocquigny. *Modelling under Risk and Uncertainty: An Introduction to Statistical, Phenomenological and Computational Methods.* Wiley, 2012. (Cited on p. 6)

[110] E. de Rocquigny, N. Devictor, and S. Tarantola, editors. *Uncertainty in Industrial Practice.* Wiley, 2008. (Cited on pp. 1, 2, 3, 168, 170, 195)

[111] A. Dean and S. Lewis, editors. *Screening: Methods for Experimentation in Industry, Drug Discovery and Genetics.* Springer, 2006. (Cited on pp. 1, 12, 19, 20, 28, 229)

[112] T. Delage, R. Sueur, and B. Iooss. Robustness analysis of epistemic uncertainties propagation studies in LOCA assessment thermal-hydraulic model. In *Proceedings of ANS Best Estimate Plus Uncertainty International Conference (BEPU 2018)*, Lucca, Italy, May 2018. (Cited on p. 174)

[113] L. Devroye. *Non-uniform Random Variate Generation.* Springer-Verlag, New York, 1986. (Cited on p. 61)

[114] L. Di Domenico, G. Pullano, C.E. Sabbatini, P.-Y. Boëlle, and V. Colizza. Expected impact of lockdown in île-de-France and possible exit strategies. *medRxiv*, 2020. (Cited on pp. 188, 189)

[115] O. Ditlevsen and H.O. Madsen, editors. *Structural Reliability Methods.* Wiley & Sons, 1996. (Cited on p. 127)

[116] F. Douarche, S. Da Veiga, M. Feraille, G. Enchéry, S. Touzanir, and R. Barsalou. Sensitivity analysis and optimization of surfactant-polymer flooding under uncertainties. *Oil & Gas Science and Technology—Revue d'IFP Energies Nouvelles*, 69(4):603–617, 2014. (Cited on p. 9)

[117] F. Douard and B. Iooss. Dealing with uncertainty in technical & economic studies of investment strategy optimization. In *Proceedings of the 7th International Conference on Sensitivity Analysis of Model Output (SAMO 2013)*, Nice, France, July 2013. (Cited on p. 7)

[118] D.J. Downing, R.H. Gardner, and F.O. Hoffman. An examination of response surface methodologies for uncertainty analysis in assessment models. *Technometrics*, 27(2):151–163, 1985. (Cited on p. 2)

[119] D. Dupuy, C. Helbert, and J. Franco. DiceDesign and DiceEval: Two R packages for design and analysis of computer experiments. *Journal of Statistical Software*, 65:1–38, 2015. (Cited on pp. 114, 230)

[120] N. Durrande, D. Ginsbourger, O. Roustant, and L. Carraro. ANOVA kernels and RKHS of zero mean functions for model-based sensitivity analysis. *Journal of Multivariate Analysis*, 155:57–67, 2013. (Cited on pp. 109, 126, 165)

[121] C. Dussert, G. Rasigni, M. Rasigni, and J. Palmari. Minimal spanning tree: A new approach for studying order and disorder. *Physical Review B*, 34(5):3528–3531, 1986. (Cited on p. 245)

[122] B. Efron and C. Stein. The jacknife estimate of variance. *Annals of Statistics*, 9:586–596, 1981. (Cited on p. 90)

[123] K. Elie-Dit-Cosaque. *Development of uncertainty measures for model risk in a context of stochastic dependence*. Thèse de l'Université Claude Bernard Lyon 1, 2020. (Cited on pp. 142, 172)

[124] J.M. Epstein, D.A.T. Cummings, S. Chakravarty, R.M. Singha, and D.S. Burke. *Toward a Containment Strategy for Smallpox Bioterror: An Individual-Based Computational Approach*. Brookings Institution Press, 2004. (Cited on p. 187)

[125] P. Etoré, C. Prieur, D.K. Pham, and L. Li. Global sensitivity analysis for models described by stochastic differential equations. In *Methodology and Computing in Applied Probability*, pages 1–29. Springer-Verlag, 2019. (Cited on p. 91)

[126] R. Faivre, B. Iooss, S. Mahévas, D. Makowski, and H. Monod, editors. *Analyse de sensibilité et exploration de modèles*. Éditions Quaé, 2013. (Cited on pp. 1, 6)

[127] K.-T. Fang, R. Li, and A. Sudjianto. *Design and Modeling for Computer Experiments*. Chapman & Hall/CRC, 2006. (Cited on pp. 1, 2, 105, 106, 108, 195, 229, 235, 242, 244)

[128] K.-T. Fang, M-Q. Liu, H. Qin, and Y.-D. Zhou. *Theory and Application of Uniform Experimental Designs*. Springer, 2018. (Cited on p. 235)

[129] H. Faure. Discrépance des suites associées à un système de numération. *Acta Arithmetica*, 41:337–351, 1982. (Cited on p. 241)

[130] J.-M. Fédou and M.-J. Rendas. Extending Morris method: Identification of the interaction graph using cycle-equitable designs. *Journal of Statistical Computation and Simulation*, 85:1398–1419, 2015. (Cited on p. 27)

[131] B. Feldman. Relative importance and value. *SSRN Electronic Journal*, 2005, doi: 10.2139/ssrn.2255827. (Cited on p. 33)

[132] M. Feraille and A. Marrel. Prediction under uncertainty on a mature field. *Oil & Gas Science and Technology—Revue d'IFP Energies Nouvelles*, 67(2):193–206, 2012. (Cited on p. 9)

[133] N. Ferguson et al. Impact of non-pharmaceutical interventions (NPIs) to reduce COVID-19 mortality and healthcare demand (Report 9). *Imperial College COVID-19 Response Team (20)*, 2020, https://www.imperial.ac.uk/mrc-global-infectious-disease-analysis/covid-19/report-9-impact-of-npis-on-covid-19/. (Cited on p. 187)

[134] F. Ferraty and P. Vieu. *Nonparametric Functional Data Analysis: Theory and Practice*. Springer, 2006. (Cited on p. 167)

[135] S. Flaxman, S. Mishra, A. Gandy, et al. Report 13: Estimating the number of infections and the impact of non-pharmaceutical interventions on COVID-19 in 11 European countries. *Technical report*, Imperial College London (30-03-2020), 2020, https://doi.org/10.25561/77731. (Cited on p. 188)

[136] A. Forrester, A. Sobester, and A. Keane, editors. *Engineering Design via Surrogate Modelling: A Practical Guide*. Wiley, 2008. (Cited on pp. 12, 111, 168)

[137] J.C. Fort, T. Klein, and A. Lagnoux. Global sensitivity analysis and Wasserstein spaces. *SIAM/ASA Journal on Uncertainty Quantification*, 9(2):880–921, 2021. (Cited on p. 91)

[138] J.C. Fort, T. Klein, A. Lagnoux, and B. Laurent. Estimation of the Sobol indices in a linear functional multidimensional model. *Journal of Statistical Planning and Inference*, 143(9):1590–1605, 2013. (Cited on pp. 84, 85, 86, 95)

[139] J.C. Fort, T. Klein, and N. Rachdi. New sensitivity analysis subordinated to a contrast. *Communications in Statistics—Theory and Methods*, 45:4349–4364, 2016. (Cited on pp. 171, 172)

[140] R. Fraiman, F. Gamboa, and L. Moreno. Sensitivity indices for output on a Riemannian manifold. *International Journal for Uncertainty Quantification*, 10:297–314, 2020. (Cited on p. 153)

[141] J. Franco, O. Vasseur, B. Corre, and M. Sergent. Minimum spanning tree: A new approach to assess the quality of the design of computer experiments. *Chemometrics and Intelligent Laboratory Systems*, 97:164–169, 2009. (Cited on pp. 245, 246, 247)

[142] H.C. Frey and S.R. Patil. Identification and review of sensitivity analysis methods. *Risk Analysis*, 22:553–578, 2002. (Cited on pp. 3, 169)

[143] J.H. Friedman and B.E. Popescu. Predictive learning via rule ensembles. *Annals of Applied Statistics*, 2(3):916–954, 2008. (Cited on p. 42)

[144] J. Fruth, O. Roustant, and S. Kuhnt. Total interaction index: A variance-based sensitivity index for second-order interaction screening. *Journal of Statistical Planning and Inference*, 147:212–223, 2014. (Cited on p. 74)

[145] J. Fruth, O. Roustant, and S. Kuhnt. Support indices: Measuring the effect of input variables over their support. *Reliability Engineering & System Safety*, 187:17–27, 2019. (Cited on pp. 36, 42, 254)

[146] F. Gamboa, P. Gremaud, T. Klein, and A. Lagnoux. Global sensitivity analysis: A new generation of mighty estimators based on rank statistics. *Preprint*, 2020, https://hal.archives-ouvertes.fr/hal-02474902/. (Cited on pp. 90, 116, 140)

[147] F. Gamboa, A. Janon, T. Klein, and A. Lagnoux. Sensitivity indices for multivariate outputs. *Comptes Rendus de l'Académie des Sciences*, 351:307–310, 2013. (Cited on pp. 52, 79, 80)

[148] F. Gamboa, A. Janon, T. Klein, and A. Lagnoux. Sensitivity analysis for multidimensional and functional outputs. *Electronic Journal of Statistics*, 8(1):575–603, 2014. (Cited on pp. 52, 80, 85)

[149] F. Gamboa, A. Janon, T. Klein, A. Lagnoux, and C. Prieur. Statistical inference for Sobol pick freeze Monte Carlo methods. *Statistics*, 50:881–902, 2016. (Cited on pp. 67, 68, 69, 71, 73)

[150] F. Gamboa, T. Klein, and A. Lagnoux. Sensitivity analysis based on Cramér–von Mises distance. *SIAM/ASA Journal on Uncertainty Quantification*, 6(2):522–548, 2018. (Cited on pp. 153, 156)

[151] F. Gamboa, T. Klein, A. Lagnoux, and L. Moreno. Sensitivity analysis in general metric spaces. *Reliability Engineering & System safety*, 212:107611, 2021. (Cited on p. 153)

[152] S.S. Garud, I. Karimi, and M. Kraft. Design of computer experiments: A review. *Computers and Chemical Engineering*, 106:71–95, 2017. (Cited on p. 230)

[153] C. Gauchy, J. Stenger, R. Sueur, and B. Iooss. An information geometry approach for robustness analysis in uncertainty quantification of computer codes. *Technometrics, In press*, 2021. (Cited on pp. 13, 175)

[154] Q. Ge and M. Menendez. Extending Morris method for qualitative global sensitivity analysis of models with dependent inputs. *Reliability Engineering & System Safety*, 162:28–39, 2017. (Cited on p. 27)

[155] A. Genizi. Decomposition of $R^2$ in multiple regression with correlated regressors. *Statistica Sinica*, 3:407–420, 1993. (Cited on p. 142)

[156] G. Geraci, P.M. Congedo, R. Abgrall, and G. Iaccarino. High-order statistics in global sensitivity analysis: Decomposition and model reduction. *Computer Methods in Applied Mechanics and Engineering*, 301:80–115, 2013. (Cited on p. 173)

[157] R. Ghanem, D. Higdon, and H. Owhadi, editors. *Springer Handbook on Uncertainty Quantification*. Springer, 2017. (Cited on p. 1)

[158] R. Ghanem and P.D. Spanos. Polynomial chaos in stochastic finite elements. *Journal of Applied Mechanics*, 57(1):197–202, 1990. (Cited on p. 107)

[159] R.G. Ghanem and P.D. Spanos. Stochastic finite element method: Response statistics. In *Stochastic Finite Elements: A Spectral Approach*, pages 101–119. Springer, 1991. (Cited on p. 107)

[160] L. Gilquin, E. Arnaud, C. Prieur, and A. Janon. Making the best use of permutations to compute sensitivity indices with replicated orthogonal arrays. *Reliability Engineering & System Safety*, 187:28–39, 2019. (Cited on pp. 66, 67, 115)

[161] L. Gilquin, C. Prieur, and A. Arnaud. Replication procedure for grouped Sobol' indices estimation in dependent uncertainty spaces. *Information and Inference: A Journal of the IMA*, 4(4):354–379, 2015. (Cited on pp. 74, 117, 120)

[162] L. Gilquin, L.A.J. Rugama, E. Arnaud, F.J. Hickernell, H. Monod, and C. Prieur. Iterative construction of replicated designs based on Sobol' sequences. *Comptes Rendus Mathematique*, 355:10–14, 2017. (Cited on p. 245)

[163] E. Giné and R. Nickl. A simple adaptive estimator of the integrated square of a density. *Bernoulli*, 14(1):47–61, 2008. (Cited on p. 153)

[164] G. Glen and K. Isaacs. Estimating Sobol sensitivity indices using correlations. *Environmental Modelling & Software*, 37:157–166, 2012. (Cited on pp. 60, 90)

[165] T. Goda. A simple algorithm for global sensitivity analysis with Shapley effects. *Reliability Engineering & System Safety*, 213:107702, 2021. (Cited on p. 139)

[166] C. Goeury, T. David, R. Ata, S. Boyaval, Y. Audouin, N. Goutal, A.-L.Popelin, M. Couplet, M. Baudin, and R. Barate. Uncertainty quantification with on a real case with TELEMAC-2D. In C. Moulinec and D.R. Emerson, editors, *Proceedings of XXIIth TELEMAC-MASCARET User Conference*, pages 44–51, Warrington, UK, 2016. Science and Technology Facilities Council, 2016. (Cited on pp. 6, 77)

[167] N. Goutal, C. Goeury, R. Ata, S. Ricciand N. El Mocayd, M. Rochoux, H. Oubanasand I. Gejadze, and P.-O. Malaterre. Uncertainty quantification for river flow simulation applied to a real test case: The Garonne Valley. In P. Gourbesville, J. Cunge, and G. Caignaert, editors, *Advances in Hydroinformatics*. Springer Water. Springer, Singapore, 2018. (Cited on pp. 6, 77)

[168] N. Goutal, J.M. Lacombe, F. Zaoui, and K. El Kadi Abderrezak. MASCARET: A 1-D open-source software for flow hydrodynamic and water quality in open channel networks. In R. Murillo Muñoz, editor, *Proceedings of the International Conference on Fluvial Hydraulics*, volume 2, pages 1169–1174, San José, Costa Rica, 2012. CRC Press, 2012. (Cited on p. 6)

[169] G. Greegar and C.S. Manohar. Global response sensitivity analysis using probability distance measures and generalization of Sobol's analysis. *Probabilistic Engineering Mechanics*, 41:21–33, 2015. (Cited on p. 149)

[170] B. Gregorutti. *Forêts aléatoires et sélection de variables: Analyse des données des enregistreurs de vol pour la sécurité aérienne*. Ph.D. thesis, Université Paris VI, France, 2015. (Cited on p. 3)

[171] A. Gretton, K.M. Borgwardt, M.J. Rasch, B. Schölkopf, and A. Smola. A kernel two-sample test. *Journal of Machine Learning Research*, 13:723–773, 2012. (Cited on p. 158)

[172] A. Gretton, O. Bousquet, A. Smola, and B. Schölkopf. Measuring statistical dependence with Hilbert-Schmidt norms. In *Proceedings Algorithmic Learning Theory*, pages 63–77. Springer-Verlag, 2005. (Cited on pp. 160, 161)

[173] A. Gretton, K. Fukumizu, C.H. Teo, L. Song, B. Schölkopf, and A.J. Smola. A kernel statistical test of independence. In *Advances in Neural Information Processing Systems*, pages 585–592. MIT Press, 2008. (Cited on pp. 36, 160, 161)

[174] A. Griewank and A. Walther. *Evaluating Derivatives: Principles and Techniques of Automatic Differentiation*, 2nd ed. SIAM, Philadelphia, 2008. (Cited on p. 37)

[175] U. Grömping. Estimators of relative importance in linear regression based on variance decomposition. *The American Statistician*, 61(2):139–147, 2007. (Cited on pp. 33, 142)

[176] J.H. Halton. On the efficiency of certain quasi-random sequences of points in evaluating multi-dimensional integrals. *Numerische Mathematik*, 2:84–90, 1960. (Cited on p. 240)

[177] J.M. Hammersley. Monte Carlo methods for solving multivariable problems. *Annals of the New York Academy of Sciences*, 86:844–874, 1960. (Cited on p. 241)

[178] O.C. Hamsici and A.M. Martinez. Rotation invariant kernels and their application to shape analysis. *IEEE Transactions on Pattern Analysis and Machine Intelligence*, 31(11):1985–1999, 2008. (Cited on p. 167)

[179] Z. Harchaoui and F. Bach. Image classification with segmentation graph kernels. In *2007 IEEE Conference on Computer Vision and Pattern Recognition*, pages 1–8. IEEE, 2007. (Cited on p. 167)

[180] J.L. Hart and P.A. Gremaud. Robustness of the Sobol' indices to marginal distribution uncertainty. *SIAM/ASA Journal on Uncertainty Quantification*, 7:1224–1244, 2019. (Cited on p. 184)

[181] J.L. Hart, A. Alexanderian, and P.A. Gremaud. Efficient computation of Sobol' indices for stochastic models. *SIAM Journal on Scientific Computing*, 39:A1514–A1530, 2017. (Cited on p. 91)

[182] J.L. Hart and P.A. Gremaud. An approximation theoretic perspective of Sobol' indices with dependent variables. *International Journal for Uncertainty Quantification*, 8(6), 2018. (Cited on pp. 122, 123)

[183] D.M. Hartley, J.G. Morris, Jr., and D.L. Smith. Hyperinfectivity: A critical element in the ability of *V. cholerae* to cause epidemics? *PLoS Medicine*, 3(1):e7, 2005. (Cited on p. 188)

[184] T. Hastie and R. Tibshirani. *Generalized Additive Models*. Chapman and Hall, London, 1990. (Cited on pp. 28, 106)

[185] T. Hastie, R. Tibshirani, and J. Friedman. *The Elements of Statistical Learning*. Springer, 2002. (Cited on pp. 3, 105)

[186] A.S. Hedayat, N.J.A. Sloane, and J. Stufken. *Orthogonal Arrays: Theory and Applications*. Springer-Verlag, New York, 1999. (Cited on pp. 66, 101)

[187] J.C. Helton. Uncertainty and sensitivity analysis techniques for use in performance assessment for radioactive waste disposal. *Reliability Engineering & System Safety*, 42:327–367, 1993. (Cited on p. 1)

[188] J.C. Helton, J.D. Johnson, W.L. Obekampf, and C.J. Salaberry. Sensitivity analysis in conjunction with evidence theory representations of epistemic uncertainty. *Reliability Engineering & System Safety*, 91:1414–1434, 2006. (Cited on p. 13)

[189] J.C. Helton, J.D. Johnson, C.J. Salaberry, and C.B. Storlie. Survey of sampling-based methods for uncertainty and sensitivity analysis. *Reliability Engineering & System Safety*, 91:1175–1209, 2006. (Cited on pp. 3, 12, 28, 33, 36)

[190] M.B. Heredia, C. Prieur, and N. Eckert. Aggregated Shapley effects: Nearest-neighbor estimation procedure and confidence intervals. Application to snow avalanche modeling. *Preprint*, 2020, https://hal.archives-ouvertes.fr/hal-02908480. (Cited on p. 140)

[191] M.B. Heredia, C. Prieur, and N. Eckert. Nonparametric estimation of aggregated Sobol' indices: Application to a depth averaged snow avalanche model. *Reliability Engineering & System Safety*, 212:107422, 2021. (Cited on p. 116)

[192] T.C. Hesterberg. Estimates and confidence intervals for importance sampling sensitivity analysis. *Mathematical and Computer Modelling*, 23:79–85, 1996. (Cited on p. 175)

[193] F.J. Hickernell. A generalized discrepancy and quadrature error bound. *Mathematics of Computation*, 67:299–322, 1998. (Cited on p. 238)

[194] F.J. Hickernell. Lattice rules: How well do they measure up? In P. Hellekalek and G. Larcher, editors, *Random and Quasi-Random Point Sets*, pages 106–166. Springer-Verlag, Berlin, New-York, 1998. (Cited on pp. 55, 235)

[195] D. Higdon, J. Gattiker, B. Williams, and M. Rightley. Computer model calibration using high-dimensional output. *Journal of the American Statistical Association*, 103:571–583, 2008. (Cited on p. 78)

[196] W. Hoeffding. A class of statistics with asymptotically normal distributions. *Annals of Mathematical Statistics*, 19:293–325, 1948. (Cited on p. 49)

[197] T. Homma and A. Saltelli. Importance measures in global sensitivity analysis of non linear models. *Reliability Engineering & System Safety*, 52:1–17, 1996. (Cited on p. 59)

[198] G. Hooker. Generalized functional ANOVA diagnostics for high-dimensional functions of dependent variables. *Journal of Computational and Graphical Statistics*, 16:709–732, 2007. (Cited on pp. 53, 118, 122, 123)

[199] C.R. Hoyt and A.B. Owen. Mean dimension of ridge functions. *SIAM Journal on Numerical Analysis*, 58:1195–1216, 2020. (Cited on p. 56)

[200] B.G.M. Husslage, G. Rennen, E.R. van Dam, and D. den Hertog. Space-filling Latin hypercube designs for computer experiments. *Optimization and Engineering*, 12:611–630, 2011. (Cited on pp. 230, 242)

[201] I.A. Ibragimov and R.Z. Has' Minskii. *Statistical Estimation: Asymptotic Theory*, volume 16 of Applications of Mathematics. Springer-Verlag, New York, 1981. (Cited on p. 68)

[202] M. Il Idrissi, V. Chabridon, and B. Iooss. Developments and applications of Shapley effects to reliability-oriented sensitivity analysis with correlated inputs. *Environmental Modelling & Software*, 143:105115, 2021. (Cited on pp. 33, 142)

[203] R.L. Iman and W.J. Conover. A distribution-free approach to inducing rank correlation among input variables. *Communications in Statistics*, 11(3):311–334, 1982. (Cited on pp. 129, 233)

[204] D.R. Insua and F. Ruggeri, editors. *Robust Bayesian Analysis*. Springer-Verlag, New York, 2000. (Cited on p. 13)

[205] B. Iooss, L. Boussouf, V. Feuillard, and A. Marrel. Numerical studies of the metamodel fitting and validation processes. *International Journal of Advances in Systems and Measurements*, 3:11–21, 2010. (Cited on pp. 106, 245)

[206] B. Iooss, S. Da Veiga, A. Janon, and G. Pujol. *sensitivity: Global Sensitivity Analysis of Model Outputs*, 2021. R package version 1.24.0. (Cited on pp. 14, 253)

[207] B. Iooss, T. Klein, and A. Lagnoux. Sobol' sensitivity analysis for stochastic numerical codes. In *Proceedings of the 8th International Conference on Sensitivity Analysis of Model Output (SAMO 2016)*, Le Tampon, Réunion Island, France, December 2016. (Cited on p. 90)

[208] B. Iooss and L. Le Gratiet. Uncertainty and sensitivity analysis of functional risk curves based on Gaussian processes. *Reliability Engineering & System Safety*, 187:58–66, 2019. (Cited on pp. 8, 13, 175)

[209] B. Iooss and P. Lemaître. A review on global sensitivity analysis methods. In C. Meloni and G. Dellino, editors, *Uncertainty Management in Simulation-Optimization of Complex Systems: Algorithms and Applications*, pages 101–122. Springer, 2015. (Cited on pp. 3, 6, 11, 33)

[210] B. Iooss and A. Marrel. Advanced methodology for uncertainty propagation in computer experiments with large number of inputs. *Nuclear Technology*, 205:1588–1606, 2019. (Cited on pp. 10, 170)

[211] B. Iooss and C. Prieur. Shapley effects for sensitivity analysis with dependent inputs: Comparisons with Sobol' indices, numerical estimation and applications. *International Journal for Uncertainty Quantification*, 9:493–514, 2019. (Cited on pp. 118, 131, 132, 133, 141)

[212] B. Iooss and M. Ribatet. Global sensitivity analysis of computer models with functional inputs. *Reliability Engineering & System Safety*, 94:1194–1204, 2009. (Cited on p. 91)

[213] B. Iooss and A. Saltelli. Introduction: Sensitivity analysis. In R. Ghanem, D. Higdon, and H. Owhadi, editors, *Springer Handbook on Uncertainty Quantification*, pages 1103–1122. Springer, 2017. (Cited on p. 3)

[214] B. Iooss, F. Van Dorpe, and N. Devictor. Response surfaces and sensitivity analyses for an environmental model of dose calculations. *Reliability Engineering & System Safety*, 91:1241–1251, 2006. (Cited on p. 109)

[215] B. Iooss, V. Vergès, and V. Larget. BEPU robustness analysis via perturbed-law based sensitivity indices. *Proceedings of the Institution of Mechanical Engineers, Part O: Journal of Risk and Reliability*, in press, 2021. (Cited on pp. 174, 175)

[216] J. Jacques, C. Lavergne, and N. Devictor. Sensitivity analysis in presence of model uncertainty and correlated inputs. *Reliability Engineering & System Safety*, 91:1126–1134, 2006. (Cited on pp. 118, 119)

[217] A. Janon. *Sensitivity analysis and model reduction: Application to oceanography*. Ph.D. thesis, Grenoble University, France, 2012. (Cited on p. 58)

[218] A. Janon, T. Klein, A. Lagnoux, M. Nodet, and C. Prieur. Asymptotic normality and efficiency of two Sobol index estimators. *ESAIM: Probability and Statistics*, 18:342–364, 2014. (Cited on pp. 59, 60, 68, 73, 74, 110, 115)

[219] A. Janon, M. Nodet, and C. Prieur. Uncertainties assessment in global sensitivity indices estimation from metamodels. *International Journal for Uncertainty Quantification*, 4:21–36, 2014. (Cited on p. 113)

[220] M.J.W. Jansen. Analysis of variance designs for model output. *Computer Physics Communications*, 117:25–43, 1999. (Cited on pp. 60, 73)

[221] Y. Jiao and J.-P. Vert. The Kendall and Mallows kernels for permutations. *IEEE Transactions on Pattern Analysis and Machine Intelligence*, 40(7):1755–1769, 2018. (Cited on p. 168)

[222] M.N. Jimenez, O.P. Le Maître, and O.M. Knio. Nonintrusive polynomial chaos expansions for sensitivity analysis in stochastic differential equations. *SIAM/ASA Journal on Uncertainty Quantification*, 5(1):378–402, 2017. (Cited on p. 91)

[223] R. Jin, W. Chen, and A. Sudjianto. An efficient algorithm for constructing optimal design of computer experiments. *Journal of Statistical Planning and Inference*, 134:268–287, 2005. (Cited on p. 242)

[224] J.W. Johnson. A heuristic method for estimating the relative weight of predictor variables in multiple regression. *Multivariate Behavioral Research*, 35:1–19, 2000. (Cited on p. 33)

[225] J.W. Johnson and J.M. LeBreton. History and use of relative importance indices in organizational research. *Organizational Research Methods*, 7:238–257, 2004. (Cited on p. 33)

[226] M.E. Johnson, L.M. Moore, and D. Ylvisaker. Minimax and maximin distance design. *Journal of Statistical Planning and Inference*, 26:131–148, 1990. (Cited on p. 236)

[227] V.R. Joseph, E. Gul, and S. Ba. Maximum projection designs for computer experiments. *Biometrika*, 102:371–380, 2015. (Cited on p. 244)

[228] A. Jourdan and J. Franco. Plans d'expériences numériques d'information de Kullback-Leibler minimale. *Journal de la Société Française de Statistique*, 150:52–64, 2009. (Cited on p. 239)

[229] A. Jourdan and J. Franco. Optimal Latin hypercube designs for the Kullback-Leibler criterion. *Advances in Statistical Analysis*, 94:341–351, 2010. (Cited on p. 239)

[230] Z. Kala. Quantile-oriented global sensitivity analysis of design resistance. *Journal of Civil Engineering and Management*, 27:297–305, 2019. (Cited on p. 172)

[231] T. Kanamori, S. Hido, and M. Sugiyama. A least-squares approach to direct importance estimation. *Journal of Machine Learning Research*, 10:1391–1445, 2009. (Cited on p. 153)

[232] T.L. Kelley. An unbiased correlation ratio measure. *Proceedings of the National Academy of Sciences of the United States of America*, 21(9):554–559, 1935. (Cited on p. 101)

[233] M.G. Kendall and A. Stuart. *The Advanced Theory of Statistics, Vol. 2: Inference and Relationship*, 2nd ed. Hafner Publishing Company, 1967. (Cited on p. 53)

[234] R.W. Kennard and L.A. Stone. Computer aided design of experiments. *Technometrics*, 11:137–148, 1969. (Cited on p. 245)

[235] R. Kennet and G. Vicario. Challenges and opportunities in simulations and computer experiments in industrial statistics: An industry 4.0 perspective. *Advances Theory and Simulations*, 4(2):2000254, 2001. (Cited on p. 1)

[236] W.O. Kermack and A.G. McKendrick. A contribution to the mathematical theory of epidemics. *Proceedings of the Royal Society of London. Series A*, 115(772):700–721, 1927. (Cited on p. 187)

[237] W.O. Kermack and A.G. McKendrick. Contributions to the mathematical theory of epidemics. II.—The problem of endemicity. *Proceedings of the Royal Society of London. Series A*, 138(834):55–83, 1932. (Cited on p. 187)

[238] W.O. Kermack and A.G. McKendrick. Contributions to the mathematical theory of epidemics. III.—Further studies of the problem of endemicity. *Proceedings of the Royal Society of London. Series A*, 141(843):94–122, 1933. (Cited on p. 187)

[239] T. Kim and J. Song. Generalized reliability importance measure (GRIM) using Gaussian mixture. *Reliability Engineering & System Safety*, 173:105–115, 2018. (Cited on p. 171)

[240] J.P.C. Kleijnen. Sensitivity analysis and related analyses: A review of some statistical techniques. *Journal of Statistical Computation and Simulation*, 57:111–142, 1997. (Cited on p. 3)

[241] J.P.C. Kleijnen. *Design and Analysis of Simulation Experiments*, 2nd ed. Springer, 2015. (Cited on pp. 1, 17, 19, 23)

[242] J.P.C. Kleijnen and J.C. Helton. Statistical analyses of scatterplots to identify important factors in large-scale simulations, 1: Review and comparison of techniques. *Reliability Engineering & System Safety*, 65:147–185, 1999. (Cited on p. 36)

[243] J.P.C. Kleijnen and R.G. Sargent. A methodology for fitting and validating metamodels in simulation. *European Journal of Operational Research*, 120:14–29, 2000. (Cited on pp. 2, 106)

[244] D.E. Knuth. *The Art of Computer Programming*, Volume 2, 3rd ed. Addison-Wesley, Boston, 1998. (Cited on p. 230)

[245] J.R. Koehler and A.B. Owen. Computer experiments. In S. Ghosh and C.R. Rao, editors, *Design and Analysis of Experiments*, volume 13 of *Handbook of Statistics*. Elsevier, 1996. (Cited on p. 229)

[246] V.Y. Korolev and I.G. Shevtsova. An upper estimate for the absolute constant in the Berry-Esseen inequality. *Teoriya Veroyatnostei i ee Primeneniya*, 54(4):671–695, 2009. (Cited on pp. 72, 73)

[247] A. Kraskov, H. Stögbauer, and P. Grassberger. Estimating mutual information. *Physical Review E*, 69(6):066138, 2004. (Cited on pp. 152, 153)

[248] B. Krzykacz-Hausmann. Epistemic sensitivity analysis based on the concept of entropy. In *Proceedings of SAMO 2001*, pages 31–35, Madrid, 2001. CIEMAT, 2001. (Cited on pp. 149, 152)

[249] S. Kucherenko and B. Iooss. Derivative-based global sensitivity measures. In R. Ghanem, D. Higdon, and H. Owhadi, editors, *Springer Handbook on Uncertainty Quantification*, pages 1241–1263. Springer, 2017. (Cited on pp. 13, 41)

[250] S. Kucherenko, O.V. Klymenko, and N. Shah. Sobol' indices for problems defined in non-rectangular domains. *Reliability Engineering & System Safety*, 167:218–231, 2017. (Cited on p. 117)

[251] S. Kucherenko, M. Rodriguez-Fernandez, C. Pantelides, and N. Shah. Monte Carlo evaluation of derivative-based global sensitivity measures. *Reliability Engineering & System Safety*, 94:1135–1148, 2009. (Cited on pp. 13, 41, 42)

[252] S. Kucherenko, S. Song, and L. Wang. Quantile-based global sensitivity measures. *Reliability Engineering & System Safety*, 185:35–48, 2019. (Cited on p. 172)

[253] S. Kucherenko, S. Tarantola, and P. Annoni. Estimation of global sensitivity indices for models with dependent variables. *Computer Physics Communications*, 183:937–946, 2012. (Cited on pp. 118, 126)

[254] F. Kuo, I. Sloan, G. Wasilkowski, and H. Woźniakowski. On decompositions of multivariate functions. *Mathematics of Computation*, 79(270):953–966, 2010. (Cited on pp. 49, 164)

[255] D. Kurowicka and R. Cooke. *Uncertainty Analysis with High Dimensional Dependence Modelling*. Wiley, 2006. (Cited on pp. 1, 12, 28, 117)

[256] G. Lacroix. *Simulation de l'écosystème pélagique de la mer Ligure à l'aide d'un modèle unidimensionnel. Etude du bilan de matière et de la variabilité saisonnière, interannuelle et spatiale.* Ph.D. thesis, Université Pierre et Marie Curie, France, 1998. (Cited on p. 4)

[257] G. Lacroix and M. Grégoire. Revisited ecosystem model (MODECOGeL) of the Ligurian sea: Seasonal and interannual variability due to atmospheric forcing. *Journal of Marine Systems*, 37(4):229–258, 2002. (Cited on p. 5)

[258] G. Lacroix and P. Nival. Influence of meteorological variability on primary production dynamics in the Ligurian Sea (NW Mediterranean sea) with a 1D hydrodynamic/biological model. *Journal of Marine Systems*, 16(1-2):23–50, 1998. (Cited on p. 5)

[259] M. Lamboni, B. Iooss, A-L. Popelin, and F. Gamboa. Derivative-based global sensitivity measures: General links with Sobol' indices and numerical tests. *Mathematics and Computers in Simulation*, 87:45–54, 2013. (Cited on pp. 41, 56, 57)

[260] M. Lamboni, D. Makowski, S. Lehuger, B. Gabrielle, and H. Monod. Multivariate global sensitivity analysis for dynamic crop models. *Fields Crop Research*, 113:312–320, 2009. (Cited on p. 52)

[261] M. Lamboni, H. Monod, and D. Makowski. Multivariate sensitivity analysis to measure global contribution of input factors in dynamic models. *Reliability Engineering & System Safety*, 96:450–459, 2011. (Cited on pp. 79, 84)

[262] B. Laurent. Efficient estimation of integral functionals of a density. *Annals of Statistics*, 24(2):659–681, 1996. (Cited on p. 153)

[263] L. Le Gratiet. Asymptotic normality of a Sobol index estimator in Gaussian process regression framework. *Working paper*, 2014, https://hal.archives-ouvertes.fr/hal-00828596. (Cited on p. 90)

[264] L. Le Gratiet, C. Cannamela, and B. Iooss. A Bayesian approach for global sensitivity analysis of (multifidelity) computer codes. *SIAM/ASA Journal on Uncertainty Quantification*, 2:336–363, 2014. (Cited on pp. 111, 113)

[265] L. Le Gratiet, B. Iooss, T. Browne, G. Blatman, S. Cordeiro, and B. Goursaud. Model assisted probability of detection curves: New statistical tools and progressive methodology. *Journal of Nondestructive Evaluation*, 36:8, 2017. (Cited on p. 8)

[266] L. Le Gratiet, S. Marelli, and B. Sudret. Metamodel-based sensitivity analysis: Polynomial chaos expansions and Gaussian processes. In R. Ghanem, D. Higdon, and H. Owhadi, editors, *Springer Handbook on Uncertainty Quantification*, pages 1289–1325. Springer, 2017. (Cited on pp. 107, 108, 111)

[267] O.P. Le Maître and O.M. Knio. PC analysis of stochastic differential equations driven by Wiener noise. *Reliability Engineering & System Safety*, 135:107–124, 2015. (Cited on p. 91)

[268] M. Ledoux. *The Concentration of Measure Phenomenon.* American Mathematical Society, 2005. (Cited on pp. 69, 71)

[269] M. Ledoux and M. Talagrand. *Probability in Banach Spaces: Isoperimetry and Processes.* Springer Science & Business Media, 2013. (Cited on p. 86)

[270] M. Lemaire, A. Chateauneuf, and J.-C. Mitteau. *Structural Reliability.* Wiley, 2009. (Cited on pp. 37, 170)

[271] P. Lemaître. *Sensitivity analysis in structural reliability.* Ph.D. thesis, Université Bordeaux I, France, 2014. (Cited on pp. 173, 174)

[272] P. Lemaître, E. Sergienko, A. Arnaud, N. Bousquet, F. Gamboa, and B. Iooss. Density modification based reliability sensitivity analysis. *Journal of Statistical Computation and Simulation*, 85:1200–1223, 2015. (Cited on pp. 13, 173, 174, 175)

[273] C. Lemieux. *Monte Carlo and quasi-Monte Carlo sampling*. Springer, 2009. (Cited on pp. 65, 230, 234, 235, 239, 240, 242)

[274] P. Lévy. *Théorie de l'addition des variables aléatoires*, volume 1. Gauthier-Villars, 1954. (Cited on p. 127)

[275] A. Lewis, R. Smith, and B. Williams. Gradient free active subspace construction using Morris screening elementary effects. *Computers and Mathematics with Applications*, 72:1603–1615, 2016. (Cited on p. 43)

[276] G. Li and H. Rabitz. General formulation of HDMR component functions with independent and correlated variables. *Journal of Mathematical Chemistry*, 50(1):99–130, 2012. (Cited on p. 125)

[277] G. Li, H. Rabitz, J. Hu, Z. Chen, and Y. Ju. Regularized random-sampling high dimensional model representation. *Journal of Mathematical Chemistry*, 43(3):6022–6032, 2008. (Cited on p. 118)

[278] G. Li, H. Rabitz, P.E. Yelvington, O.O. Oluwole, F. Bacon, C.E. Kolb, and J. Schoendorf. Global sensitivity analysis for systems with independent and/or correlated inputs. *Journal of Physical Chemistry*, 114:6022–6032, 2010. (Cited on pp. 118, 125)

[279] L. Li, Z. Lu, J. Feng, and B. Wang. Moment-independent importance measure of basic variable and its state dependent parameter solution. *Structural Safety*, 38:40–47, 2012. (Cited on p. 168)

[280] W. Li, L. Lu, X. Xie, and M. Yang. A novel extension algorithm for optimized Latin hypercube sampling. *Journal of Statistical Computation and Simulation*, 87:2549–2559, 2017. (Cited on p. 245)

[281] M. Lim and T. Hastie. Learning interactions via hierarchical group-lasso regularization. *Journal of Computational and Graphical Statistics*, 24(3):627–654, 2015. (Cited on p. 34)

[282] C.D. Lin and B. Tang. Latin hypercubes and space-filling designs. In A. Dean, M. Morris, J. Stufken, and D. Bingham, editors, *Handbook of Design and Analysis of Experiments*, Handbooks of Modern Statistical Methods. Chapman & Hall/CRC, 2015. (Cited on pp. 232, 234)

[283] Y. Lin and H.H. Zhang. Component selection and smoothing in smoothing spline analysis of variance models. *Annals of Statistics*, 34(5):2272–2297, 2006. (Cited on pp. 34, 109)

[284] R.H. Lindeman, P.F. Merenda, and R.Z. Gold. *Introduction to Bivariate and Multivariate Analysis*. Scott Foresman and Company, Glenview, IL, 1980. (Cited on pp. 33, 142)

[285] C. Linkletter, D. Bingham, N. Hengartner, D. Higdon, and K.Q. Ye. Variable selection for Gaussian process models in computer experiments. *Technometrics*, 48:478–490, 2006. (Cited on p. 252)

[286] H. Liu, W. Chen, and A. Sudjianto. Relative entropy based method for probabilistic sensitivity analysis in engineering design. *ASME Journal of Mechanical Design*, 128:326–336, 2006. (Cited on pp. 149, 152)

[287] R. Liu and A.B. Owen. Estimating mean dimensionality of analysis of variance decompositions. *Journal of the American Statistical Association*, 101(474):712–721, 2006. (Cited on pp. 55, 74)

[288] Z. Liu, P. Magal, O. Seydi, and G. Webb. Understanding unreported cases in the COVID-19 epidemic outbreak in Wuhan, China, and the importance of major public health interventions. *Biology*, 9(3):50, 2020. (Cited on pp. 188, 190)

[289] Z. Liu, P. Magal, and G.F. Webb. Predicting the number of reported and unreported cases for the COVID-19 epidemics in China, South Korea, Italy, France, Germany and United Kingdom. *Journal of Theoretical Biology*, 509(110501), 2021. (Cited on p. 188)

[290] S. Lo Piano, F. Feretti, A. Puy, D. Albrecht, and A. Saltelli. Variance-based sensitivity analysis: The quest for better estimators between explorativity and economy. *Reliability Engineering & System Safety*, 206(107300), 2021. (Cited on p. 90)

[291] J. Lonchampt and K. Fessart. Investments portfolio optimal planning, a tool for an integrated life management. In *Proceedings of ANS Embedded Meeting on Risk Management for Complex Socio-technical Systems (RM4CSS)*, Washington DC, 2013. (Cited on p. 6)

[292] A. Lopez-Benito and R. Bolado-Lavin. A case study on global sensitivity analysis with dependent inputs: The natural gas transmission model. *Reliability Engineering & System Safety*, 165:11–21, 2017. (Cited on p. 117)

[293] X. Lu and E. Borgonovo. Is time to intervention in the COVID-19 outbreak really important? A global sensitivity analysis approach. *Preprint*, 2020, arXiv:2005.01833. (Cited on p. 188)

[294] S. Lundberg and S.-I. Lee. A unified approach to interpreting model predictions. *Advances in Neural Information Processing Systems*, 30:4765–4774, 2017. (Cited on p. 142)

[295] Y. Maday, A.T. Patera, and D.V. Rovas. A blackbox reduced-basis output bound method for non-coercive linear problems. In D. Cioranescu and J.-L. Lions, editors, *Nonlinear Partial Differential Equations and Their Applications Collège de France Seminar Volume XIV*, Studies in Mathematics and Its Applications. Elsevier, 2002. (Cited on p. 112)

[296] H.O. Madsen. Omission sensitivity factors. *Structural Safety*, 5.1:35–45, 1988. (Cited on p. 171)

[297] H.O. Madsen, S. Krenk, and N.C. Lind, editors. *Methods of Structural Safety*. Prentice Hall, 1986. (Cited on pp. 170, 171)

[298] W.R. Madych and S.A. Nelson. Bounds on multivariate polynomials and exponential error estimates for multiquadric interpolation. *Journal of Approximation Theory*, 70(1):94–114, 1992. (Cited on p. 112)

[299] S. Mak and V.R. Joseph. Minimax and minimax projection designs using clustering. *Journal of Computational and Graphical Statistics*, 27:166–178, 2018. (Cited on p. 237)

[300] S. Mak and V.R. Joseph. Support points. *Annals of Statistics*, 46:2562–2592, 2018. (Cited on p. 239)

[301] T. Mara. Extension of the RBD-FAST method to the computation of global sensitivity indices. *Reliability Engineering & System Safety*, 94:1274–1281, 2009. (Cited on p. 96)

[302] T. Mara and O.R. Joseph. Comparison of some efficient methods to evaluate the main effect of computer model factors. *Journal of Statistical Computation and Simulation*, 78:167–178, 2008. (Cited on pp. 62, 74)

[303] T. Mara and S. Tarantola. Variance-based sensitivity indices for models with dependent inputs. *Reliability Engineering & System Safety*, 107:115–121, 2012. (Cited on pp. 118, 128)

[304] T. Mara, S. Tarantola, and P. Annoni. Non-parametric methods for global sensitivity analysis of model output with dependent inputs. *Environmental Modeling & Software*, 72:173–183, 2015. (Cited on pp. 118, 126, 127, 128, 129)

[305] A. Marrel. *Mise en oeuvre et exploitation du métamodèle processus gaussien pour l'analyse de modèles numériques—Application à un code de transport hydrogéologique*. Ph.D. thesis, INSA Toulouse, France, 2008. (Cited on p. 109)

[306] A. Marrel and V. Chabridon. Statistical developments for target and conditional sensitivity analysis: Application on safety studies for nuclear reactor. *Reliability Engineering & System Safety*, 214:107711, 2021. (Cited on p. 168)

[307] A. Marrel, B. Iooss, S. Da Veiga, and M. Ribatet. Global sensitivity analysis of stochastic computer models with joint metamodels. *Statistics and Computing*, 22:833–847, 2012. (Cited on p. 91)

[308] A. Marrel, B. Iooss, M. Jullien, B. Laurent, and E. Volkova. Global sensitivity analysis for models with spatially dependent outputs. *Environmetrics*, 22:383–397, 2011. (Cited on p. 78)

[309] A. Marrel, B. Iooss, B. Laurent, and O. Roustant. Calculations of the Sobol indices for the Gaussian process metamodel. *Reliability Engineering & System Safety*, 94:742–751, 2009. (Cited on p. 109)

[310] A. Marrel, B. Iooss, F. Van Dorpe, and E. Volkova. An efficient methodology for modeling complex computer codes with Gaussian processes. *Computational Statistics and Data Analysis*, 52:4731–4744, 2008. (Cited on pp. 109, 111)

[311] A. Marrel, N. Perot, and C. Mottet. Development of a surrogate model and sensitivity analysis for spatio-temporal numerical simulators. *Stochastic Environmental Research and Risk Assessment*, 29:959–974, 2015. (Cited on p. 78)

[312] A. Marrel, N. Saint-Geours, and M. De Lozzo. Sensitivity analysis of spatial and/or temporal phenomena. In R. Ghanem, D. Higdon, and H. Owhadi, editors, *Springer Handbook on Uncertainty Quantification*, pages 1327–1357. Springer, 2017. (Cited on pp. 76, 77)

[313] M. Marseguerra, R. Masini, E. Zio, and G. Cojazzi. Variance decomposition-based sensitivity analysis via neural networks. *Reliability Engineering & System Safety*, 79:229–238, 2003. (Cited on p. 109)

[314] P. Massart. *Concentration Inequalities and Model Selection*. Springer, 2007. (Cited on p. 171)

[315] G. Matheron. Principles of geostatistics. *Economic Geology*, 58:1246–1266, 1963. (Cited on p. 108)

[316] V. Maume-Deschamps and I. Niang. Estimation of quantile-oriented sensitivity indices. *Statistics and Probability Letters*, 134:122–127, 2018. (Cited on p. 172)

[317] P. Mazgaj, J.-L. Vacher, and S. Carnevali. Comparison of CATHARE results with the experimental results of cold leg intermediate break LOCA obtained during ROSA-2/LSTF test 7. *European Journal of Physics—Nuclear Sciences & Technology (EPJ-N)*, 2:1, 2016. (Cited on p. 10)

[318] G. Mazo. An optimal tradeoff between explorations and repetitions in global sensitivity analysis for stochastic computer models. *Preprint*, 2020, https://hal.archives-ouvertes.fr/hal-02113448v2/. (Cited on p. 91)

[319] M.D. McKay. Evaluating prediction uncertainty. Technical report NUREG/CR-6311, U.S. Nuclear Regulatory Commission and Los Alamos National Laboratory, 1995. (Cited on pp. 53, 61, 62)

[320] M.D. McKay, R.J. Beckman, and W.J. Conover. A comparison of three methods for selecting values of input variables in the analysis of output from a computer code. *Technometrics*, 21:239–245, 1979. (Cited on p. 232)

[321] L. Meier, S. Van de Geer, and P. Bühlmann. High-dimensional additive modeling. *Annals of Statistics*, 37(6B):3779–3821, 2009. (Cited on p. 34)

[322] M. Menz, S. Dubreuil, J. Morio, C. Gogu, N. Bartoli, and M. Chiron. Variance based sensitivity analysis for Monte Carlo and importance sampling reliability assessment with Gaussian processes. *Preprint*, 2021, arXiv:2011.15001. (Cited on p. 111)

[323] A. Meynaoui, A. Marrel, and B. Laurent. New statistical methodology for second level global sensitivity analysis. *Preprint*, 2019, https://hal.archives-ouvertes.fr/hal-02019412. (Cited on pp. 177, 179)

[324] H. Monod, C. Naud, and D. Makowski. Uncertainty and sensitivity analysis for crop models. In D. Wallach, D. Makowski, and J. W. Jones, editors, *Working with Dynamic Crop Models: Evaluation, Analysis, Parameterization, and Applications*, pages 55–99. Elsevier, 2006. (Cited on pp. 68, 74, 115)

[325] D.C. Montgomery. *Design and Analysis of Experiments* 6th ed. John Wiley & Sons, 2004. (Cited on pp. 17, 18, 19, 229)

[326] J.J. Moré. The Levenberg-Marquardt algorithm: Implementation and theory. In *Numerical Analysis*, pages 105–116. Springer, 1978. (Cited on p. 211)

[327] J. Morio. Influence of input pdf parameters of a model on a failure probability estimation. *Simulation Modelling Practice and Theory*, 19:2244–2255, 2011. (Cited on p. 13)

[328] J. Morio and M. Balesdent. *Estimation of rare event probabilities in complex aerospace and other systems*. Woodhead Publishing, 2016. (Cited on pp. 168, 170)

[329] M.D. Morris. Factorial sampling plans for preliminary computational experiments. *Technometrics*, 33:161–174, 1991. (Cited on pp. 23, 24, 252)

[330] M.D. Morris and T.J. Mitchell. Exploratory designs for computational experiments. *Journal of Statistical Planning and Inference*, 43:381–402, 1995. (Cited on p. 242)

[331] M.D. Morris, L.M. Moore, and M.D. McKay. Sampling plans based on balanced incomplete block designs for evaluating the importance of computer model inputs. *Journal of Statistical Planning and Inference*, 136:3203–3220, 2006. (Cited on p. 59)

[332] M.D. Morris, L.M. Moore, and M.D. McKay. Using orthogonal arrays in the sensitivity analysis of computer models. *Technometrics*, 50:205–215, 2008. (Cited on p. 62)

[333] M.D. Morris and M. Moore. Design of computer experiments: Introduction and background. In A. Dean, M. Morris, J. Stufken, and D. Bingham, editors, *Handbook of Design and Analysis of Experiments*, Handbooks of Modern Statistical Methods. Chapman & Hall/CRC, 2015. (Cited on pp. 229, 230, 232)

[334] V. Moutoussamy, S. Nanty, and B. Pauwels. Emulators for stochastic simulation codes. *ESAIM: Proceedings and Surveys*, 48:116–155, 2015. (Cited on pp. 7, 91)

[335] A. Müller. Integral probability metrics and their generating classes of functions. *Advances in Applied Probability*, 29(2):429–443, 1997. (Cited on p. 156)

[336] W.G. Müller. *Collecting Spatial Data*, 3rd ed. Springer, 2007. (Cited on pp. 229, 245)

[337] B.L. Nelson. "Some tactical problems in digital simulation" for the next 10 years. *Journal of Simulation*, 10:2–11, 2016. (Cited on p. 256)

[338] H. Niederreiter. *Random Number Generation and Quasi-Monte Carlo Methods*. SIAM, Philadelphia, 1992. (Cited on pp. 240, 241)

[339] J.C.J. Nihoul and S. Djenidi. Perspectives in three-dimensional modelling of the marine system. In J.C.J. Nihoul and B.M. Jamart, editors, *Three-Dimensional Models of Marine and Estuarine Dynamics*, pages 1–34. Elsevier, Amsterdam, 1987. (Cited on p. 5)

[340] M. Nodet and A. Vidard. Variational methods. In R. Ghanem, D. Higdon, and H. Owhadi, editors, *Springer Handbook on Uncertainty Quantification*, pages 1123–1142. Springer, 2017. (Cited on p. 37)

[341] E. Nouy and A. de Crécy. Quantification of the uncertainty of physical models integrated into system thermohydraulic codes. *Nuclear Engineering and Design*, 321:278–287, 2017. (Cited on p. 10)

[342] W.T. Nutt and G.B. Wallis. Evaluation of nuclear safety from the outputs of computer codes in the presence of uncertainties. *Reliability Engineering & System Safety*, 83:57–77, 2004. (Cited on p. 10)

[343] J.E. Oakley and A. O'Hagan. Probabilistic sensitivity analysis of complex models: A Bayesian approach. *Journal of the Royal Statistical Society, Series B*, 66:751–769, 2004. (Cited on p. 109)

[344] A.B. Owen. Lattice sampling revisited: Monte Carlo variance of means randomized orthogonal arrays. *Annals of Statistics*, 22(2):930–945, 1994. (Cited on pp. 90, 102)

[345] A.B. Owen. A central limit theorem for Latin hypercube sampling. *Journal of the Royal Statistical Society. Series B*, 16:140–149, 1992. (Cited on p. 65)

[346] A.B. Owen. Orthogonal arrays for computer experiments, integration and visualization. *Statistica Sinica*, 2:439–452, 1992. (Cited on p. 90)

[347] A.B. Owen. Variance and discrepancy with alternative scramblings. *ACM Transactions on Modeling and Computer Simulation*, 13:363–378, 1993. (Cited on p. 242)

[348] A.B. Owen. The dimension distribution and quadrature test functions. *Statistica Sinica*, 13:1–17, 2003. (Cited on p. 55)

[349] A.B. Owen. Better estimation of small Sobol' sensitivity indices. *ACM Transactions on Modeling and Computer Simulation*, 23:11, 2013. (Cited on pp. 59, 60, 74)

[350] A.B. Owen. Sobol' indices and Shapley value. *SIAM/ASA Journal on Uncertainty Quantification*, 2:245–251, 2014. (Cited on pp. 118, 132, 133, 134)

[351] A.B. Owen. Monte Carlo theory, methods and examples. http://statweb.stanford.edu/ owen/mc/, unpublished, 2013. (Cited on p. 56)

[352] A.B. Owen, J. Dick, and S. Chen. Higher order Sobol' indices. *Information and Inference: A Journal of the IMA*, 3:59–81, 2014. (Cited on p. 173)

[353] A.B. Owen and C. Prieur. On Shapley value for measuring importance of dependent inputs. *SIAM/ASA Journal on Uncertainty Quantification*, 5:986–1002, 2017. (Cited on pp. 118, 132, 142)

[354] H. Owhadi, C. Scovel, T.J. Sullivan, M. McKerns, and M. Ortiz. Optimal uncertainty quantification. *SIAM Review*, 55:271–345, 2013. (Cited on p. 185)

[355] S. Pannier and W. Graf. Sectional global sensitivity measures. *Reliability Engineering & System Safety*, 134:110–117, 2015. (Cited on p. 36)

[356] I. Papaioannou and D. Straub. Variance-based reliability sensitivity analysis and the FORM $\alpha$-factors. *Reliability Engineering & System Safety*. 206(107496), 2021. (Cited on p. 171)

[357] F. Pappenberger, M. Ratto, and V. Vandenberghe. Review of sensitivity analysis methods. In P.A. Vanrolleghem, editor, *Modelling Aspects of Water Framework Directive Implementation*, pages 191–265. IWA Publishing, 2010. (Cited on p. 3)

[358] C.K. Park and K.I. Ahn. A new approach for measuring uncertainty importance and distributional sensitivity in probabilistic safety assessment. *Reliability Engineering & System Safety*, 46:253–261, 1994. (Cited on p. 152)

[359] K.S. Park and L. Xu. *Data Assimilation for Atmospheric, Oceanic and Hydrologic Applications*. Springer, 2008. (Cited on p. 37)

[360] A. Pasanisi and A. Dutfoy. An industrial viewpoint on uncertainty quantification in simulation: Stakes, methods, tools, examples. In A.M. Dienstfrey and R.F. Boisvert, editors, *Uncertainty Quantification in Scientific Computing—10th IFIP WG 2.5 Working Conference, WoCoUQ 2011, Boulder, CO, August 1–4, 2011*, volume 377 of IFIP Advances in Information and Communication Technology, pages 27–45. Springer, Berlin, 2012. (Cited on p. 255)

[361] E. Patelli, H.J. Pradlwarter, and G.I. Schuëller. Global sensitivity of structural variability by random sampling. *Computer Physics Communications*, 181:2072–2081, 2010. (Cited on p. 41)

[362] H. Peng, F. Long, and C. Ding. Feature selection based on mutual information criteria of max-dependency, max-relevance, and min-redundancy. *IEEE Transactions on Pattern Analysis*, 27(8):1226–1238, 2005. (Cited on pp. 34, 35)

[363] G. Perrin and G. Defaux. Efficient estimation of reliability-oriented sensitivity indices. *Journal of Scientific Computing*, 80(3), 2019. (Cited on pp. 13, 169, 174, 175)

[364] T.V.E. Perrin, O. Roustant, J. Rohmer, O. Alata, J.-P. Naulin, D. Idier, R. Pedreros, D. Moncoulon, and P. Tinard. Functional principal component analysis for global sensitivity analysis of model with spatial output. *Reliability Engineering & System Safety*, 211:107522, 2021. (Cited on p. 78)

[365] S.M. Pesenti, P. Millossovich, and A. Tsanakas. Cascade sensitivity measures. https://doi.org/10.2139/ssrn.3270839, 2019. (Cited on p. 185)

[366] M. Petelet, B. Iooss, O. Asserin, and A. Loredo. Latin hypercube sampling with inequality constraints. *Advances in Statistical Analysis*, 94:325–339, 2010. (Cited on pp. 117, 233)

[367] S. Petit, C. Goeury, N. Goutal, A-L. Popelin, and F. Zaoui. Couplage entre indices à base de dérivées et mode adjoint pour l'estimation d'indices de Sobol. Application sur le code Mascaret. *Working paper*, 2016, https://hal.archives-ouvertes.fr/hal-01373535. (Cited on p. 6)

[368] F. Pianosi and T. Wagener. A simple and efficient method for global sensitivity analysis based on cumulative distribution functions. *Environmental Modelling & Software*, 67:1–11, 2015. (Cited on p. 149)

[369] I. Pinelis and R. Molzon. Berry-Esseen bounds for general nonlinear statistics, with applications to Pearson's and non-central Student's and Hotelling's. *Working paper*, 2009, https://arxiv.org/abs/0906.0177v5. (Cited on p. 72)

[370] E. Plischke. An effective algorithm for computing global sensitivity indices (EASI). *Reliability Engineering & System Safety*, 95:354–360, 2010. (Cited on pp. 101, 105, 115)

[371] E. Plischke. An adaptive correlation ratio using the cumulative sum of reordered output. *Reliability Engineering & System Safety*, 107:149–156, 2012. (Cited on p. 29)

[372] E. Plischke, G. Rabitti, and E. Borgonovo. Computing Shapley effects for sensitivity analysis. *SIAM/ASA Journal on Uncertainty Quantification, In press*, 2021. (Cited on p. 138)

[373] C. Prieur and S. Tarantola. Variance-based sensitivity analysis: Theory and estimation algorithms. In R. Ghanem, D. Higdon, and H. Owhadi, editors, *Springer Handbook on Uncertainty Quantification*, pages 1217–1239. Springer, 2017. (Cited on pp. 13, 90)

[374] C. Prieur, L. Viry, E. Blayo, and J.-M. Brankart. A global sensitivity analysis approach for marine biogeochemical modeling. *Ocean Modelling*, 139:101402, 2019. (Cited on p. 5)

[375] L. Pronzato. Minimax and maximin space-filling designs: Some properties and methods for construction. *Journal de la Societe Française de Statistique*, 158:7–36, 2017. (Cited on pp. 230, 237)

[376] L. Pronzato. Sensitivity analysis via Karhunen-Loève expansion of a random field model: Estimation of Sobol' indices and experimental design. *Reliability Engineering & System Safety*, 187:93–109, 2019. (Cited on p. 115)

[377] L. Pronzato and W. Müller. Design of computer experiments: Space filling and beyond. *Statistics and Computing*, 22:681–701, 2012. (Cited on pp. 230, 237)

[378] L. Pronzato and A. Zhigljavsky. Bayesian quadrature, energy minimization, and space-filling design. *SIAM/ASA Journal on Uncertainty Quantification*, 8:959–1011, 2020. (Cited on p. 245)

[379] A. Prosek and B. Mavko. The state-of-the-art theory and applications of best-estimate plus uncertainty methods. *Nuclear Technology*, 158:69–79, 2007. (Cited on p. 10)

[380] G. Pujol. Simplex-based screening designs for estimating metamodels. *Reliability Engineering & System Safety*, 94:1156–1160, 2009. (Cited on pp. 27, 230)

[381] P.Z.G. Qian. Nested Latin hypercube designs. *Biometrika*, 96:957–970, 2009. (Cited on p. 245)

[382] P.Z.G. Qian. Sliced Latin hypercube designs. *Journal of the American Statistical Association*, 497:393–399, 2012. (Cited on p. 245)

[383] P.Z.G. Qian, M. Ai, and C.F.J. Wu. Construction of nested space-filling designs. *Annals of Statistics*, 37:3616–3643, 2009. (Cited on p. 245)

[384] P.Z.G. Qian and C.F.J. Wu. Sliced space filling designs. *Biometrika*, 96:945–956, 2009. (Cited on p. 245)

[385] R Development Core Team. *R: A Language and Environment for Statistical Computing*. R Development Core Team, 2006. (Cited on p. 14)

[386] G. Rabitti and E. Borgonovo. A Shapley–Owen index for interaction quantification. *SIAM/ASA Journal on Uncertainty Quantification*, 7:1060–1075, 2019. (Cited on pp. 118, 142)

[387] H. Raguet and A. Marrel. Target and conditional sensitivity analysis with emphasis on dependence measures. *Working paper*, 2018, https://arxiv.org/abs/1801.10047. (Cited on p. 168)

[388] S. Rahman. A generalized ANOVA dimensional decomposition for dependent probability measures. *SIAM/ASA Journal on Uncertainty Quantification*, 2:670–697, 2014. (Cited on p. 118)

[389] S. Rahman. The $f$-sensitivity index. *SIAM/ASA Journal on Uncertainty Quantification*, 4:130–162, 2016. (Cited on p. 152)

[390] S. Rahman. Mathematical properties of polynomial dimensional decomposition. *SIAM/ASA Journal on Uncertainty Quantification*, 6(2):816–844, 2018. (Cited on p. 118)

[391] O. Rakovec, M.C. Hill, M.P. Clark, A.H. Weerts, A.J. Teuling, and R. Uijlenhoet. Distributed evaluation of local sensitivity analysis (DELSA), with application to hydrologic models. *Water Resources Research*, 50:1–18, 2014. (Cited on p. 39)

[392] C.E. Rasmussen and C.K.I. Williams. *Gaussian Processes for Machine Learning*. MIT Press, 2006. (Cited on p. 108)

[393] M. Ratto and A. Pagano. Using recursive algorithms for the efficient identification of smoothing spline ANOVA models. *Advances in Statistical Analysis*, 94:367–388, 2010. (Cited on p. 113)

[394] P. Ravikumar, J. Lafferty, H. Liu, and L. Wasserman. Sparse additive models. *Journal of the Royal Statistical Society. Series B*, 71(5):1009–1030, 2009. (Cited on p. 34)

[395] S. Razavi, A. Jakeman, A. Saltelli, C. Prieur, B. Iooss, E. Borgonovo, E. Plischke, S. Lo Piano, T. Iwanaga, W. Becker, S. Tarantola, J.H.A. Guillaume, J. Jakeman, H. Gupta, N. Melillo, G. Rabiti, V. Chabridon, Q. Duan, X. Sun, S. Smith, R. Sheikholeslami, N. Hosseini, M. Asadzadeh, A. Puy, S. Kucherenko, and H.R. Maier. The future of sensitivity analysis: An essential discipline for systems modelling and policy making. *Environmental Modelling and Software*, 137(104954), 2020. (Cited on pp. 1, 3)

[396] A. Ribés, J. Pouderoux, and B. Iooss. A visual sensitivity analysis for parameter-augmented ensembles of curves. *Journal of Verification, Validation and Uncertainty Quantification*, 4:041007, 2019. (Cited on pp. 78, 257)

[397] R.T. Rockafellar and J.O. Royset. Engineering decisions under risk averseness. *Journal of Risk and Uncertainty in Engineering Systems. Part A: Civil Engineering*, 1(2), 2015. (Cited on p. 168)

[398] M. Rosenblatt. Remarks on a multivariate transformation. *Annals of Mathematical Statistics*, 23(3):470–472, 1952. (Cited on p. 127)

[399] B.C. Ross. Mutual information between discrete and continuous data sets. *PLoS One*, 9(2):e87357, 2014. (Cited on p. 43)

[400] G.-C. Rota. On the foundations of combinatorial theory I. Theory of Möbius functions. *Zeitschrift für Wahrscheinlichkeitstheorie und verwandte Gebiete*, 2(4):340–368, 1964. (Cited on pp. 51, 52, 138)

[401] O. Roustant, F. Barthe, and B. Iooss. Poincaré inequalities on intervals: Application to sensitivity analysis. *Electronic Journal of Statistics*, 2:3081–3119, 2017. (Cited on pp. 6, 41, 42, 56, 57)

[402] O. Roustant, J. Fruth, B. Iooss, and S. Kuhnt. Crossed-derivative-based sensitivity measures for interaction screening. *Mathematics and Computers in Simulation*, 105:105–118, 2014. (Cited on pp. 41, 42)

[403] O. Roustant, F. Gamboa, and B. Iooss. Parseval inequalities and lower bounds for variance-based sensitivity indices. *Electronic Journal of Statistics*, 14:386–412, 2020. (Cited on p. 115)

[404] O. Roustant, D. Ginsbourger, and Y. Deville. DiceKriging, DiceOptim: Two R packages for the analysis of computer experiments by kriging-based metamodeling and optimization. *Journal of Statistical Software*, 21:1–55, 2012. (Cited on p. 113)

[405] R.Y. Rubinstein. The score function approach for sensitivity analysis of computer simulation models. *Mathematics and Computers in Simulation*, 28:351–379, 1986. (Cited on p. 171)

[406] J. Sacks, W.J. Welch, T.J. Mitchell, and H.P. Wynn. Design and analysis of computer experiments. *Statistical Science*, 4:409–435, 1989. (Cited on pp. 2, 108)

[407] H. Salje, C.T. Kiem, N. Lefrancq, N. Courtejoie, P. Bosetti, J. Paireau, A. Andronico, N. Hozé, J. Richet, C.-L. Dubost, et al. Estimating the burden of SARS-CoV-2 in France. *Science*, 369:208–211, 2020. (Cited on p. 188)

[408] *Salome: The Open Source Integration Platform for Numerical Simulation.* OPEN CASCADE, SAS, 2006. (Cited on p. 256)

[409] A. Saltelli. Making best use of model evaluations to compute sensitivity indices. *Computer Physics Communications*, 145:280–297, 2002. (Cited on pp. 58, 59, 64, 66, 67, 73, 115)

[410] A. Saltelli. Sensitivity analysis for importance assessment. *Risk Analysis*, 22:1–12, 2002. (Cited on p. 90)

[411] A. Saltelli, T.H. Andres, and T. Homma. Sensitivity analysis of model output: An investigation of new techniques. *Computational Statistics and Data Analysis*, 15:211–238, 1993. (Cited on p. 59)

[412] A. Saltelli, P. Annoni, I. Azzini, F. Campolongo, M. Ratto, and S. Tarantola. Variance based sensitivity analysis of model output. Design and estimator for the total sensitivity index. *Computer Physics Communications*, 181:259–270, 2010. (Cited on pp. 59, 90)

[413] A. Saltelli, G. Bammer, I. Bruno, E. Charters, M. Di Fiore, et al. Five ways to ensure that models serve society: A manifesto (short comments). *Nature*, 582:482–484, 2020. (Cited on p. 188)

[414] A. Saltelli, K. Chan, and E.M. Scott, editors. *Sensitivity Analysis*, Wiley Series in Probability and Statistics. Wiley, 2000. (Cited on pp. 1, 28, 32, 33, 90, 169, 195, 251)

[415] A. Saltelli, M. Ratto, T. Andres, F. Campolongo, J. Cariboni, D. Gatelli, M. Salsana, and S. Tarantola. *Global Sensitivity Analysis: The Primer.* Wiley, 2008. (Cited on p. 1)

[416] A. Saltelli and S. Tarantola. On the relative importance of input factors in mathematical models: Safety assessment for nuclear waste disposal. *Journal of American Statistical Association*, 97:702–709, 2002. (Cited on pp. 4, 117)

[417] A. Saltelli, S. Tarantola, F. Campolongo, and M. Ratto. *Sensitivity Analysis in Practice: A Guide to Assessing Scientific Models*. Wiley, 2004. (Cited on pp. 3, 4, 13, 26)

[418] A. Saltelli, S. Tarantola, and K. Chan. A quantitative, model-independent method for global sensitivity analysis of model output. *Technometrics*, 41:39–56, 1999. (Cited on pp. 95, 96)

[419] J. Santiago, M. Claeys-Bruno, and M. Sergent. Construction of space-filling designs using WSP algorithm for high dimensional spaces. *Chemometrics and Intelligent Laboratory Systems*, 113:26–31, 2012. (Cited on p. 246)

[420] J. Santiago, B. Corre, M. Claeys-Bruno, and M. Sergent. Improved sensitivity through Morris extension. *Chemometrics and Intelligent Laboratory Systems*, 113:52–57, 2012. (Cited on p. 27)

[421] T. Santner, B. Williams, and W. Notz. *The Design and Analysis of Computer Experiments*. Springer, 2003. (Cited on pp. 1, 106, 108, 111)

[422] G. Sarazin, P. Derennes, and J. Morio. Estimation of high-order moment-independent importance measures for Shapley value analysis. *Applied Mathematical Modelling*, 88:396–417, 2020. (Cited on p. 142)

[423] G. Sarazin, J. Morio, A. Lagnoux, M. Balesdent, and L. Brevault. Reliability-oriented sensitivity analysis in presence of data-driven epistemic uncertainty. *Reliability Engineering & System Safety*, 215:107733, 2021. (Cited on p. 169)

[424] R. Schaback. Mathematical results concerning kernel techniques. In *Proc. 13th IFAC Symposium on System Identification, Rotterdam*, pages 1814–1819. Citeseer, 2003. (Cited on pp. 111, 112)

[425] J.H. Schaibly and K. Shuler. Study of the sensitivity of coupled reaction systems to uncertainties in rate coefficients: Applications. *Journal of Chemical Physics*, 59:3879–3888, 1973. (Cited on p. 96)

[426] M. Schartau, P. Wallhead, J. Hemmings, U. Löptien, I. Kriest, S. Krishna, B.A. Ward, T. Slawig, and A. Oschlies. Reviews and syntheses: Parameter identification in marine planktonic ecosystem modelling. *Biogeosciences (BG)*, 14(6):1647–1701, 2017. (Cited on p. 4)

[427] H. Scheffe. *The Analysis of Variance*. John Wiley & Sons, 1999. (Cited on pp. 46, 51)

[428] M. Scheuerer, R. Schaback, and M. Schlather. Interpolation of spatial data—a stochastic or a deterministic problem? *European Journal of Applied Mathematics*, 24(04):601–629, 2013. (Cited on p. 111)

[429] D. Sejdinovic, B. Sriperumbudur, A. Gretton, and K. Fukumizu. Equivalence of distance-based and RKHS-based statistics in hypothesis testing. *Annals of Statistics*, 41(5):2263–2291, 2013. (Cited on p. 239)

[430] B. Shang and D.W. Apley. Fully-sequential space-filling design algorithms for computer experiments. *Journal of Quality Technology*, 53:173–196, 2020. (Cited on p. 245)

[431] L.S. Shapley. A value for n-persons game. In H.W. Kuhn and A.W. Tucker, editors, *Contributions to the Theory of Games* II, volume 28 of Annals of Mathematics Studies, pages 307–317. Princeton University Press, Princeton, NJ, 1953. (Cited on pp. 118, 133)

[432] R. Sheikholeslami and S. Razavi. Progressive Latin hypercube sampling: An efficient approach for robust sampling-based analysis of environmental models. *Environmental Modelling & Software*, 93:109–126, 2017. (Cited on p. 245)

[433] W. Shi and J.P.C. Kleijnen. Testing the assumptions of sequential bifurcation for factor screening. *Simulation Modelling Practice and Theory*, 81:85–99, 2018. (Cited on p. 23)

[434] W. Shi, J.P.C. Kleijnen, and Z. Liu. Factor screening for simulation with multiple responses: Sequential bifurcation. *European Journal of Operational Research*, 237:136–147, 2014. (Cited on p. 23)

[435] B. Shölkopf, R. Herbrich, and A. Smola. A generalized representer theorem. In *Proceedings of the Annual Conference of Computational Learning Theory*, pages 416–426. Springer, London, 2001. (Cited on p. 250)

[436] T.W. Simpson, J.D. Peplinski, P.N. Kock, and J.K. Allen. Metamodel for computer-based engineering designs: Survey and recommandations. *Engineering with Computers*, 17:129–150, 2001. (Cited on pp. 105, 106)

[437] Y. G. Sinai. *Introduction to Ergodic Theory.* Princeton University Press, 1977. (Cited on p. 95)

[438] R.C. Smith. *Uncertainty Quantification.* SIAM, Philadelphia, 2014. (Cited on pp. 1, 2, 76, 188, 208, 213, 255, 256)

[439] A. Smola, A. Gretton, L. Song, and B. Schölkopf. A Hilbert space embedding for distributions. In *International Conference on Algorithmic Learning Theory*, pages 13–31. Springer, 2007. (Cited on pp. 157, 238)

[440] I.M. Sobol'. On the distributions of points in a cube and the approximate evaluation of integrals. *USSR Computational Mathematics and Mathematical Physics*, 7:86–112, 1967. (Cited on p. 241)

[441] I.M. Sobol'. Sensitivity estimates for non linear mathematical models (in Russian). *Matematicheskoe Modelirovanie*, 2:112–118, 1990. (Cited on pp. 53, 90)

[442] I.M. Sobol'. Sensitivity estimates for non linear mathematical models. *Mathematical Modelling and Computational Experiments*, 1:407–414, 1993. (Cited on pp. 1, 13, 53, 58, 59, 73, 90, 252)

[443] I.M. Sobol' and A. Gershman. On an alternative global sensitivity estimators. In *Proceedings of SAMO 1995*, pages 40–42, Belgirate, Italy, 1995. (Cited on p. 41)

[444] I.M. Sobol' and S. Kucherenko. Derivative based global sensitivity measures and their links with global sensitivity indices. *Mathematics and Computers in Simulation*, 79:3009–3017, 2009. (Cited on pp. 41, 42, 56)

[445] I.M. Sobol' and S. Kucherenko. A new derivative based importance criterion for groups of variables and its link with the global sensitivity indices. *Computer Physics Communications*, 181:1212 – 1217, 2010. (Cited on p. 41)

[446] I.M. Sobol', S. Tarantola, D. Gatelli, S.S. Kucherenko, and W. Mauntz. Estimating the approximation errors when fixing unessential factors in global sensitivity analysis. *Reliability Engineering & System Safety*, 92:957–960, 2007. (Cited on p. 73)

[447] E. Song, B.L. Nelson, and J. Staum. Shapley effects for global sensitivity analysis: Theory and computation. *SIAM/ASA Journal on Uncertainty Quantification*, 4:1060–1083, 2016. (Cited on pp. 118, 132, 133, 134, 135, 136, 137, 138, 139)

[448] L. Song, A. Smola, A. Gretton, J. Bedo, and K. Borgwardt. Feature selection via dependence maximization. *Journal of Machine Learning Research*, 13:1393–1434, 2012. (Cited on pp. 35, 160, 161, 167)

[449] A Spagnol. *Kernel-based sensitivity indices for high-dimensional optimization problems.* Ph.D. thesis, University Lyon, France, 2020. (Cited on p. 169)

[450] A. Spagnol, R. Le Riche, and S. Da Veiga. Global sensitivity analysis for optimization with variable selection. *SIAM/ASA Journal on Uncertainty Quantification*, 7(2):417–443, 2019. (Cited on pp. 168, 169)

[451] R. Spear and G. Hornberger. Eutrophication in peel inlet II. Identification of critical uncertainties via generalized sensitivity analysis. *Water Research*, 14:43–49, 2019. (Cited on p. 169)

[452] B.K. Sriperumbudur, K. Fukumizu, A. Gretton, B. Schölkopf, and G.R. Lanckriet. On the empirical estimation of integral probability metrics. *Electronic Journal of Statistics*, 6:1550–1599, 2012. (Cited on p. 157)

[453] B.K. Sriperumbudur, A. Gretton, K. Fukumizu, B. Schölkopf, and G.R. Lanckriet. Hilbert space embeddings and metrics on probability measures. *Journal of Machine Learning Research*, 11:1517–1561, 2010. (Cited on p. 167)

[454] M. Stein. Large sample properties of simulations using Latin hypercube sampling. *Technometrics*, 29:143–151, 1987. (Cited on p. 232)

[455] M.L. Stein. *Interpolation of Spatial Data*. Springer, 1999. (Cited on p. 108)

[456] I. Steinwart, D. Hush, and C. Scovel. An explicit description of the reproducing kernel Hilbert spaces of Gaussian RBF kernels. *IEEE Transactions on Information Theory*, 52(10):4635–4643, 2006. (Cited on p. 164)

[457] J. Stenger, F. Gamboa, and M. Keller. Optimization of quasi-convex function over product measure sets. *SIAM Journal on Optimization*, 31:425–447, 2021. (Cited on p. 185)

[458] J. Stenger, F. Gamboa, M. Keller, and B. Iooss. Optimal uncertainty quantification of a risk measurement from a thermal-hydraulic code using canonical moments. *International Journal for Uncertainty Quantification*, 10:35–53, 2020. (Cited on p. 185)

[459] C. J. Stone. The use of polynomial splines and their tensor products in multivariate function estimation. *Annals of Statistics*, 22(1):118–184, 1994. (Cited on pp. 118, 122, 123)

[460] C.B. Storlie, H.D. Bondell, B.J. Reich, and H.H. Zhang. Surface estimation, variable selection, and the nonparametric oracle property. *Statistica Sinica*, 21(2):679, 2011. (Cited on p. 43)

[461] C.B. Storlie and J.C. Helton. Multiple predictor smoothing methods for sensitivity analysis: Description of techniques. *Reliability Engineering & System Safety*, 93:28–54, 2008. (Cited on pp. 12, 28, 106)

[462] C.B. Storlie, L.P. Swiler, J.C. Helton, and C.J. Salaberry. Implementation and evaluation of nonparametric regression procedures for sensitivity analysis of computationally demanding models. *Reliability Engineering & System Safety*, 94:1735–1763, 2009. (Cited on p. 110)

[463] B. Sudret. Global sensitivity analysis using polynomial chaos expansion. *Reliability Engineering & System Safety*, 93:964–979, 2008. (Cited on p. 107)

[464] B. Sudret and Y. Caniou. Analysis of covariance (ANCOVA) using polynomial chaos expansions. In G. Deodatis, editor, *Proceedings of 11th International Conference on Structural Safety and Reliability (ICOSSAR 2013)*, New York, July 2013. (Cited on pp. 118, 125)

[465] B. Sudret and C.-V. Mai. Computing derivative-based global sensitivity measures using polynomial chaos expansions. *Reliability Engineering & System Safety*, 134:241–250, 2015. (Cited on p. 105)

[466] R. Sueur, N. Bousquet, B. Iooss, and J. Bect. Perturbed-law based sensitivity indices for sensitivity analysis in structural reliability. In *Proceedings of the 8th International Conference on Sensitivity Analysis of Model Output (SAMO 2016)*, Le Tampon, Réunion Island, France, December 2016. (Cited on pp. 174, 175)

[467] R. Sueur, B. Iooss, and T. Delage. Sensitivity analysis using perturbed-law based indices for quantiles and application to an industrial case. In *Proceedings of the 10th International Conference on Mathematical Methods in Reliability (MMR 2017)*, Grenoble, France, 2017. (Cited on pp. 10, 13, 174, 175)

[468] M. Sugiyama, T. Suzuki, and T. Kanamori. *Density Ratio Estimation in Machine Learning*. Cambridge University Press, 2012. (Cited on p. 153)

[469] X. Sun, B. Croke, S. Roberts, and A. Jakeman. Comparing methods of randomizing Sobol' sequences for improving uncertainty of metrics in variance-based global sensitivity estimation. *Reliability Engineering & System Safety*, 210: 107499, 2021. (Cited on p. 90)

[470] T. Suzuki, M. Sugiyama, J. Sese, and T. Kanamori. Approximating mutual information by maximum likelihood density ratio estimation. *Journal of Machine Learning Research-Proceedings Track*, 4:5–20, 2008. (Cited on p. 153)

[471] G.J. Székely and M.L. Rizzo. Testing for equal distributions in high dimension. *InterStat*, 5:1–6, 2004. (Cited on p. 239)

[472] G.J. Székely and M.L. Rizzo. Energy statistics: A class of statistics based on distances. *Journal of Statistical Planning and Inference*, 143:1249–1272, 2013. (Cited on p. 239)

[473] G.J. Székely, M.L. Rizzo, and N.K. Bakirov. Measuring and testing dependence by correlation of distances. *Annals of Statistics*, 25:2769–2794, 2007. (Cited on p. 36)

[474] B. Tang. Orthogonal array-based Latin hypercubes. *Journal of the American Statistical Association*, 88:1392–1397, 1993. (Cited on p. 234)

[475] K. Tang, P.M. Congedo, and R. Abgrall. Sensitivity analysis using anchored ANOVA expansion and high-order moments computation. *International Journal for Numerical Methods in Engineering*, 102(9):1554–1584, 2015. (Cited on p. 49)

[476] S. Tarantola, D. Gatelli, and T. Mara. Random balance designs for the estimation of first order global sensitivity indices. *Reliability Engineering & System Safety*, 91:717–727, 2006. (Cited on pp. 97, 100)

[477] S. Tarantola, E. Plischke, and T. Mara. Quasi-random balance designs for sensitivity analysis. In *Eleventh International Conference on Monte Carlo and Quasi-Monte Carlo Methods in Scientific Computing*, 2014. (Cited on p. 100)

[478] H.J. ter Horst. On Stieltjes integration in Euclidean space. *Journal of Mathematical Analysis and Applications*, 114(1):57–74, 1986. (Cited on p. 155)

[479] T. Terraz, A. Ribes, Y. Fournier, B. Iooss, and B. Raffin. Large scale in transit global sensitivity analysis avoiding intermediate files. In *Proceedings the International Conference for High Performance Computing, Networking, Storage and Analysis (Supercomputing)*, Denver, CO, November 2017. (Cited on p. 77)

[480] L. Tierney. Markov chains for exploring posterior distributions. *Annals of Statistics*, 22(4), 1701–1728, 1994. (Cited on p. 215)

[481] J.-Y. Tissot. *On ANOVA decomposition and Sobol' indices estimation. Application to a marine ecosystem model.* Ph.D. thesis, Grenoble University, France, 2012. (Cited on pp. 103, 104, 105)

[482] J.-Y. Tissot and C. Prieur. A bias correction method for the estimation of sensitivity indices based on random balance designs. *Reliability Engineering & System Safety*, 107:205–213, 2012. (Cited on p. 105)

[483] J.-Y. Tissot and C. Prieur. Variance-based sensitivity analysis using harmonic analysis. *Working paper*, 2012, http://hal.archives-ouvertes.fr/hal-00680725. (Cited on pp. 98, 100, 103, 104, 105, 115)

[484] J.-Y. Tissot and C. Prieur. A randomized orthogonal array-based procedure for the estimation of first- and second-order Sobol' indices. *Journal of Statistical Computation and Simulation*, 85:1358–1381, 2015. (Cited on pp. 62, 65, 66, 69, 74)

[485] T. Tjur. *Conditional Probability Distributions*, volume 2. Institute of Mathematical Statistics, University of Copenhagen, 1974. (Cited on pp. 47, 48)

[486] S. Touzani. *Response surface methods based on functional variance decomposition with application to sensitivity analysis*. Ph.D. thesis, Grenoble University, France, 2011. (Cited on p. 109)

[487] T. Turanyi. Sensitivity analysis for complex kinetic system, tools and applications. *Journal of Mathematical Chemistry*, 5:203–248, 1990. (Cited on pp. 37, 39)

[488] A.W. van der Vaart. *Asymptotic Statistics*. Cambridge University Press, 1998. (Cited on pp. 68, 69, 72, 83, 87)

[489] F.A.C. Viana. A tutorial on Latin hypercube design of experiments. *Quality Reliability Engineering International*, 32:1975–1985, 2016. (Cited on pp. 230, 242)

[490] G. Vignat, G. Taliercio, J. Lamouroux, S. Da Veiga, N. Savary, and P. Duchaine. Analysis of performance sensitivity to geometrical variations of a modern helicopter engine combustor using LES simulations. In *ASME Turbo Expo 2017: Turbomachinery Technical Conference and Exposition*, pages V04BT04A003–V04BT04A003. American Society of Mechanical Engineers, 2017. (Cited on p. 11)

[491] E. Volkova, B. Iooss, and F. Van Dorpe. Global sensitivity analysis for a numerical model of radionuclide migration from the RRC "Kurchatov Institute" radwaste disposal site. *Stochastic Environmental Research and Risk Assessment*, 22:17–31, 2008. (Cited on p. 77)

[492] G. Wahba, Y. Wang, C. Gu, R. Klein, B. Klein, et al. Smoothing spline ANOVA for exponential families, with application to the Wisconsin Epidemiological Study of Diabetic Retinopathy: The 1994 Neyman Memorial Lecture. *Annals of Statistics*, 23(6):1865–1895, 1995. (Cited on p. 165)

[493] C. Walder and O. Chapelle. Learning with transformation invariant kernels. In *Advances in Neural Information Processing Systems*, pages 1561–1568. MIT Press, 2008. (Cited on p. 167)

[494] H. Wan, B. Ankenman, and B.L. Nelson. Controlled sequential bifurcation: A new factor-screening method for discrete-event simulation. *Operations Research*, 54:743–755, 2006. (Cited on p. 23)

[495] H. Wan, B. Ankenman, and B.L. Nelson. Improving the efficiency and efficacy of controlled sequential bifurcation for simulation factor screening. *INFORMS Journal on Computing*, 22:482–492, 2010. (Cited on p. 23)

[496] J. Wang, R. Faivre, H. Richard, and H. Monod. mtk (Mexico Toolkit): An R package for uncertainty and sensitivity analyses of numerical experiments. *The R Journal*, 7:206–226, 2015. (Cited on p. 255)

[497] P. Wei, Z. Lu, W. Hao, J. Feng, and B. Wang. Efficient sampling methods for global reliability sensitivity analysis. *Computer Physics Communications*, 183:1728–1743, 2012. (Cited on p. 168)

[498] P. Wei, Z. Lu, and J. Song. A comprehensive comparison of two variable importance analysis techniques in high dimensions: Application to an environmental multi-indicators system. *Environmental Modelling & Software*, 70:178–190, 2015. (Cited on p. 109)

[499] P. Wei, Z. Lu, and J. Song. Variable importance analysis: A comprehensive review. *Reliability Engineering & System Safety*, 142:399–432, 2015. (Cited on p. 3)

[500] H. Weyl. Über die gleichverteilung von zahlen mod. eins. *Mathematische Annalen*, 77:313–352, 1916. (Cited on p. 95)

[501] H. Weyl. Mean motion. *American Journal of Mathematics*, 60:889–896, 1938. (Cited on p. 95)

[502] N. Wiener. The homogeneous chaos. *American Journal of Mathematics*, 60(4):897–936, 1938. (Cited on p. 107)

[503] B. Williamson and J. Feng. Efficient nonparametric statistical inference on population feature importance using Shapley values. In *International Conference on Machine Learning*, pages 10282–10291. PMLR, 2020. (Cited on p. 142)

[504] E. Winter. The Shapley value. In R.J. Aumann and S. Hart, editors, *Handbook of Game Theory with Economic Applications. Vol.* 3, pages 2025–2054. Elsevier Science, 2002. (Cited on p. 133)

[505] C.Y. Wong, P. Seshadri, and G. Parks. Extremum sensitivity analysis with polynomial Monte Carlo filtering. *Reliability Engineering & System Safety*, 212:107609, 2021. (Cited on p. 173)

[506] D. Woods and S. Lewis. Design of experiments for screening. In R. Ghanem, D. Higdon, and H. Owhadi, editors, *Springer Handbook on Uncertainty Quantification*, pages 1143–1185. Springer, 2017. (Cited on pp. 12, 19)

[507] C. Xu and G. Gertner. Extending a global sensitivity analysis technique to models with correlated parameters. *Computational Statistics and Data Analysis*, 51:5579–5590, 2007. (Cited on p. 125)

[508] C. Xu and G. Gertner. Uncertainty and sensitivity analysis for models with correlated parameters. *Reliability Engineering & System Safety*, 93:1563–1573, 2008. (Cited on pp. 118, 125)

[509] C. Xu and G. Gertner. Understanding and comparisons of different sampling approaches for the Fourier Amplitudes Sensitivity Test (FAST). *Computational Statistics and Data Analysis*, 55(1):184–198, 2011. (Cited on p. 100)

[510] M. Yamada, W. Jitkrittum, L. Sigal, E.P. Xing, and M. Sugiyama. High-dimensional feature selection by feature-wise non-linear lasso. *Neural Computation*, 26:185–207, 2014. (Cited on pp. 36, 161, 167)

[511] M. Yuan and Y. Lin. Model selection and estimation in regression with grouped variables. *Journal of the Royal Statistical Society. Series B*, 68(1):49–67, 2006. (Cited on p. 43)

[512] O. Zahm, P.G. Constantine, C. Prieur, and Y.M. Marzouk. Gradient-based dimension reduction of multivariate vector-valued functions. *SIAM Journal on Scientific Computing*, 42:A534–A558, 2020. (Cited on p. 88)

[513] Q. Zhang, S. Filippi, A. Gretton, and D. Sejdinovic. Large-scale kernel methods for independence testing. *Statistics and Computing*, 28(1):113–130, 2018. (Cited on p. 36)

[514] Y. Zhou, Z. Lu, S. Xiao, and W. Yun. Distance correlation-based sensitivity analysis of models with dependent inputs. *Structural and Multidisciplinary Optimization*, 60:1189–1207, 2019. (Cited on p. 36)

[515] X. Zhu and B. Sudret. Global sensitivity analysis for stochastic simulators based on generalized lambda surrogate models. *Reliability Engineering & System Safety*, 214:107815, 2021. (Cited on p. 91)

[516] E. Zio and N. Pedroni. Monte Carlo simulation-based sensitivity analysis of the model of a thermal-hydraulic passive system. *Reliability Engineering & System Safety*, 2012:90–106, 2008. (Cited on p. 171)

# Index


active subspaces, 43
additive model, 28, 34, 105, 125
adjoint-based method, 37
adjoint model, 42
aggregated Sobol' indices, 79
aggregation procedure
  random permutation, 137
  subset, 138
analysis of variance, 46, 118
  kernels, 109, 126
ANOVA, *see* analysis of variance

Bernoulli, 46
Berry–Esseen-like results, 69, 71
block-additive models, 139
bootstrap, 32, 106, 110, 137, 175
bootstrap-based confidence intervals, 67, 75
Bratley function, 120, 251

central limit theorem, 68, 69, 86, 175, 232
circular permutation, 127
CLT, *see* central limit theorem
cobweb plots, 28
coefficient of determination $R^2$, 32, 33, 142
computer experiments, 1, 3, 106, 117, 229
concentration inequalities, 69
conditional SA, 168
contrast function, 171
correlation, 117, 119, 141
correlation ratio, 53
covariance decomposition, 122
covariance function, 108
covariance kernel, 108, 126
covariance matrix, 68, 79
covariance operator, 84
Cramér–von Mises distance, 153, 155
Cramér–von Mises indices, 155
cross-validation, 34, 106
Csiszár $f$-divergences, 150, 152, 156
curse of dimensionality, 126, 153, 156, 238
CUSONORO plots, 29

DELSA, 39
Delta method, 69, 72, 83, 87
dependence, 2, 117, 122, 125, 161, 177, 231
dependence measure, 35, 152
  kernel-based, 35, 152
dependent inputs, 185
derivative-based global sensitivity measure, 41, 56, 88, 105, 115
derivatives, 38
design
  fractional factorial, 18, 19
  full factorial, 18
  one-at-a-time, 17, 18, 23
design of experiments, 17, 18, 43, 105, 195, 229, 235
DGSM, *see* derivative-based global sensitivity measure
dimension reduction, 123
discrepancy
  centered $\mathbb{L}^2$, 235
  wraparound $\mathbb{L}^2$, 235
discrepancy measure, 235
dissimilarity measure, 142, 150, 151, 157, 160
DOE, *see* design of experiments

EASI, 101, 115
effective dimension, 54, 55, 103
elasticities, 37

factor fixing, 4, 7, 10, 12, 141, 161, 188, 199
factor prioritization, 4, 6, 8, 9, 11, 12, 42, 141, 188, 202, 223
failure probability, 170, 174, 175
FANOVA, *see* functional analysis of variance
FAST, *see* Fourier amplitude sensitivity test
feature selection, 17, 34, 43, 160
first-/second-order reliability methods, 170
FORM/SORM, *see* first-/second-order reliability methods
Fourier amplitude sensitivity test, 93, 95, 101
functional analysis of variance, 45, 52, 53, 90, 94, 118, 119, 156, 162, 163
functional output, 78, 84, 167

Gaussian process, 105, 108, 111, 113, 115, 126
global sensitivity measures, 12, 24, 36
  quantile-based, 172
goal-oriented sensitivity analysis, 168
GOSA, *see* goal-oriented sensitivity analysis
group screening, 20, 28

HDMR, *see* high-dimensional model representation
Hessian-based bounds, 87

HeterDisc function, 28, 32, 39, 251
high-dimensional model representation, 118
high-order quantiles, 10
Hilbert–Schmidt independence criterion, 160, 164, 169, 177, 182, 188, 201
Hoeffding decomposition, 49, 57, 98, 109, 118, 122, 154
  generalized, 124
HSIC, *see* Hilbert–Schmidt independence criterion
HSIC-Shapley effect, 165

importance measures, 41
  variance-based, 53
interaction, 3, 12, 18, 19, 23, 29, 34, 39, 42, 51, 54, 64, 66, 106, 117, 133, 141, 142, 230
interaction effects, 134
Ishigami function, 54, 75, 113, 145, 150, 153, 176, 179, 251

KL, *see* Kullback–Leibler
kriging, 105, 108
Kullback–Leibler divergence, 151, 153, 160, 174, 239

lasso, 34, 43, 125, 161
Latin hypercube sample, 61, 105, 232–234, 236, 242
  centered $\mathbb{L}^2$-discrepancy, 244
  design, 233
  maximin, 112, 244
  optimized, 246
  replicated, 62, 65
LHS, *see* Latin hypercube sample
linear regression, 31
linearization, 38
Linkletter decreasing function, 21, 252
local approach, 37
local sensitivity, 37

machine learning, 3, 34, 142, 158, 167
max-dependency, 35
maximum mean discrepancy, 157–169, 183, 238
  kernel-based, 238
maximum projection criterion, 244
mean dimension, 55
Mercer's theorem, 159, 162, 164
metamodel, 2, 3, 7, 12, 91, 105, 107–109, 115, 133, 170
  error, 110, 112
metamodeling, 105
minimum spanning tree, 245
MMD, *see* maximum mean discrepancy
MMD-Shapley effect, 165
Möbius inverses, 138
model exploration, 3, 4
moment-independent measures, 149
moment-independent sensitivity analysis, 185
moment-independent sensitivity indices, 159, 162
Monte Carlo, 3, 28, 106, 110, 135, 137, 156, 159, 167, 175, 229, 231
  error, 110
  estimator, 110
  filtering method, 169
  integration, 111
  method, 240
  sample, 12
  sampling, 39
  simulation, 6
Morris
  design, 24, 25
  function, 26, 38, 252
  method, 23–27, 41, 43
MST, *see* minimum spanning tree

nearest neighbors, 116, 139, 144, 153
  algorithm, 223
  approach, 90

OAT, *see* design, one-at-a-time
orthogonal array, 66, 101, 234

Parseval's identity, 94, 96, 97
partial correlation coefficient, 33
partial derivative, 37
PCA, *see* principal component analysis
PCC, *see* partial correlation coefficient
permutation importance measure, 109
permuted column sampling, 61, 109
perturbed law–based indices, 13, 173
pick-freeze, 58, 59, 64, 73, 82, 90, 138, 159
  estimator, 62, 67
  sample, 111
PLI, *see* perturbed law–based indices
Poincaré constants, 57
Poincaré inequality, 56, 57
polynomial chaos, 34, 105, 107, 125
predictivity coefficient, 32
principal component analysis, 84

QMC, *see* quasi-Monte Carlo
QRBD, *see* quasi-random balance design
qualitative variables, 43
quantile, 29, 172, 174, 175
quantile-oriented SA, 172
quasi-Monte Carlo, 42, 55, 61, 195, 239
quasi-random balance design, 100

random balance design, 62, 95, 101
random permutation, 97, 100, 133, 137, 233
randomized orthogonal array, 65
rank correlation, 129
RB, *see* reduced basis
RBD, *see* random balance design
reduced basis, 112
replicated designs, 61, 66
replicated sampling plan based on randomized orthogonal array, 66
reproducing kernel Hilbert space, 108, 111, 115, 157, 164, 178, 238
RKHS, *see* reproducing kernel Hilbert space
robustness, 4, 8, 10, 173
robustness sensitivity measure, 178

Rosenblatt transform, 127, 129, 174

SA, *see* sensitivity analysis
sampling-based, 28, 43
SB, *see* sequential bifurcation
scatterplot, 28, 39, 63
screening, 12, 17, 19, 36, 42, 43, 57, 161, 188, 199, 229, 230
sensitivity analysis
  conditional, 168
  covariance-based, 133
  goal-oriented, 149
  quantile-oriented, 172
  target, 168
  ubiquitous, 77
  variance-based, 4, 52
sensitivity indices
  covariance-based, 121, 123
  variance-based, 42
sequential bifurcation, 20
Shapley effects, 129, 133, 165, 223
  decomposition, 134
  quantile-oriented, 142
Shapley values, 132, 133
simulated annealing, 242, 244
smoothing, 12, 113
smoothing techniques, 28, 106
Sobol' decomposition, 49
Sobol' $g$-function, 35, 54, 74, 76, 166, 169, 252
Sobol' index, 1, 45–91, 107, 108, 119
  closed, 53
  first-order, 53, 150
  second-order, 64–67
  total, 53
space-filling, 115
  criteria, 229
  design, 28, 61, 112, 230
  property, 106
SRC, *see* standard regression coefficient
SRRC, *see* standard rank regression coefficient
standard rank regression coefficient, 32
standard regression coefficient, 31
statistical learning, 3, 171
stochastic computer code, 90
superquantile, 174
supersaturated designs, 28
support points, 239

target SA, 168
total variance theorem, 53

ubiquitous SA, 77
uncertainty quantification, 1, 9, 168, 184, 187, 188, 197
unnormalized first-order HSIC, 161
UQ, *see* uncertainty quantification

variance-based SA, 4, 52
vectorial output, 78